EDEXCEL A LEVEL

BIOLOGY

2

Ed Lees
Martin Rowland
C. J. Clegg

Although every effort has been made to ensure that website addresses are correct at time of going to press, Hodder Education cannot be held responsible for the content of any website mentioned in this book. It is sometimes possible to find a relocated web page by typing in the address of the home page for a website in the URL window of your browser.

Orders: please contact Hachette UK Distribution, Hely Hutchinson Centre, Milton Road, Didcot, Oxfordshire, OX11 7HH. Telephone: +44 (0)1235 827827. Email education@hachette.co.uk Lines are open from 9 a.m. to 5 p.m., Monday to Friday.
You can also order through our website: www.hoddereducation.com

First published in 2015 by
Hodder Education,
An Hachette UK Company
Carmelite House, 50 Victoria Embankment
London EC4Y 0DZ

Impression number 5

Year 2022

Cover photo © Gaschwald - Fotolia

Typeset in 11/13 pts., Bembo Std by Aptara Inc.

Printed by CPI Group (UK) Ltd, Croydon CR0 4YY

A catalogue record for this title is available from the British Library.

ISBN 9781471807374

Contents

Get the most from this book

Welcome to the **Edexcel A level Biology 2 Student's Book**. This book covers Year 2 of the Edexcel A level Biology specification.

The following features have been included to help you get the most from this book.

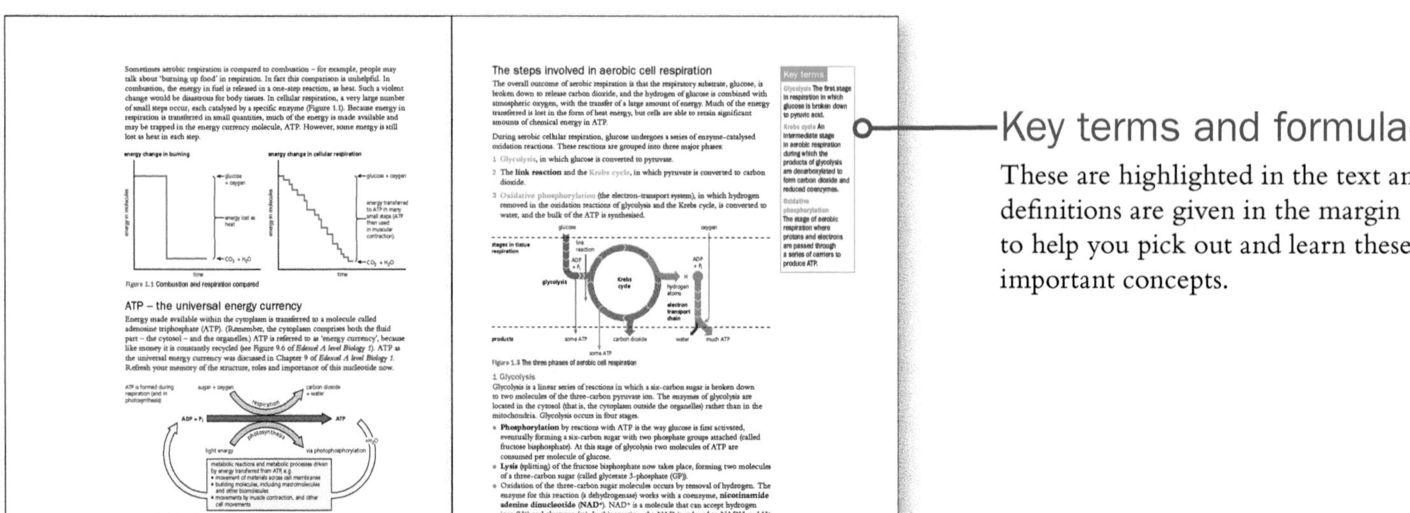

Key terms and formulae

These are highlighted in the text and definitions are given in the margin to help you pick out and learn these important concepts.

Examples

Examples of questions and calculations feature full workings and sample answers.

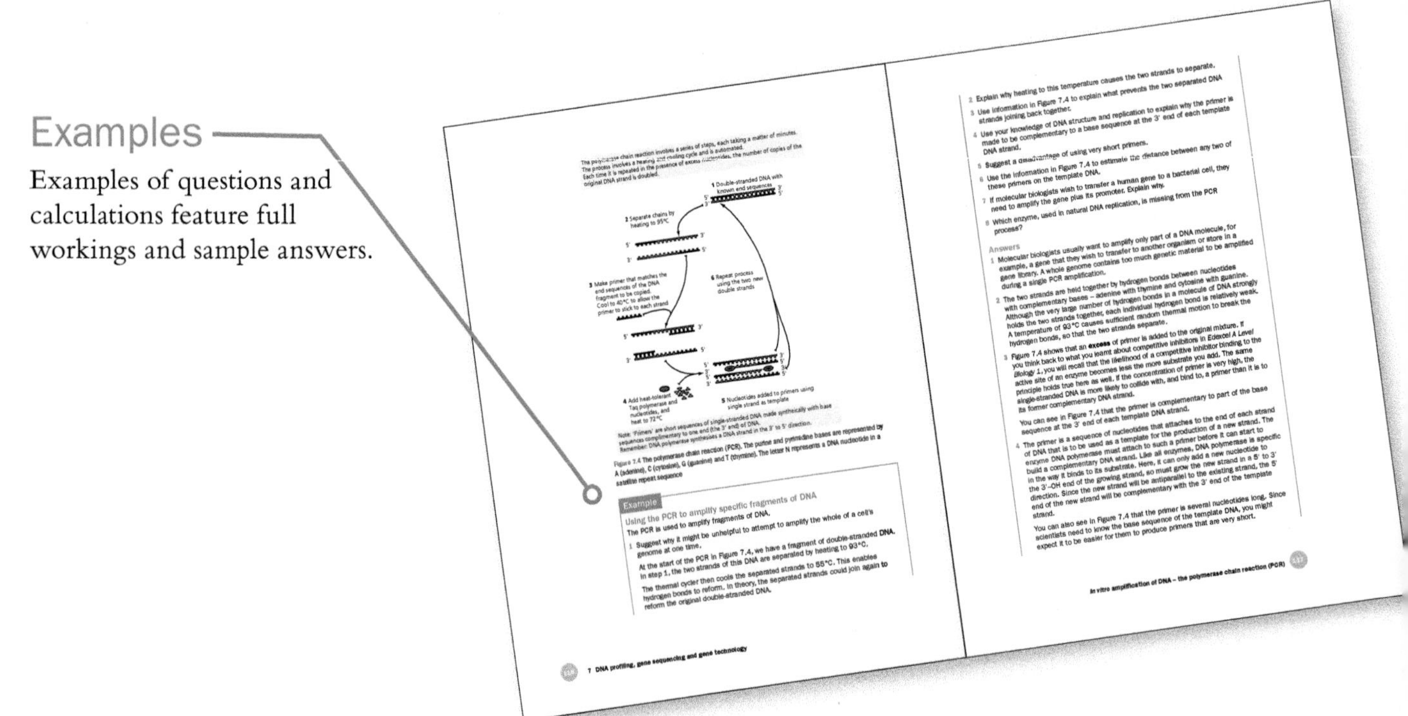

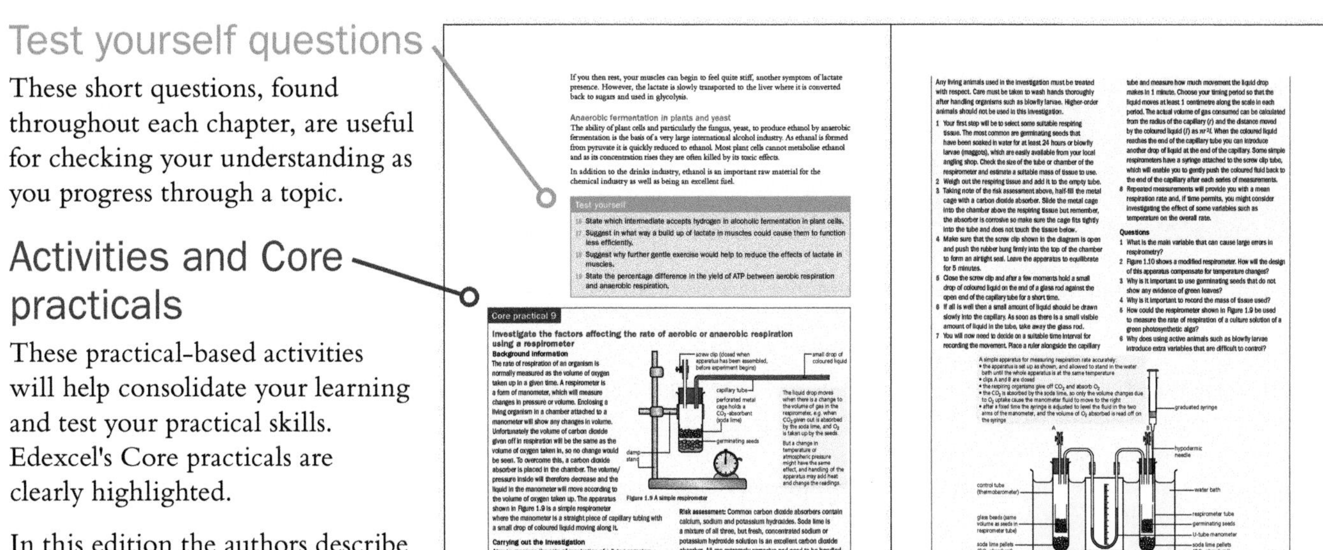

Test yourself questions

These short questions, found throughout each chapter, are useful for checking your understanding as you progress through a topic.

Activities and Core practicals

These practical-based activities will help consolidate your learning and test your practical skills. Edexcel's Core practicals are clearly highlighted.

In this edition the authors describe many important experimental procedures to conform to recent changes in the A level curriculum. Teachers should be aware that, although there is enough information to inform students of techniques and many observations for exam purposes, there is not enough information for teachers to replicate the experiments themselves, or with students, without recourse to CLEAPSS Hazcards or Laboratory worksheets that have undergone a risk assessment procedure.

Exam practice questions

1 A coenzyme can be described as:

A a protein catalyst

B a non-protein catalyst

C a protein essential for some enzyme activity

D a non-protein essential for some enzyme activity (1)

2 During chemiosmosis in mitochondria, energy is used to pump hydrogen ions:

A across the outer membrane into the cytoplasm

B across the outer membrane into the space between membranes

C across the inner membrane into the matrix

D across the inner membrane into the space between membranes (1)

Tip Question 2 is a straightforward recall question but it does illustrate the level of detail you must aim at learning in your revision.

3 The end product of the link reaction in aerobic respiration is:

A acetyl coenzyme A

B pyruvic acid

C citric acid

D glycerate-3-phosphate (1)

4 The diagram shows an outline of part of the biochemical pathways of anaerobic respiration in plant and animal cells.

Name the following parts of this diagram:

a) the stage of respiration labelled A (1)

b) the coenzyme formed at B (1)

c) the compound labelled C (1)

d) the compound labelled D (1)

e) the 1C molecule labelled E (1)

5 A student investigated the rate of respiration of germinating seeds at two different temperatures using the apparatus shown in the diagram.

Tip Question 5 contains typical elements of a core practical-based question. Some parts are designed to test your understanding of the experimental procedure itself and others test your application of Level 2 maths or interpretation of data.

1 Cellular respiration

Exam practice questions

You will find Exam practice questions at the end of every chapter. These follow the style of the different types of questions you might see in your examination and are colour coded to highlight the level of difficulty. Test your understanding even further with Maths questions and Stretch and challenge questions.

Tips

These highlight important facts, common misconceptions and signpost you towards other relevant topics.

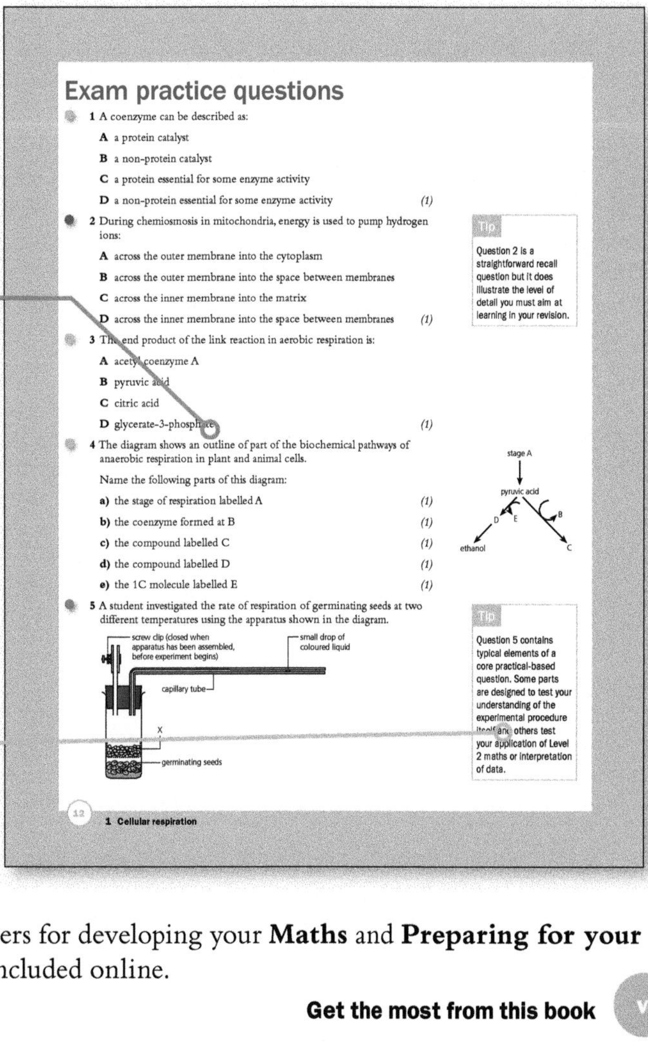

Dedicated chapters for developing your **Maths** and **Preparing for your exam** are also included online.

Acknowledgements

The Publisher would like to thank the following for permission to reproduce copyright material.

Photo credits:

p.20 © Dr Kari Lounatmaa/Science Photo Library; **p.21** © Dr Kenneth R. Miller/Science Photo Library; **p.35** © Dr Jeremy Burgess/Science Photo Library; **p.38** © Dr Linda Stannard, Uct/Science Photo Library; **p.39** © Doncaster And Bassetlaw Hospitals/Science Photo Library; **p.40** *both* © Trevor Clifford Photography/Science Photo Library; **p.42** © John Durham/Science Photo Library; **p.46** © Amelie-Benoist/Bsip/Science Photo Library; **p.47** © Martyn F. Chillmaid/Science Photo Library; **p.50** © Amelie-Benoist/Bsip/Science Photo Library; **p.58** © Kwangshin Kim/Science Photo Library; **p.59** *t* © Juergen Berger/Science Photo Library, *b* © Cnri/Science Photo Library; **p.60** *t* © Eye Of Science/Science Photo Library, *b* © Simon Fraser/Science Photo Library; **p.65** © Dr Linda Stannard, Uct/Science Photo Library; **p.68** © Yue Jin/Us Department Of Agriculture/Science Photo Library; **p.69** © Nick Gregory / Alamy; **p.71** © Dr. Pete Billingsley, University Of Aberdeen/Sinclair Stammers/Science Photo Library; **p.79** © Steve Gschmeissner/Science Photo Library; **p.80** © Steve Gschmeissner/Science Photo Library; **p.86** © Steve Gschmeissner/Science Photo Library; **p.96** *tl* and *tr* © Gary Meszaros/Science Photo Library, *bl* © Gilles Mermet/Science Photo Library; **p.103** © Biophoto Associates/Science Photo Library; **p.115** © Uk Crown Copyright Courtesy Of Fera/Science Photo Library; **p.122** © David Parker/Science Photo Library; **p.131** © Matt Meadows/Photo library/Getty Images; **p.133** © Nigel Cattlin/Science Photo Library; **p.135** © Janice and Nolan Braud / Alamy; **p.148** © Solvin Zankl/Visuals Unlimited, Inc. /Science Photo Library; **p.165** © Imagestate Media (John Foxx); **p.166** © Dave Watts / Alamy; **p.169** © Peter Falkner/Science Photo Library; **p.174** © Gene Cox Estate; **p.175** © Gene Cox Estate; **p.180** © Gene Cox Estate; **p.187** © Prof S. Cinti/Science Photo Library; **p.190** © NHPA/Photoshot; **p.194** *tl* © Vladimir Melnik – Fotolia, *tm* © Nazzu – Fotolia, *tr* © ZambeziShark/iStock/Thinkstock, *bl* © Marcelo Sanchez – Fotolia, *br* © Faultier/iStock/Thinkstock; **p.202** © Ross Koning; **p.213** © Gene Cox Estate; **p.226** © P&R Fotos - www.superstock.com; **p.233** © Jim Zipp/Science Photo Library; **p.239** © Noaa Pmel Vents Program/Science Photo Library; **p.248** © Christian GUY/imageBROKER - www.superstock.com; **p.256** *l* © G A Matthews/Science Photo Library, *r* © Glenis Moore/Science Photo Library; **p.263** © Marshall Ikonography / Alamy; **p.267** *tl* © Marvin Dembinsky Photo Associates / Alamy, *tr* © Martin Harvey / Alamy, *bl* © Premaphotos / Alamy, *br* © RM Floral / Alamy; **p.268** © Biophoto Associates/Science Photo Library; **p.278** *l* © Greg Arthur, *r* © Nature Picture Library / Alamy; **p.279** *t* © Frederick William Wallace/National Geographic Creative/Corbis, *b* © Pierre Gleizes / Greenpeace; **p.280** © Greenpeace; **p.281** © Marine Stewardship Council; **p.286** *all* © Dr Ian Fuller, Massey University; **p.287** © Georgette Douwma/Science Photo Library.

t = top, *b* = bottom, *l* = left, *r* = right, *m* = middle

Every effort has been made to trace all copyright holders, but if any have been inadvertently overlooked, the Publisher will be pleased to make the necessary arrangements at the earliest opportunity.

1 Cellular respiration

Prior knowledge

In this chapter you will need to recall that:

- → chemical reduction involves the gain of electrons
- → chemical oxidation involves the loss of electrons
- → hydrogen atoms have one proton and one electron
- → hydrogen ions have just one proton and hence are positively charged
- → ATP and ADP molecules are used to transfer energy within cells
- → mitochondria are the sites of ATP formation
- → glucose is the most common respiratory substrate
- → energy is neither created nor destroyed but simply transformed from one form into another
- → aerobic respiration yields more ATP than anaerobic respiration
- → the end products of aerobic respiration are carbon dioxide, water and ATP.

Test yourself on prior knowledge

1 State why reactions are referred to as REDOX reactions rather than simply oxidation or reduction.

2 We very rarely find hydrogen atoms moving around freely. Why is this?

3 Explain why ATP is described as the 'energy currency' of the cell.

4 Galactose is a 6-carbon monosaccharide with the formula $C_6H_{12}O_6$. Whilst this is the same as glucose, many organisms cannot use galactose as a respiratory substrate. Explain why this is the case.

5 State the most common form of energy, apart from chemical energy, found in an actively respiring cell.

6 State what other compounds form respiratory substrates in animals apart from glucose.

Aerobic respiration

Key term

Aerobic respiration The chemical breakdown of substrate molecules in cells to release energy in the form of ATP when oxygen is present.

Organisms require energy to maintain all their living cells and to carry out their activities and functions. Respiration is the process by which that energy is transferred in usable form. Cellular respiration that involves oxygen is described as **aerobic respiration**. Most animals and plants and very many microorganisms respire aerobically most, if not all of the time.

In **aerobic respiration**, sugar is oxidised to carbon dioxide and water and much energy is made available. The steps involved in aerobic respiration can be summarised by a single equation. Note that this equation is equivalent to a balance sheet of inputs (the raw materials) and outputs (the products), but it tells us nothing about the steps.

glucose + oxygen → carbon dioxide + water + energy

$$C_6H_{12}O_6 + 6O_2 \rightarrow 6CO_2 + 6H_2O + \text{energy}$$

Sometimes aerobic respiration is compared to combustion – for example, people may talk about 'burning up food' in respiration. In fact this comparison is unhelpful. In combustion, the energy in fuel is released in a one-step reaction, as heat. Such a violent change would be disastrous for body tissues. In cellular respiration, a very large number of small steps occur, each catalysed by a specific enzyme (Figure 1.1). Because energy in respiration is transferred in small quantities, much of the energy is made available and may be trapped in the energy currency molecule, ATP. However, some energy is still lost as heat in each step.

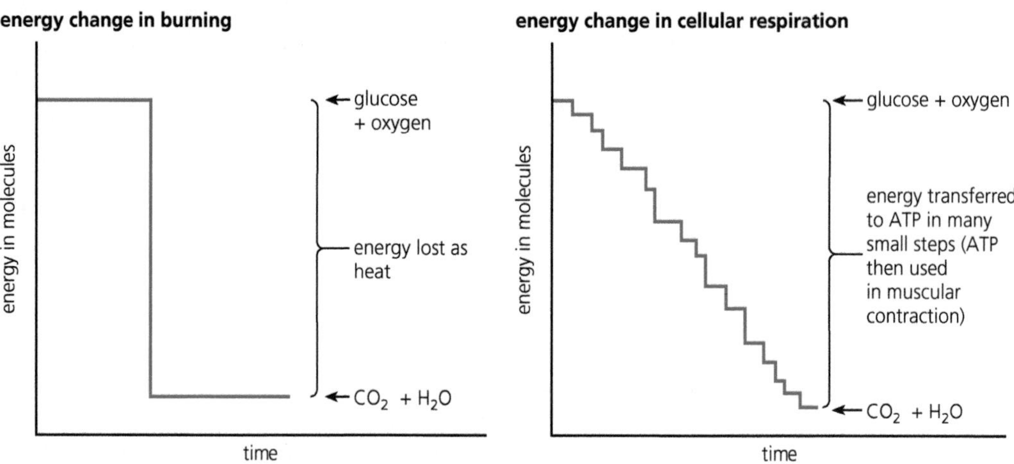

Figure 1.1 Combustion and respiration compared

ATP – the universal energy currency

Energy made available within the cytoplasm is transferred to a molecule called adenosine triphosphate (ATP). (Remember, the cytoplasm comprises both the fluid part – the cytosol – and the organelles.) ATP is referred to as 'energy currency', because like money it is constantly recycled (see Figure 9.6 of *Edexcel A level Biology 1*). ATP as the universal energy currency was discussed in Chapter 9 of *Edexcel A level Biology 1*. Refresh your memory of the structure, roles and importance of this nucleotide now.

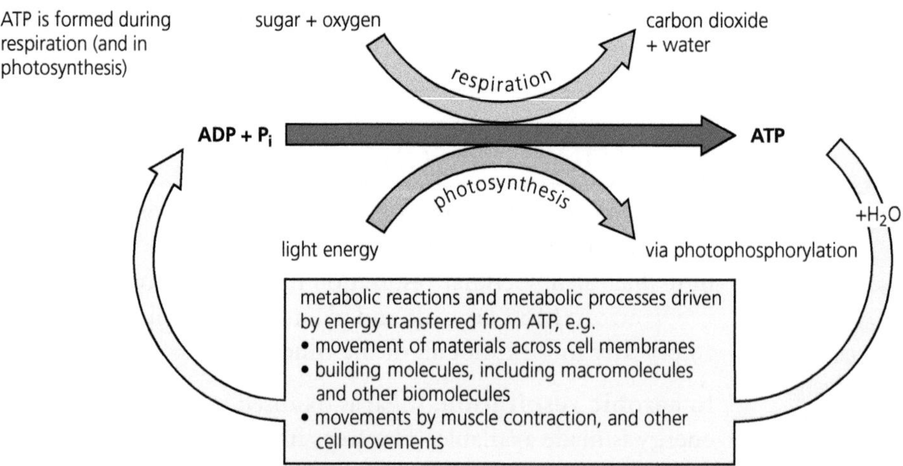

Figure 1.2 The ATP $\longrightarrow$ ADP + P_i cycle

The steps involved in aerobic cell respiration

The overall outcome of aerobic respiration is that the respiratory substrate, glucose, is broken down to release carbon dioxide, and the hydrogen of glucose is combined with atmospheric oxygen, with the transfer of a large amount of energy. Much of the energy transferred is lost in the form of heat energy, but cells are able to retain significant amounts of chemical energy in ATP.

During aerobic cellular respiration, glucose undergoes a series of enzyme-catalysed oxidation reactions. These reactions are grouped into three major phases:

1. Glycolysis, in which glucose is converted to pyruvate.
2. The **link reaction** and the Krebs cycle, in which pyruvate is converted to carbon dioxide.
3. Oxidative phosphorylation (the electron-transport system), in which hydrogen removed in the oxidation reactions of glycolysis and the Krebs cycle, is converted to water, and the bulk of the ATP is synthesised.

> **Key terms**
>
> **Glycolysis** The first stage in respiration in which glucose is broken down to pyruvic acid.
>
> **Krebs cycle** An intermediate stage in aerobic respiration during which the products of glycolysis are decarboxylated to form carbon dioxide and reduced coenzymes.
>
> **Oxidative phosphorylation** The stage of aerobic respiration where protons and electrons are passed through a series of carriers to produce ATP.

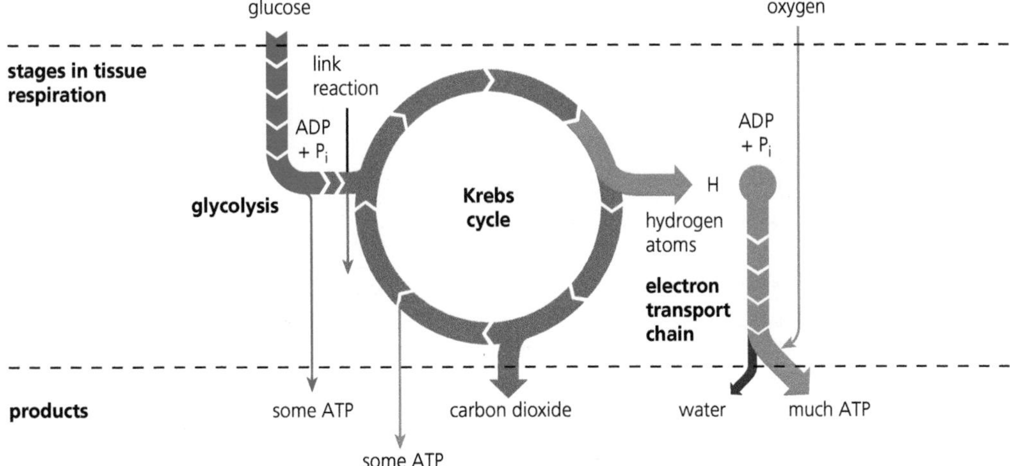

Figure 1.3 The three phases of aerobic cell respiration

1 Glycolysis

Glycolysis is a linear series of reactions in which a six-carbon sugar is broken down to two molecules of the three-carbon pyruvate ion. The enzymes of glycolysis are located in the cytosol (that is, the cytoplasm outside the organelles) rather than in the mitochondria. Glycolysis occurs in four stages.

- **Phosphorylation** by reactions with ATP is the way glucose is first activated, eventually forming a six-carbon sugar with two phosphate groups attached (called fructose bisphosphate). At this stage of glycolysis two molecules of ATP are consumed per molecule of glucose.
- **Lysis** (splitting) of the fructose bisphosphate now takes place, forming two molecules of a three-carbon sugar (called glycerate 3-phosphate (GP)).
- Oxidation of the three-carbon sugar molecules occurs by removal of hydrogen. The enzyme for this reaction (a dehydrogenase) works with a coenzyme, **nicotinamide adenine dinucleotide (NAD^+)**. NAD^+ is a molecule that can accept hydrogen ions (H^+) and electrons (e^-). In this reaction, the NAD is reduced to NADH and H^+ (known as reduced NAD):

  $NAD^+ + 2H^+ + 2e^- \longrightarrow NADH + H^+$ (sometimes represented as $NADH_2$)

  (Reduced NAD can pass hydrogen ions and electrons on to other acceptor molecules, as described below, and when it does it becomes oxidised back to NAD.)

- **ATP formation** occurs twice in the reactions, by which each triose phosphate molecule is converted to pyruvate. This form of ATP synthesis is referred to as being 'at substrate level' in order to differentiate it from the bulk of ATP synthesis that occurs later in cell respiration, during operation of the electron transport chain (see below). At this stage of glycolysis as two molecules of glycerate 3-phosphate (GP) are converted to pyruvate, four molecules of ATP are synthesised. So, in total, there is a net gain of two ATPs in glycolysis.

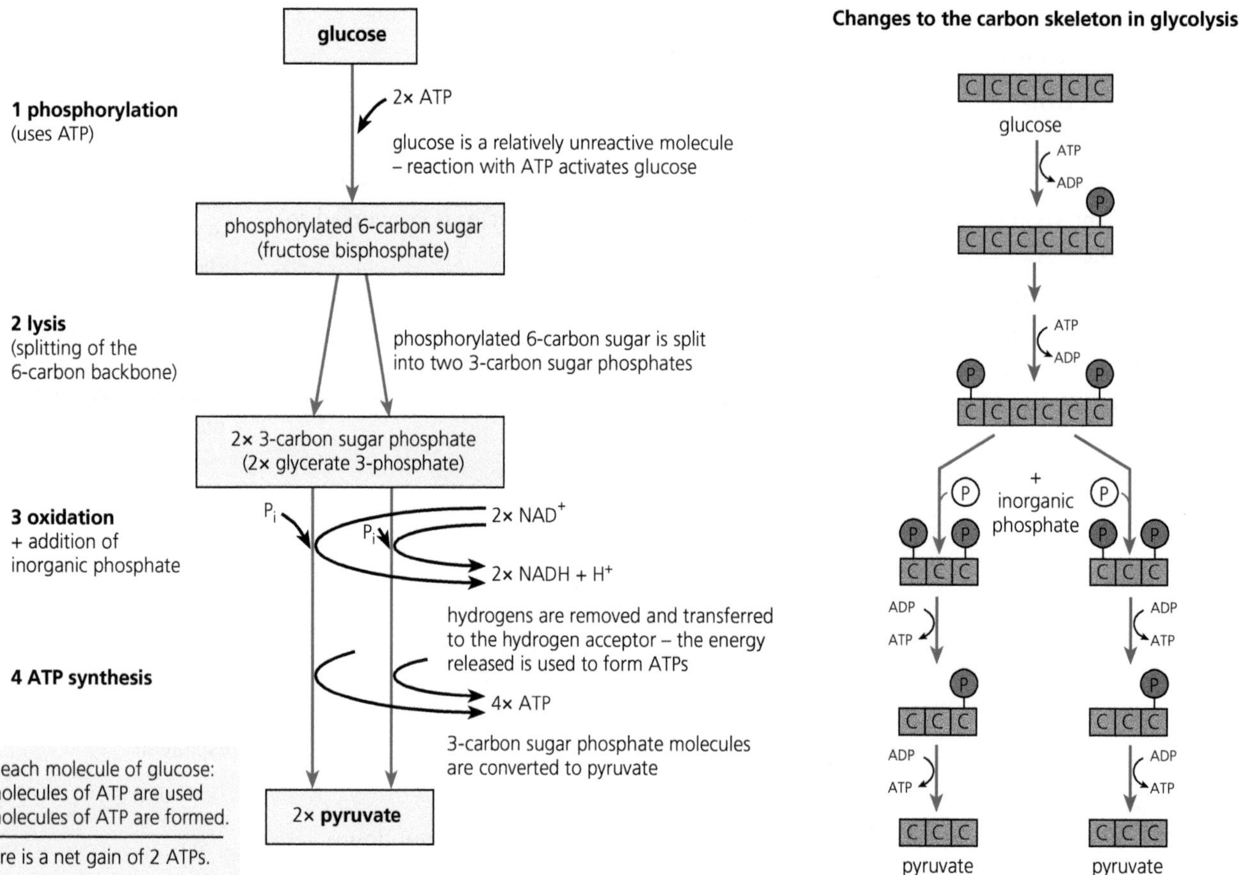

Figure 1.4 A summary of glycolysis

Test yourself

1. Explain why two molecules of ATP are used to phosphorylate glucose at the start of glycolysis.
2. State the name of the three-carbon sugar formed by the splitting of fructose bisphosphate.
3. State the end products of glycolysis.
4. Describe where in the cell the reactions of glycolysis take place.

2 The link reaction and Krebs cycle

The subsequent steps in aerobic respiration occur in the organelles known as mitochondria.

In the link reaction, pyruvate diffuses into the matrix of the mitochondrion as it forms, and is metabolised there. First, the three-carbon pyruvate is decarboxylated by removal of carbon dioxide and, at the same time, oxidised by removal of hydrogen. Reduced NAD is formed. The product of this oxidative decarboxylation reaction is an acetyl group – a two-carbon fragment. This acetyl group is then combined with a **coenzyme** called coenzyme A, forming acetyl coenzyme A.

The production of acetyl coenzyme A from pyruvate is known as the link reaction because it connects glycolysis to reactions of the Krebs cycle, details of which now follow.

Key term

Coenzyme An organic non-protein molecule that binds to an enzyme to allow it to carry out its function.

In the Krebs cycle, acetyl coenzyme A reacts with a four-carbon organic acid (oxaloacetate, OAA). The products of this reaction are a six-carbon acid (citrate) and, of course, coenzyme A. This latter, on release, is re-used in the link reaction.

The Krebs cycle is named after Hans Krebs who discovered it, but it is also sometimes referred to as the citric acid cycle, after the first intermediate acid formed.

Then the citrate is converted back to the four-carbon acid (an acceptor molecule, in effect) by the reactions of the Krebs cycle. These involve the following changes:

- Two molecules of carbon dioxide are given off in separate decarboxylation reactions.
- A molecule of ATP is formed as part 1 of the reactions of the cycle – as with glycolysis, this ATP synthesis is 'at substrate level' too.
- Three molecules of reduced NAD are formed.
- One molecule of another hydrogen accepter – FAD (flavin adenine dinucleotide) is reduced. (NAD is the chief hydrogen-carrying coenzyme of respiration but FAD is another coenzyme with this role in the Krebs cycle).

Because glucose is converted to two molecules of pyruvate in glycolysis, the whole Krebs cycle sequence of reactions 'turns' twice for every molecule of glucose that is metabolised by aerobic cellular respiration.

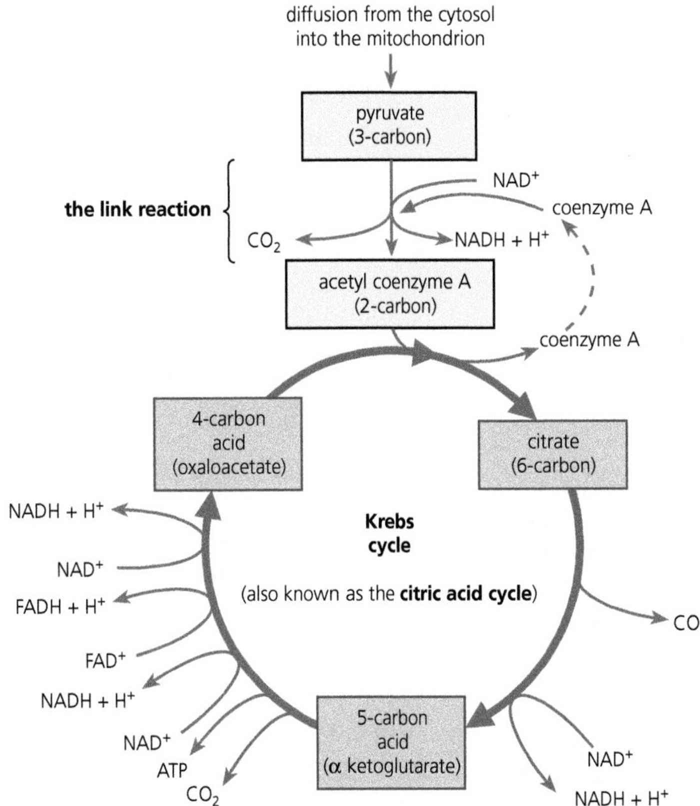

Figure 1.5 A summary of the Krebs cycle

This gives us the products shown in Table 1.1.

Table 1.1 Net products of aerobic respiration of glucose at the end of the Krebs cycle

Step	Product			
	CO_2	ATP	Reduced NAD	Reduced FAD
Glycolysis	0	2	2	0
Link reaction (pyruvate → acetyl CoA)	2	0	2	0
Krebs cycle	4	2	6	2
Totals	6	4	10	2

As you can see, this table shows a very small yield of ATP but a large yield of reduced NAD and some reduced FAD. The final stage of aerobic respiration explains how these reduced coenzymes can be converted into ATP and exactly how the remaining product, water, is formed.

Test yourself

5 State which final product of respiration is generated in the link reaction and Krebs cycle.
6 Describe where exactly in the cell the reactions of the link reaction and the Krebs cycle take place.
7 Describe how reduced NAD is formed in the link reaction.
8 State which product of the link reaction enters the Krebs cycle.
9 Explain why NAD and FAD are known as coenzymes.
10 The Krebs cycle completes the breakdown of glucose molecules. Explain why the ATP yield is so low.

3 Oxidative phosphorylation and terminal oxidation

This section depends on your knowledge of mitochondrial structure, which you met in Chapter 4 of *Edexcel A level Biology 1*, Figure 4.6. It is important that you look back at your notes before continuing.

Tip

Before attempting to understand this section, make sure you are very clear about hydrogen ions and atoms. Hydrogen atoms lose their only electron so a hydrogen ion is left as a proton. This is why you will see H^+ and proton as alternatives in the diagram. The electrons move down a chain of carriers.

The removal of pairs of hydrogen atoms from various intermediates of the respiratory pathway is a feature of several of the steps in glycolysis and the Krebs cycle. On most occasions, oxidised NAD is converted to reduced NAD, but in the Krebs cycle it is an alternative hydrogen-acceptor coenzyme known as FAD that is reduced.

In this final stage of aerobic respiration, the hydrogen atoms (or their electrons) are transported along a series of carriers, from the reduced NAD (or FAD), to be combined with oxygen to form water. Hence the role of oxygen in this process is that of the **final hydrogen acceptor**. Note that this final reaction to form water only occurs after the energy level has been lowered by a series of transfers between carriers, bringing about the gradual transfer of energy shown in Figure 1.1.

As electrons are passed between the carriers in the series, energy is transferred. Transfer of energy in this manner is controlled and can be used by the cell. The energy is transferred to ADP and P_i, forming ATP. Normally, for every molecule of reduced NAD that is oxidised (that is, for every pair of hydrogens) approximately three molecules of ATP are produced.

The process is summarised in Figure 1.6. The total yield from aerobic respiration is **about 38 ATPs per molecule of glucose** respired. This is obviously much more than from glycolysis. There are other pathways associated with this process, which means the overall yield is given as 'about' 38 ATPs.

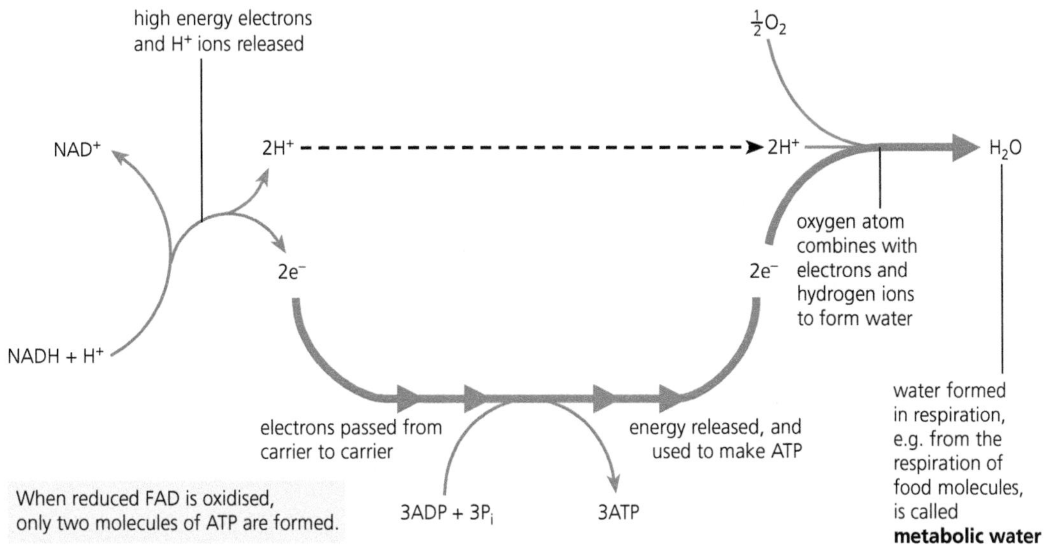

Figure 1.6 Terminal oxidation and formation of ATP

ATP formation by chemiosmosis

In Book 1 we discussed how scientists suggested new ideas or models that formed the basis of predictions that could be tested experimentally. Only when sufficient experimental data accumulate does the model gradually become accepted.

The 'chemiosmotic model' grew out of studies of bacterial metabolism carried out by biochemist Peter Mitchell in 1961. Gradually, over the next decade and more, scientists were able to confirm the presence of the correct gradients of H^+ ions and the necessary protein pumps in the mitochondrial membranes. The model stood up well to these investigations and in 1975 Mitchell was awarded the Nobel Prize for his work, confirming that this had indeed become the accepted model.

Chemiosmosis is a process by which the synthesis of ATP is coupled to electron transport via the movement of protons as shown in Figure 1.7. **Electron-carrier proteins** are arranged in the inner mitochondrial membrane in a highly ordered way. These carrier proteins oxidise the reduced coenzymes, and energy from this process is used to pump hydrogen ions (protons) from the matrix of the mitochondrion into the space between inner and outer mitochondrial membranes.

Key term

Chemiosmosis The process by which the movement of protons across the inner mitochondrial membrane is coupled to the synthesis of ATP.

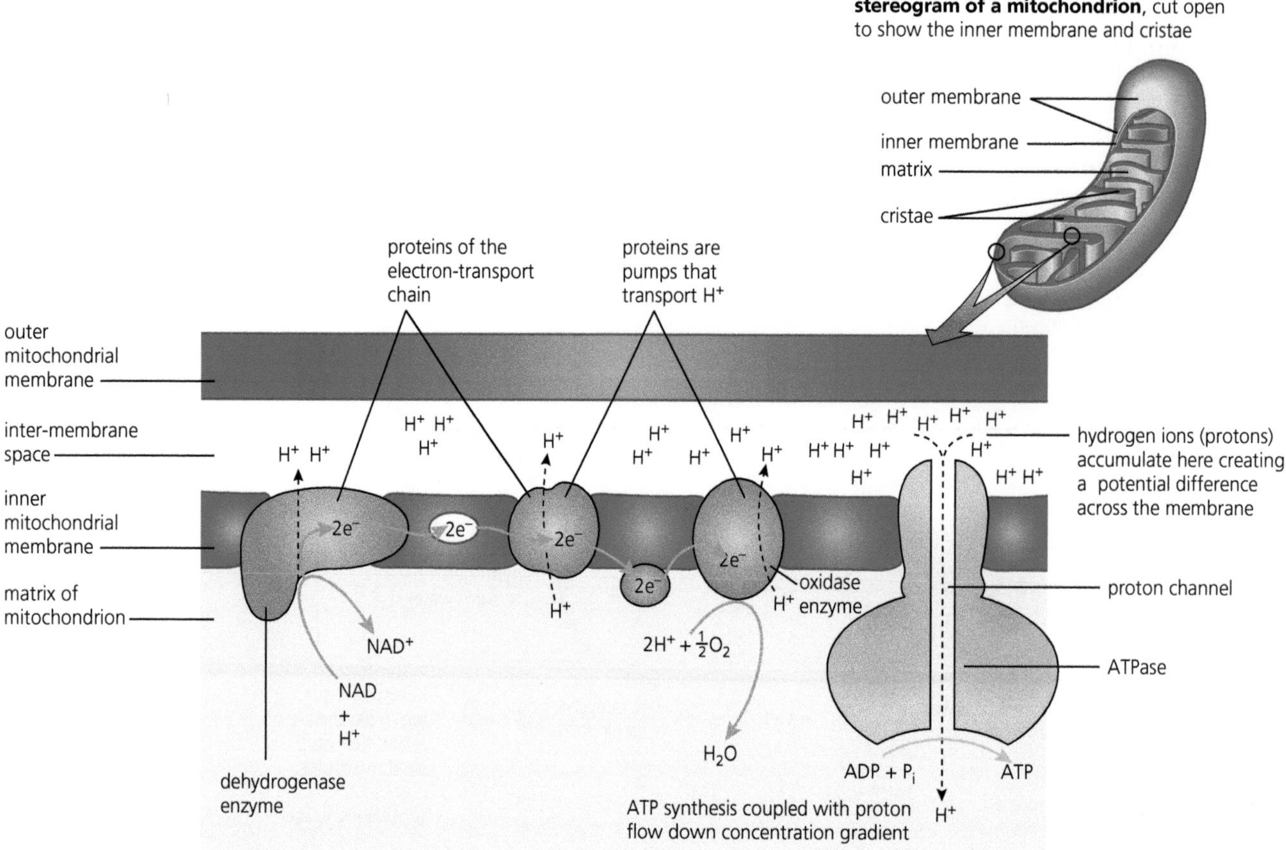

Figure 1.7 Mitchell's chemiosmotic theory

> **Key term**
>
> **ATP synthetase** An enzyme found in the inner mitochondrial membrane, which channels hydrogen ions through the membrane during ATP synthesis by chemiosmosis.

Here the H^+ ions accumulate – incidentally, causing the pH to drop. Because the inner membrane is largely impermeable to ions, a significant difference in hydrogen ion concentration builds up, generating an electrochemical gradient across the inner membrane – a store of potential energy.

Eventually, the protons do flow back into the matrix, via channels in **ATP synthetase** enzymes, also found in the inner mitochondrial membrane. As the protons flow down their concentration gradient through the enzyme, the energy is transferred and ATP synthesis occurs.

Test yourself

11 During oxidative phosphorylation electrons are passed from one carrier to another. Explain why they do this.

12 Explain why hydrogen is combined with oxygen to form water only at the end of the chain of carriers.

13 a) State where in the mitochondria there is a high concentration of H^+ ions.

b) Describe how a high concentration of H^+ ions is built up in this region.

14 Explain why a high H^+ ion concentration means a drop in pH.

15 In what ways is ATP synthetase unlike a normal enzyme?

Anaerobic respiration or fermentation

In the absence of oxygen, many organisms (and sometimes certain tissues in organisms when deprived of sufficient oxygen) will continue to respire by a process known as fermentation or anaerobic respiration, at least for a short time.

A knowledge of aerobic respiration shows us the effect that a lack of oxygen will have. If oxygen is the final hydrogen acceptor, then without it the carriers of oxidative phosphorylation will all become reduced and the flow of electrons and protons will cease. This will also mean that the supply of NAD^+ will be halted and the Krebs cycle will also come to a stop.

With only glycolysis operating, you can see that there is a net gain of only 2ATPs per glucose molecule compared with about 38ATPs from complete aerobic respiration. Therefore, whilst anaerobic respiration can continue without oxygen it is a very inefficient process.

A second consequence of a lack of oxygen is that the end-product of glycolysis, pyruvic acid, will begin to accumulate. As the concentration of pyruvic acid increases it is channelled into other biochemical pathways, as shown in Figure 1.8. In animal cells this results in the formation of **lactate** (lactic acid) as the pyruvate acts as the acceptor for reduced NAD, whilst in plant cells **ethanal** acts as this acceptor and this results in the formation of **ethanol**. In effect these compounds are replacing oxygen as the final hydrogen acceptor to allow glycolysis to continue.

Both lactate and ethanol contain large quantities of chemical energy, indicating that the glucose has only been partially broken down and explaining the low ATP yield.

> **Key term**
>
> **Anaerobic respiration** The process by which substrate molecules are broken down in cells to release energy in the form of ATP in the absence of oxygen.

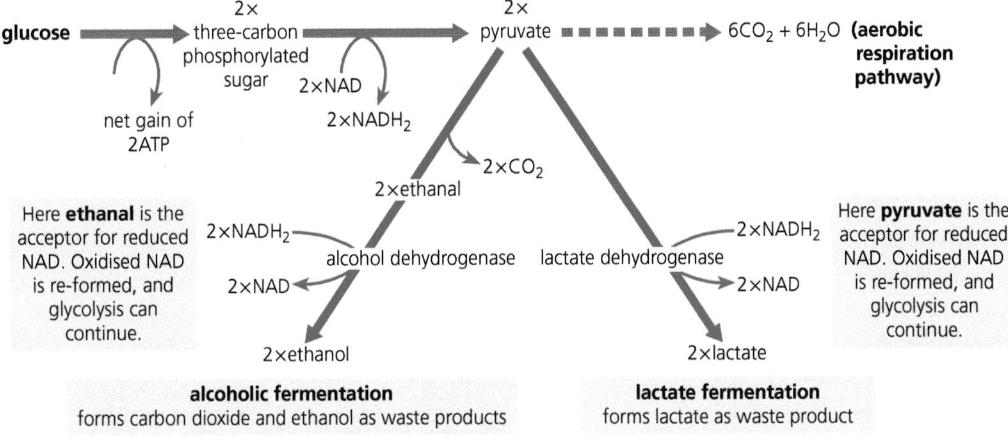

Figure 1.8 The respiratory pathways of anaerobic respiration

Anaerobic respiration in vertebrate muscle

Active vertebrate muscles have a high oxygen demand. If they continue to work at a high rate their demand for oxygen quickly exceeds the maximum rate of supply. In order to continue working, the muscles need to respire anaerobically. This means that the concentration of lactate will begin to rise. The effect of this lactate is to gradually inhibit the muscle contractions.

You are probably familiar with this effect. If you begin to sprint, you will quickly begin to experience lactate build up in your muscles and the effect we know as fatigue. No matter how much you try to run faster, your muscles will not respond, and then begin to feel painful.

If you then rest, your muscles can begin to feel quite stiff, another symptom of lactate presence. However, the lactate is slowly transported to the liver where it is converted back to sugars and used in glycolysis.

Anaerobic fermentation in plants and yeast

The ability of plant cells and particularly the fungus, yeast, to produce ethanol by anaerobic fermentation is the basis of a very large international alcohol industry. As ethanal is formed from pyruvate it is quickly reduced to ethanol. Most plant cells cannot metabolise ethanol and as its concentration rises they are often killed by its toxic effects.

In addition to the drinks industry, ethanol is an important raw material for the chemical industry as well as being an excellent fuel.

Test yourself

16 State which intermediate accepts hydrogen in alcoholic fermentation in plant cells.

17 Suggest in what way a build up of lactate in muscles could cause them to function less efficiently.

18 Suggest why further gentle exercise would help to reduce the effects of lactate in muscles.

19 State the percentage difference in the yield of ATP between aerobic respiration and anaerobic respiration.

Core practical 9

Investigate the factors affecting the rate of aerobic or anaerobic respiration using a respirometer

Background information

The rate of respiration of an organism is normally measured as the volume of oxygen taken up in a given time. A respirometer is a form of manometer, which will measure changes in pressure or volume. Enclosing a living organism in a chamber attached to a manometer will show any changes in volume. Unfortunately the volume of carbon dioxide given off in respiration will be the same as the volume of oxygen taken in, so no change would be seen. To overcome this, a carbon dioxide absorber is placed in the chamber. The volume/pressure inside will therefore decrease and the liquid in the manometer will move according to the volume of oxygen taken up. The apparatus shown in Figure 1.9 is a simple respirometer where the manometer is a straight piece of capillary tubing with a small drop of coloured liquid moving along it.

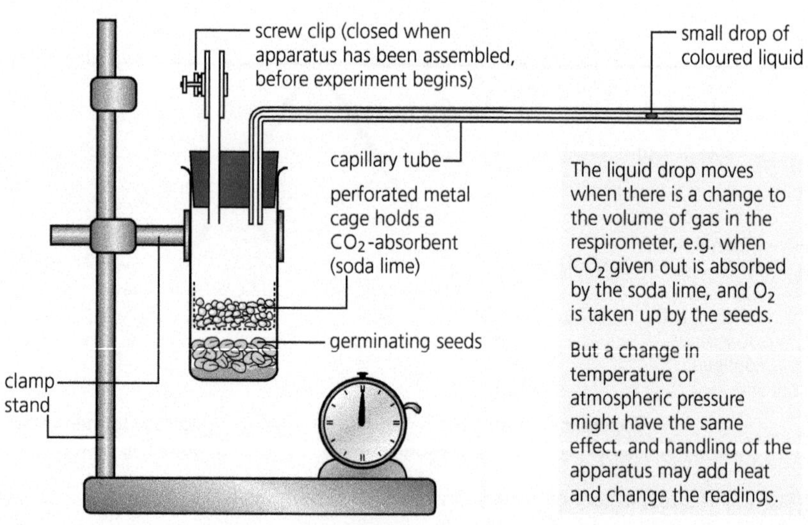

Figure 1.9 A simple respirometer

Carrying out the investigation

Aim: to measure the rate of respiration of a living organism.

Risk assessment: Common carbon dioxide absorbers contain calcium, sodium and potassium hydroxides. Soda lime is a mixture of all three, but fresh, concentrated sodium or potassium hydroxide solution is an excellent carbon dioxide absorber. All are extremely corrosive and eye protection should be worn at all times. If you are handling the chemicals yourself, you must wear chemically resistant (nitrile) gloves and goggles. Your teacher will show you the safe procedure and supervise closely.

Your teacher may decide to provide place the alkaline solutions/solids in the respirometer before the lesson, in which case you will not need to handle the chemicals directly.

Any living animals used in the investigation must be treated with respect. Care must be taken to wash hands thoroughly after handling organisms such as blowfly larvae. Higher-order animals should not be used in this investigation.

1 Your first step will be to select some suitable respiring tissue. The most common are germinating seeds that have been soaked in water for at least 24 hours or blowfly larvae (maggots), which are easily available from your local angling shop. Check the size of the tube or chamber of the respirometer and estimate a suitable mass of tissue to use.
2 Weigh out the respiring tissue and add it to the empty tube.
3 Taking note of the risk assessment above, half-fill the metal cage with a carbon dioxide absorber. Slide the metal cage into the chamber above the respiring tissue but remember, the absorber is corrosive so make sure the cage fits tightly into the tube and does not touch the tissue below.
4 Make sure that the screw clip shown in the diagram is open and push the rubber bung firmly into the top of the chamber to form an airtight seal. Leave the apparatus to equilibrate for 5 minutes.
5 Close the screw clip and after a few moments hold a small drop of coloured liquid on the end of a glass rod against the open end of the capillary tube for a short time.
6 If all is well then a small amount of liquid should be drawn slowly into the capillary. As soon as there is a small visible amount of liquid in the tube, take away the glass rod.
7 You will now need to decide on a suitable time interval for recording the movement. Place a ruler alongside the capillary tube and measure how much movement the liquid drop makes in 1 minute. Choose your timing period so that the liquid moves at least 1 centimetre along the scale in each period. The actual volume of gas consumed can be calculated from the radius of the capillary (r) and the distance moved by the coloured liquid (l) as $\pi r^2 l$. When the coloured liquid reaches the end of the capillary tube you can introduce another drop of liquid at the end of the capillary. Some simple respirometers have a syringe attached to the screw clip tube, which will enable you to gently push the coloured fluid back to the end of the capillary after each series of measurements.
8 Repeated measurements will provide you with a mean respiration rate and, if time permits, you might consider investigating the effect of some variables such as temperature on the overall rate.

Questions

1 What is the main variable that can cause large errors in respirometry?
2 Figure 1.10 shows a modified respirometer. How will the design of this apparatus compensate for temperature changes?
3 Why is it important to use germinating seeds that do not show any evidence of green leaves?
4 Why is it important to record the mass of tissue used?
5 How could the respirometer shown in Figure 1.9 be used to measure the rate of respiration of a culture solution of a green photosynthetic alga?
6 Why does using active animals such as blowfly larvae introduce extra variables that are difficult to control?

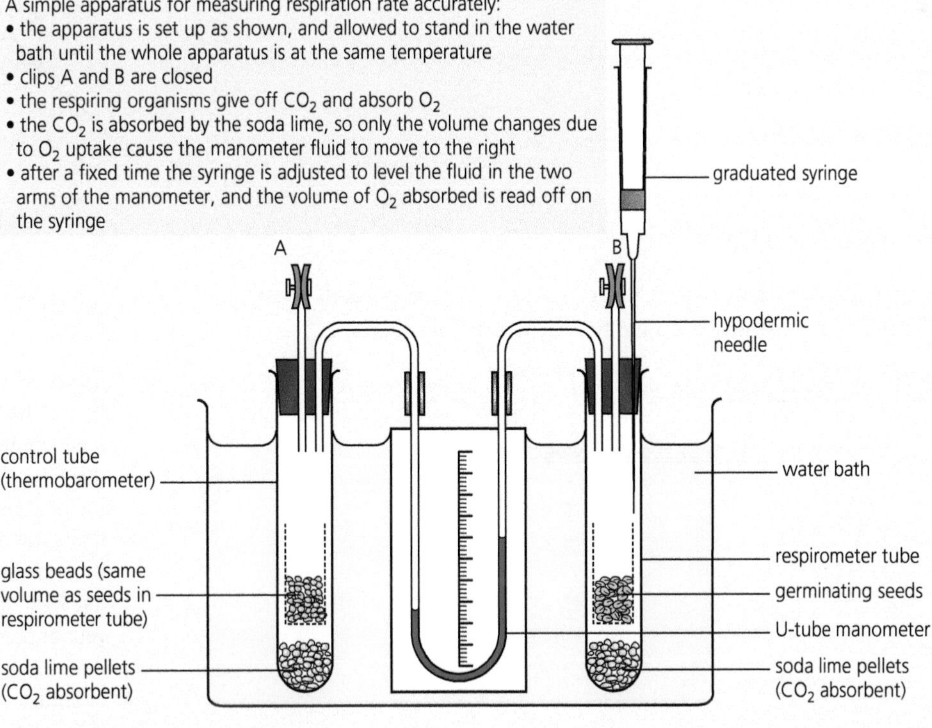

Figure 1.10 A more advanced respirometer

Exam practice questions

1 A coenzyme can be described as:

A a protein catalyst

B a non-protein catalyst

C a protein essential for some enzyme activity

D a non-protein essential for some enzyme activity *(1)*

2 During chemiosmosis in mitochondria, energy is used to pump hydrogen ions:

A across the outer membrane into the cytoplasm

B across the outer membrane into the space between membranes

C across the inner membrane into the matrix

D across the inner membrane into the space between membranes *(1)*

Tip

Question 2 is a straightforward recall question but it does illustrate the level of detail you must aim at learning in your revision.

3 The end product of the link reaction in aerobic respiration is:

A acetyl coenzyme A

B pyruvic acid

C citric acid

D glycerate-3-phosphate *(1)*

4 The diagram shows an outline of part of the biochemical pathways of anaerobic respiration in plant and animal cells.

Name the following parts of this diagram:

a) the stage of respiration labelled A *(1)*

b) the coenzyme formed at B *(1)*

c) the compound labelled C *(1)*

d) the compound labelled D *(1)*

e) the 1C molecule labelled E *(1)*

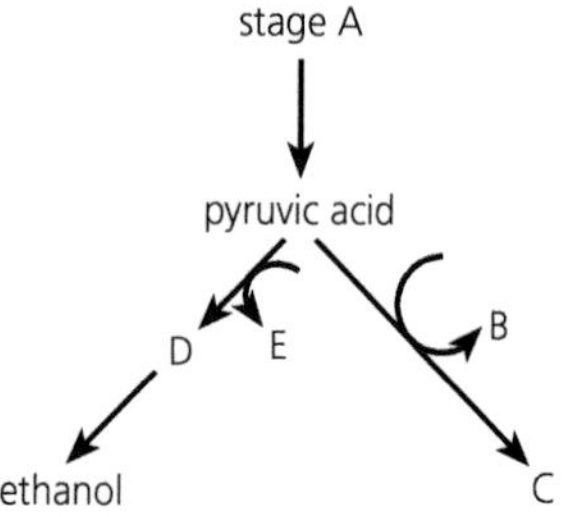

5 A student investigated the rate of respiration of germinating seeds at two different temperatures using the apparatus shown in the diagram.

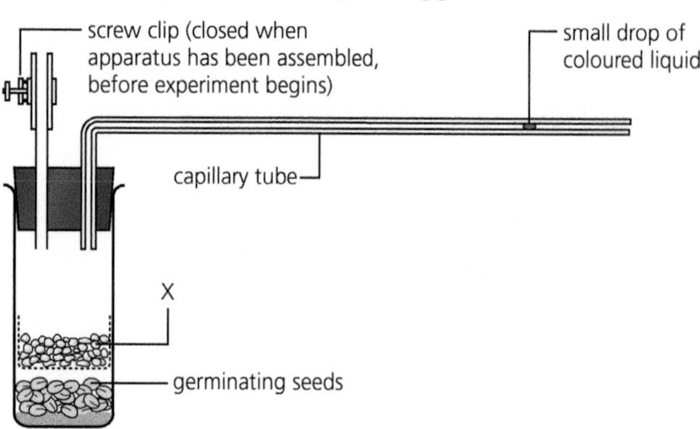

Tip

Question 5 contains typical elements of a core practical-based question. Some parts are designed to test your understanding of the experimental procedure itself and others test your application of Level 2 maths or interpretation of data.

a) Name **one** compound that would be added to the wire cage X and explain its purpose. *(2)*

b) The student assembled the apparatus as shown in the diagram by adding 8 g of germinating wheat seeds to the tube, which was then immersed in a water bath at 20 °C for 10 minutes. The screw clip was then closed and readings taken of the movement of the coloured liquid each minute for 10 minutes. This was then repeated using a water bath maintained at 30 °C.

Explain why the screw clip was left open for 10 minutes before the investigation was started. *(2)*

c) The graph shows the results of this investigation.

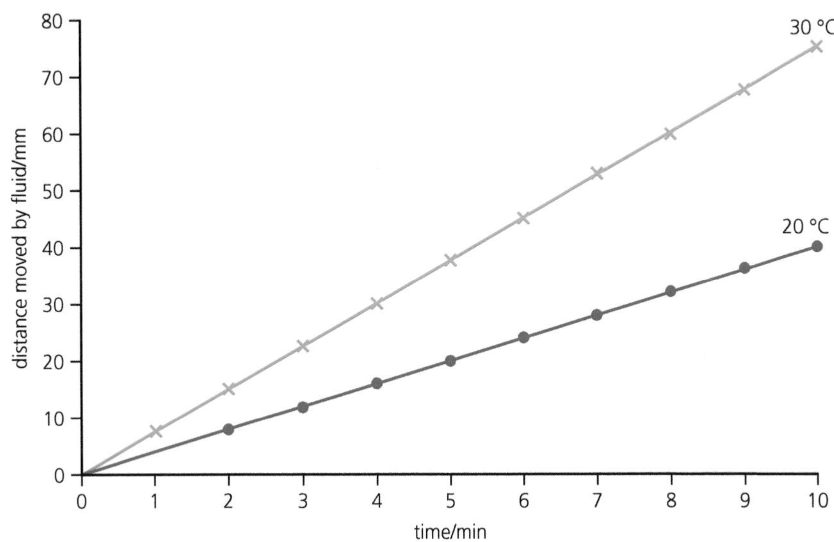

i) Temperature or pressure fluctuations often cause problems when using respirometers. Explain what the graph shows about the effectiveness of variable control in this investigation. *(2)*

ii) The diameter of the capillary tube was 3 mm. Calculate the rate of oxygen uptake of the germinating seeds at each temperature during the course of this investigation. *(4)*

iii) Calculate the percentage increase in respiration rate caused by raising the temperature from 20 °C to 30 °C. *(2)*

6 Explain how the structure and properties of cell membranes are essential for the production of ATP in aerobic respiration. *(7)*

Stretch and challenge

7 Most organisms use other respiratory substrates in addition to glucose. Both lipids and proteins can be channelled into respiration by various biochemical pathways.

a) What is meant by a 'biochemical pathway'?

b) Many plants and animals use lipids as energy storage molecules. Suggest what the advantages of this would be to a groundnut seed and a mammal such as a fox.

Tip

Question 6 is a typical synoptic question that you will find in A level papers. It asks you to show knowledge and understanding of membrane structure and the biochemistry of aerobic respiration, and then to apply your knowledge to explain some selected detail of the process (AO2). As this is a slightly extended prose question, take care to keep to the rubric without straying into lots of details of respiration or mitochondria, which are not relevant.

c) If germinating seeds with a very high lipid content are used in a respirometer, the results show that a much greater volume of oxygen is required for the respiration of the fatty acids they contain compared to respiration of glucose.

The summary equation for the respiration of fatty acids is shown below.

$$\underset{\text{fatty acid}}{C_{18}H_{36}O_2} + 26O_2 \rightarrow 18CO_2 + 18H_2O$$

i) Explain how this equation shows that more oxygen will be needed to oxidise one molecule of fatty acid compared to one molecule of glucose.

ii) How do the proportions of hydrogen and oxygen in glucose and fatty acid help to explain this difference?

8 Training programmes for endurance sports such as long-distance running and cycling involve careful monitoring of lactate levels in the blood by taking blood samples.

The graph below shows the blood lactate levels (Blood [La]) of an athlete at increasing workloads, measured in Watts (W) achieved by increasing speeds on a treadmill. LT indicates the lactate threshold.

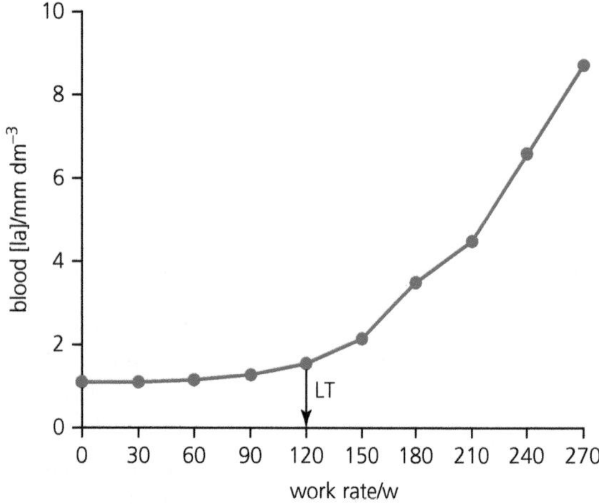

a) Explain the changes in lactate levels at work rates below the threshold and the changes at work rates above the threshold.

b) Suggest how the results of such monitoring might be used to design individual training programmes for endurance athletes.

2 Photosynthesis

Prior knowledge

In this chapter you will need to recall that:

- ➜ green plants are able to manufacture sugars using light energy, carbon dioxide and water
- ➜ oxygen is a waste product of photosynthesis
- ➜ several stages of photosynthesis take place in cell organelles called chloroplasts
- ➜ light is absorbed by plant cells using pigments such as chlorophyll
- ➜ colours of light are determined by different wavelengths
- ➜ photosynthesis using sunlight energy is the starting point for the large majority of food chains
- ➜ plants need to respire to produce ATP for cellular processes; photosynthesis produces the respiratory substrates required.

Test yourself on prior knowledge

1 State where chlorophyll molecules are attached in chloroplasts.

2 Explain why a solution of ink will appear blue when white light is shone through it.

3 If photosynthesis produces sugars that are respired to produce ATP, suggest why many plants accumulate large food stores.

4 Explain why many plant cells have food stores consisting of starch rather than sugars.

5 Name the organisms that cause the Earth's oceans to be a major contributor to atmospheric oxygen.

Photosynthetic pigments

Key term

Accessory pigments Light-absorbing molecules that pass on the energy they absorb to chlorophyll molecules at the start of photosynthesis.

The purpose of photosynthetic pigments is to absorb light energy and convert it into chemical energy. Higher plants have two types of pigments – chlorophylls and carotenoids. The most important pigments are the chlorophylls, which play a direct role in the first stages of photosynthesis. Carotenoids absorb light energy, which is passed on to chlorophyll, so they are called **accessory pigments**. Other plants such as marine algae (seaweeds) have other accessory pigments. Some common photosynthetic pigments are listed in Table 2.1.

Table 2.1 Common photosynthetic pigments

Chlorophylls	Carotenes	Others
Chlorophyll *a* (green)	β-carotene (orange)	Phycoerithrin (red)
Chlorophyll *b* (green)	Xanthophyll (yellow)	Fucoxanthin (brown)

the structure of chlorophyll *a*

chlorophyll *b* has an aldehyde group (—CHO) in place of this —CH_3 group

conjugated protein head containing magnesium (hydrophilic and associated with the proteins in the membranes of the grana)

hydrocarbon tail (hydrophobic and occurs folded, associated with the lipid of the membranes)

Figure 2.1 The structure of chlorophyll

The structure of chlorophyll

The chlorophyll molecule has two parts. The head of the molecule is a **hydrophilic** ring structure with a magnesium atom at its centre. Attached to this is a long **hydrophobic** hydrocarbon tail. This arrangement means that chlorophyll molecules are attached to the membranes of the chloroplast by their long tails whilst the heads lie flat on the membrane surface to absorb the maximum amount of light. The structure of chlorophyll can be seen in Figure 2.1. You will see how this is linked to the first stage of photosynthesis later in this chapter.

Absorption and action spectra

To investigate the roles of photosynthetic pigments it is important to know more about their absorption of light. You can do this by simply shining light of different wavelengths through a solution of the pigment and measuring how much is absorbed. The graph of the amount of light absorbed at each wavelength, as shown in Figure 2.2, is known as an **absorption spectrum**.

To provide evidence that these pigments do play an important part in the process of photosynthesis, you can produce an **action spectrum** by plotting a graph of the rate of photosynthesis of a green plant (measured as explained in Core practical 10) at different wavelengths. As you can see in Figure 2.2, the action spectrum for photosynthesis shows a very similar pattern to that of the absorption spectrum for the pigments. This provides good evidence to support the model of light being trapped by these pigments and used in photosynthesis.

Key terms

Hydrophilic Having an affinity for water and soluble in water.

Hydrophobic Substances that repel water. They are mainly insoluble in water but can be soluble in lipids.

absorption spectra

the amount of light absorbed by each pigment, measured at each wavele

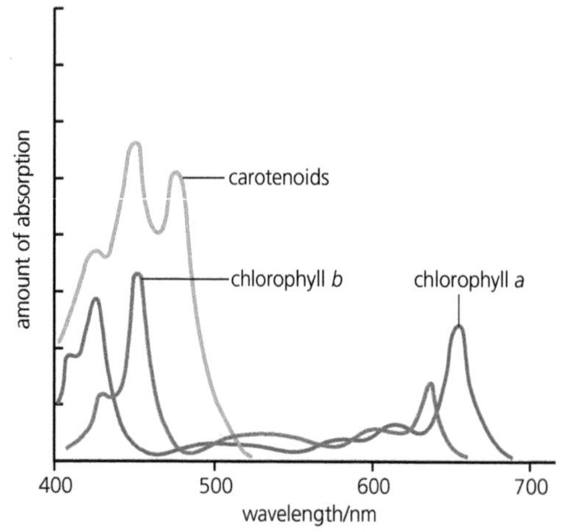

action spectrum

the amount of photosynthesis occurring at each wavelength

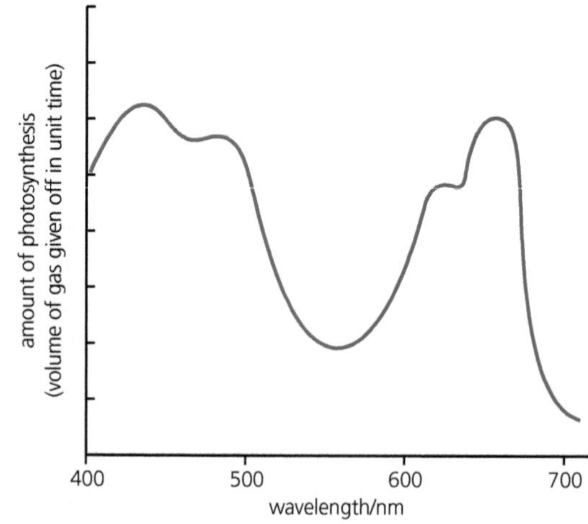

Figure 2.2 Absorption and action spectra

Why do plants need more than one photosynthetic pigment?

If you look carefully at the absorption spectra in Figure 2.2 you can see that the absorption of chlorophyll *a* and *b* is not quite the same. Both chlorophyll *a* and *b* absorb strongly in the blue and red ends of the spectrum but in the blue end of the spectrum chlorophyll *a* absorbs most strongly at about 430 nm, whilst chlorophyll *b* absorbs most strongly at 470 nm. In this way a combination of two or more pigments means that a greater range of wavelengths can be absorbed efficiently. You can see a very similar pattern at the red end of the spectrum. Almost all plants use chlorophylls as their primary pigments at the start of photosynthesis but the range of accessory pigments can vary considerably depending upon their habitat.

Key terms

Absorption spectrum A graph of the amount of light absorbed at different wavelengths by a pigment.

Action spectrum A graph of the rate of photosynthesis of a plant at different wavelengths of light.

Test yourself

1 Suggest why the hydrocarbon chains of a chlorophyll molecule will attach themselves easily to chloroplast membranes.

2 Describe how the 'heads' of chlorophyll molecules are adapted to capture the maximum number of photons.

3 State which colour of light in the visible spectrum has the longest wavelength.

4 Describe what happens to the energy trapped by accessory pigments.

Core practical 11

Investigate the presence of different chloroplast pigments using chromatography

Background information

Chromatography is a very common technique used in biochemistry to separate and identify small quantities of different compounds. There are many different variations but all work on similar principles.

The process involves a stationary phase – the chromatogram, which can be absorptive paper in paper chromatography, a powdered solid in column chromatography or a thin film of dried solid on a glass or plastic sheet in thin-layer chromatography. In each case the mixture to be separated is loaded on to the stationary phase as a small spot and allowed to dry. The edge of the paper or other medium is then dipped in a solvent, which will be drawn up through the spot by capillarity. The different compounds in the mixture will have different solubilities in the solvent and will interact with the stationary phase in different ways, so will move up the paper at different rates. As part of this interaction it is usually necessary to ensure the chromatogram is surrounded by solvent vapour by sealing it in a container.

If the compounds are coloured then it is easy to find the bands or spots formed on the complete chromatogram; if they are not then some other treatment will be necessary such as staining or viewing them under UV light.

Carrying out the investigation

Aim: To identify the pigments present in an extract of plant leaves.

Risk assessment: The solvents used in this investigation, propanone and petroleum ether, produce highly flammable heavy vapours. These can spread unseen along benches so there should be no other naked flames in the laboratory during their use. Both solvents produce potentially harmful vapours, which should not be inhaled. Filling of chromatography vessels should be carried out in a fume cupboard. In general, you should wear eye protection. Gloves may be advisable if you have sensitive skin.

1 To prepare the extract for chromatography you will need to collect leaves of a plant that have a dark green colour. Soft leaves such as spinach, which are easier to crush, are ideal.
2 Cut the leaves into small pieces with scissors and place them into a small mortar with a sprinkling of fine sand.
3 Use a pestle to grind the leaves into a fine paste. Add a small amount of propanone to the paste as you grind. Keep the volume of propanone as low as you can to make sure your extract is as concentrated as possible.
4 Allow the mixture to settle and then draw off the dark green chlorophyll solution with a small pipette. Place this in a small, sealed tube and allow any solids to settle. The mixture can be filtered but you will need to add even more propanone as some is absorbed by the filter paper.
5 To prepare the chromatogram for loading, cut a piece of chromatography paper to fit the container you are to use. Use a pencil to draw a line about 1 cm from the bottom of your paper as shown in Figure 2.3, on the next page.
6 Assemble your chromatogram inside the container and make a mark on the outside to indicate the level of solvent you will need to add. This should be sufficient to cover the end of the paper but it must not touch the spot of extract you are about to load onto it.

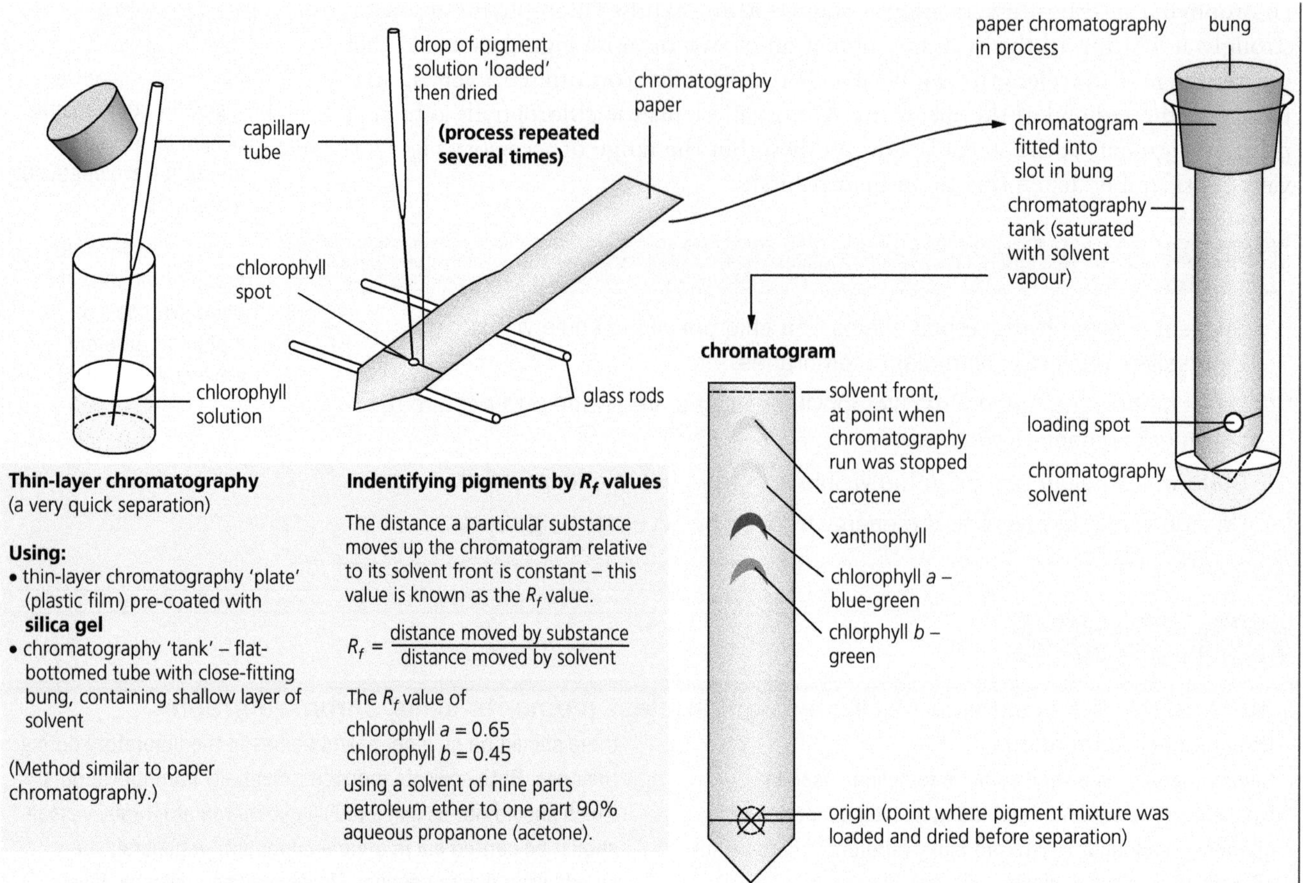

Figure 2.3 Chromatographic separation of chlorophyll pigments

7 To load the extract on the paper (the stationary phase), you will need to support the paper so that the origin does not touch any surface. Then use a very fine paintbrush or a short length of fine capillary tubing. Dip this into the extract and just touch a tiny spot onto the origin line you have marked on the paper. At this stage it is crucial to use a little patience. Your spot must be as small and as concentrated as you can make it, so you need to make several additions to it, but you must allow each addition to dry before adding the next. The final spot should not be more than about 2 mm in diameter.
The solvent you will use is made up of nine parts petroleum ether to one part propanone. (CAUTION! see Risk assessment.)

8 Add the solvent to your container until it reaches the mark you made. Attach your loaded paper to the bung and lower it carefully into the solvent so that the bung fits tightly and the bottom of the paper touches the solvent. The paper must hang freely and not touch the sides of the container. Keep the tube in a shaded position, or a dark cupboard, as several of the coloured pigments fade quickly.

9 Allow the solvent to rise almost to the top of the paper but do not allow it to continue any longer. When removing the completed chromatogram you will need to act quickly. Make sure you have a sharp pencil; quickly mark the final level of the solvent and draw around any coloured spots you see, noting their colour. Avoid inhaling the solvent vapour.

10 To help identify the pigments, measure the distances shown in Figure 2.3 and calculate their R_f values.

Questions

1 Why is it necessary to use powerful organic solvents in this investigation?
2 Why do you need to act quickly to mark the solvent front and the pigments?
3 How can the pigments be identified from their R_f values?
4 Why is it important not to allow the solvent to touch the origin spot at the start of the investigation?
5 Why must the loading spot be as small as possible?
6 Why is it important to remove the chromatogram before the solvent front reaches the top of the paper?

Photosynthesis

Green plants use the energy from sunlight to produce sugars from the inorganic raw materials, carbon dioxide and water, by a process called photosynthesis. The waste product is oxygen. Photosynthesis occurs in plant cells containing chloroplasts – typically, these are found mainly in the leaves of green plants. Here, light energy is trapped by the green pigment chlorophyll, and becomes the chemical energy in molecules such as glucose and ATP. (Note that we say light energy is transferred to organic compounds in photosynthesis, rather than talking of the 'conversion' of energy, although the latter term was used widely at one time.) This mode of nutrition is known as **autotrophic** as large organic molecules are built up from simple inorganic ingredients such as carbon dioxide and water.

> **Key term**
>
> **Autotrophic nutrition** The synthesis of larger organic molecules from simpler inorganic compounds as carried out by plants during photosynthesis.

Sugar formed in photosynthesis may temporarily be stored as starch, but sooner or later most is used in metabolism. For example, plants manufacture other carbohydrates, together with lipids, proteins, growth factors, and all the other metabolites they require. For this they need, in addition, certain mineral ions, which are absorbed from the soil solution. Figure 2.4 is a summary of photosynthesis and its place in plant metabolism.

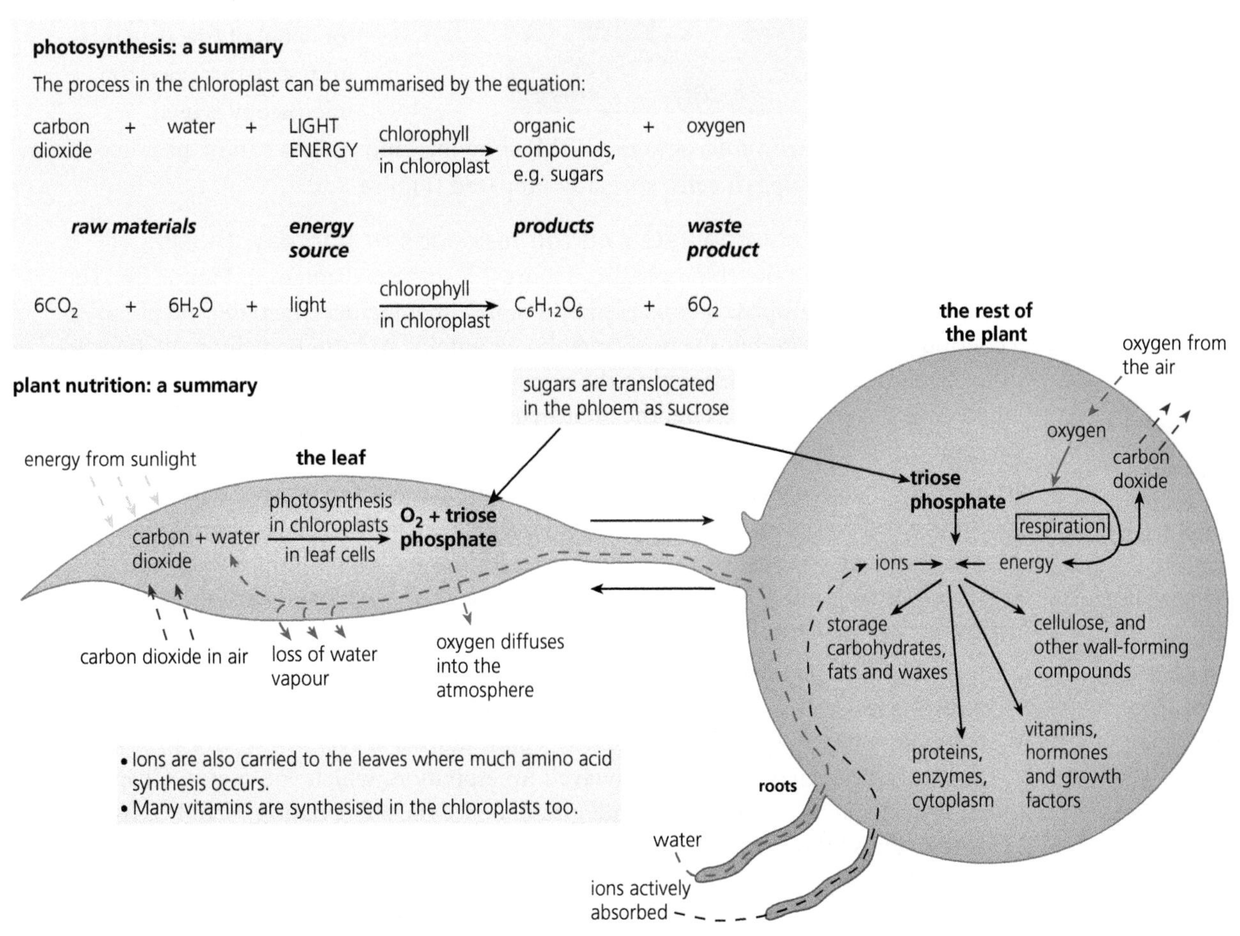

Figure 2.4 Photosynthesis and its place in plant nutrition

Chloroplasts – site of photosynthesis

Just as the mitochondria are the site of many of the reactions of respiration, as we explained in Chapter 1, so the chloroplasts are the organelles where the reactions of photosynthesis occur. Remember, chloroplasts are members of a group of organelles called plastids. (Amyloplasts, where starch is stored, are also plastids.)

the process

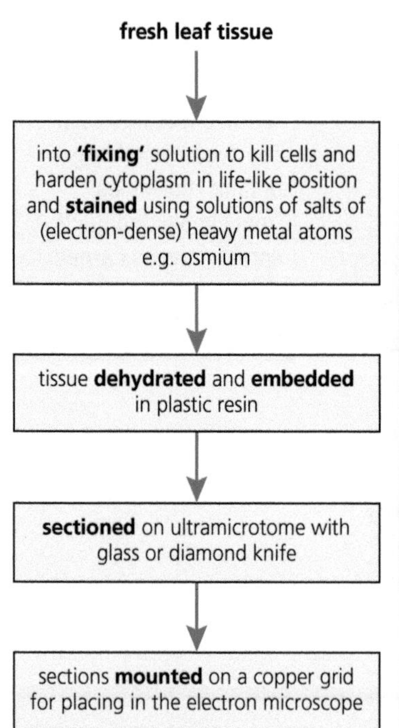

TEM of thin section of chloroplasts (×22 000)

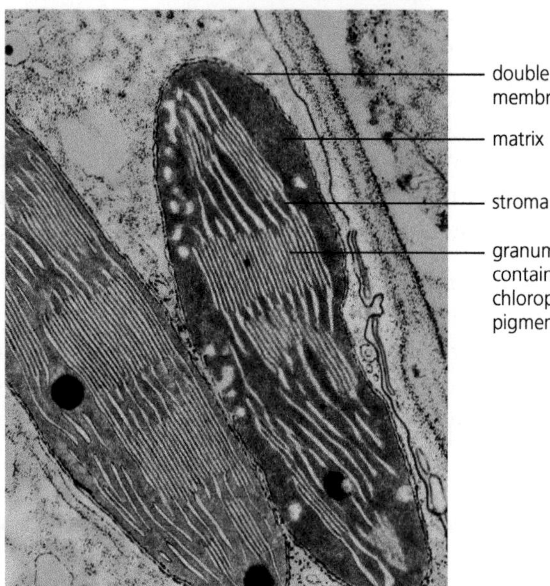

Figure 2.5 The production of a TEM of chloroplasts

The chloroplast is one of the larger organelles found in plant cells, yet typically measures only 4–10 μm long and 2–3 μm wide. (A micrometre or micron, μm, is one-millionth of a millimetre.) Consequently, while chloroplasts can be seen in outline by light microscopy, for detail of fine structure (ultrastructure) electron microscopy is used.

A transmission electron micrograph (TEM) showing chloroplasts can be produced from thin sections of mesophyll cells, specially prepared (Figure 2.5).

Ultrastructure of chloroplasts and the reactions of photosynthesis

Examine the TEM of the chloroplasts in Figure 2.5, and the diagram in Figure 2.6. You will see that the chloroplast is contained by a double membrane. The outer membrane is a continuous boundary, but the inner membrane 'infolds' to form branching membranes called lamellae or thylakoids within the organelle. Some of the thylakoids are arranged in circular piles called grana. Here, the photosynthetic pigment, chlorophyll, is held. Between the grana, the lamellae are loosely arranged in an aqueous matrix, forming the stroma.

It turns out that photosynthesis consists of a complex set of reactions, which take place in illuminated chloroplasts (unsurprisingly). Biochemical studies by several teams of scientists have established that the many reactions by which light energy brings about the production of sugars, using the raw materials water and carbon dioxide, fall naturally into two interconnected stages (Figure 2.7).

- In the **light-dependent reactions**, light energy is used directly to split water (a process known as 'photolysis', for obvious reasons). Hydrogen is then removed and retained by the photosynthetic-specific hydrogen acceptor, known as $NADP^+$. ($NADP^+$ is very similar to the coenzyme NAD^+ involved in respiration, which you met in Chapter 1, but it carries an additional phosphate group, hence the abbreviation NADP). At the same time, ATP is generated from ADP and phosphate, also using energy from light. This is known as photophosphorylation. Oxygen is given off as a waste product of the light-dependent reactions. This stage occurs in the grana of the chloroplasts.
- In the **light-independent reactions**, sugars are built up using carbon dioxide. This stage occurs in the stroma of the chloroplast. Of course, the light-independent reactions require a continuous supply of the products of the light-dependent reactions (ATP and reduced hydrogen acceptor $NADPH + H^+$), but do not directly involve light energy (hence the name). Names can be misleading, however, because sugar production is an integral part of photosynthesis, and photosynthesis is a process that is powered by transfer of light energy.

Key terms

Grana Circular piles of membrane-bound vesicles called thylakoids found in a chloroplast.

Stroma The aqueous matrix found inside chloroplasts.

Thylakoids Infolds of the inner membranes of chloroplasts that carry photosynthetic pigments.

Photophosphorylation The production of ATP from ADP using energy from light during photosynthesis.

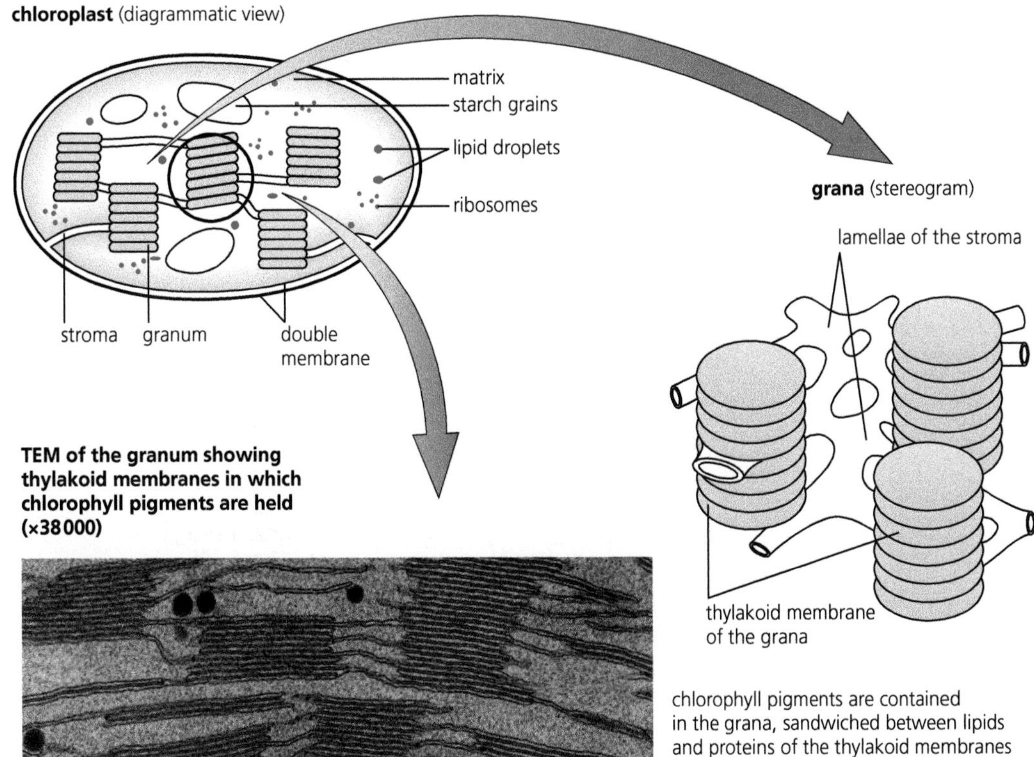

Figure 2.6 The ultrastructure of a chloroplast

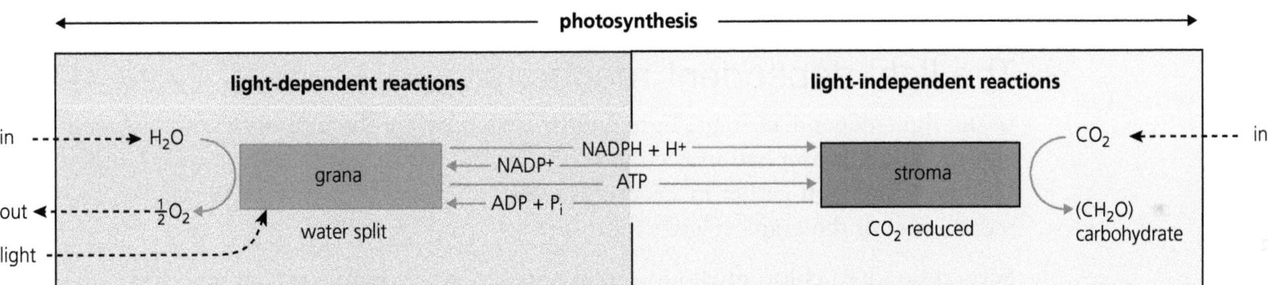

Figure 2.7 The two reactions of photosynthesis

In the next section we shall consider each stage in turn, in order to understand more about how these complex changes are brought about.

Test yourself

5 Explain what is meant by *transmission electron microscopy.*

6 Give the general term used to describe compounds such as the hydrogen acceptor $NADP^+$.

7 State in what form carbon dioxide enters the cytoplasm and chloroplasts of a plant cell.

8 Name the stage of photosynthesis during which oxygen is given off.

9 Describe the difference between the main hydrogen acceptors in photosynthesis and respiration.

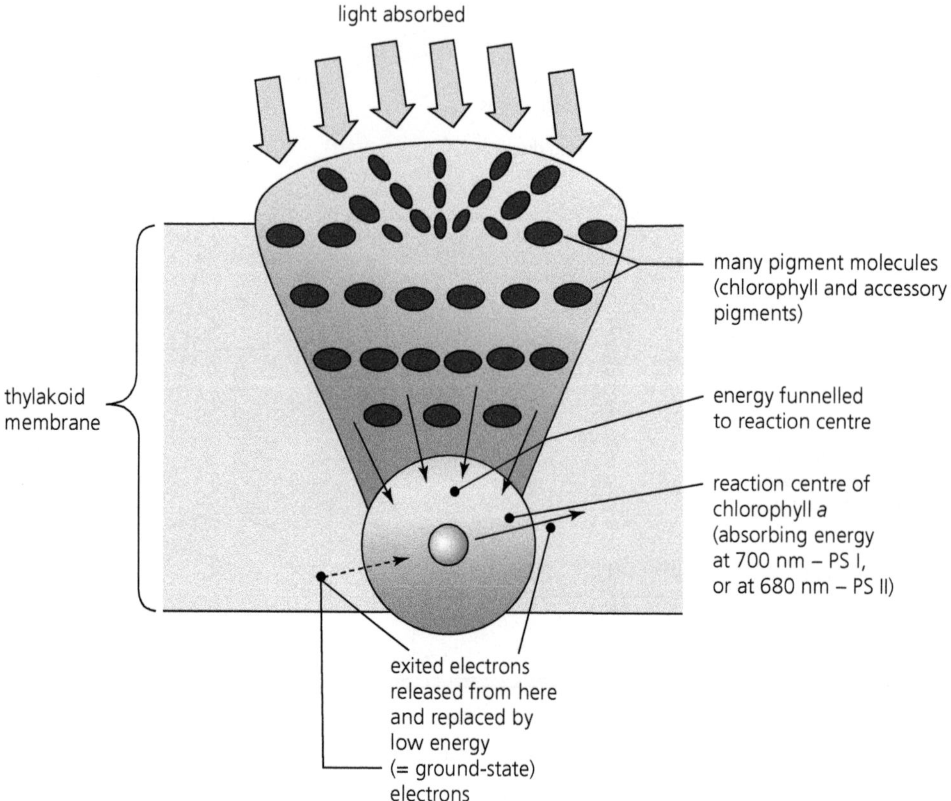

Figure 2.8 The structure of photosystems in the chloroplast membranes

The light-dependent reactions

In the light-dependent stage, light energy is trapped by the photosynthetic pigment, chlorophyll. Chlorophyll molecules do not occur haphazardly in the grana. Rather, they are grouped together in structures called photosystems, held in the thylakoid membranes of the grana (Figure 2.8).

Several hundred chlorophyll molecules plus accessory pigments (carotene and xanthophylls) are arranged in each photosystem. All these pigment molecules harvest light energy, and they funnel the energy to a single chlorophyll molecule in the photosystem, known as the reaction centre. The different pigments around the reaction centres absorb light energy of slightly different wavelengths.

There are two types of photosystem present in the thylakoid membranes of the grana, identified by the wavelength of light that the chlorophyll of the reaction centre absorbs.

- **Photosystem I** has a reaction centre activated by light of wavelength 700 nm. This reaction centre is also referred to as P700.
- **Photosystem II** has a reaction centre activated by light of wavelength 680 nm. This reaction centre is also referred to as P680.

Key term

Photosystems I and II
Two chains of electron carriers responsible for the capture of light energy in the first stages of photosynthesis.

Photosystems I and II have differing roles, as you shall see shortly. However, they occur grouped together in the thylakoid membranes of the grana, along with certain proteins that function quite specifically in one of the following roles:

1 Enzymes catalysing the splitting of water into hydrogen ions, electrons and oxygen atoms.

2 Enzymes catalysing the formation of ATP from ADP and phosphate (P_i).

3 Enzymes catalysing the conversion of oxidised H-carrier ($NADP^+$) to reduced carrier ($NADPH + H^+$).

4 Electron-carrier molecules (these are large proteins).

When light energy reaches a reaction centre, 'ground-state' electrons in the key chlorophyll molecule are raised to an 'excited' state by the light energy received. As a result, high-energy electrons are released from this chlorophyll molecule, and these electrons bring about the biochemical changes of the light-dependent reactions (Figure 2.9). The spaces vacated by the high-energy (excited) electrons are continuously refilled by non-excited or 'ground-state' electrons.

We will examine this sequence of reactions in the two photosystems next.

- Firstly, the excited electrons from photosystem II are picked up by, and passed along, a chain of electron-carriers. As these excited electrons pass, some of the energy causes the pumping of hydrogen ions (protons) from the chloroplast's matrix into the thylakoid spaces. Here they accumulate – incidentally, causing the pH to drop. The result is a proton gradient that is created across the thylakoid membrane, and which sustains the synthesis of ATP. This is an example of chemiosmosis, which is described in detail in Chapter 1.

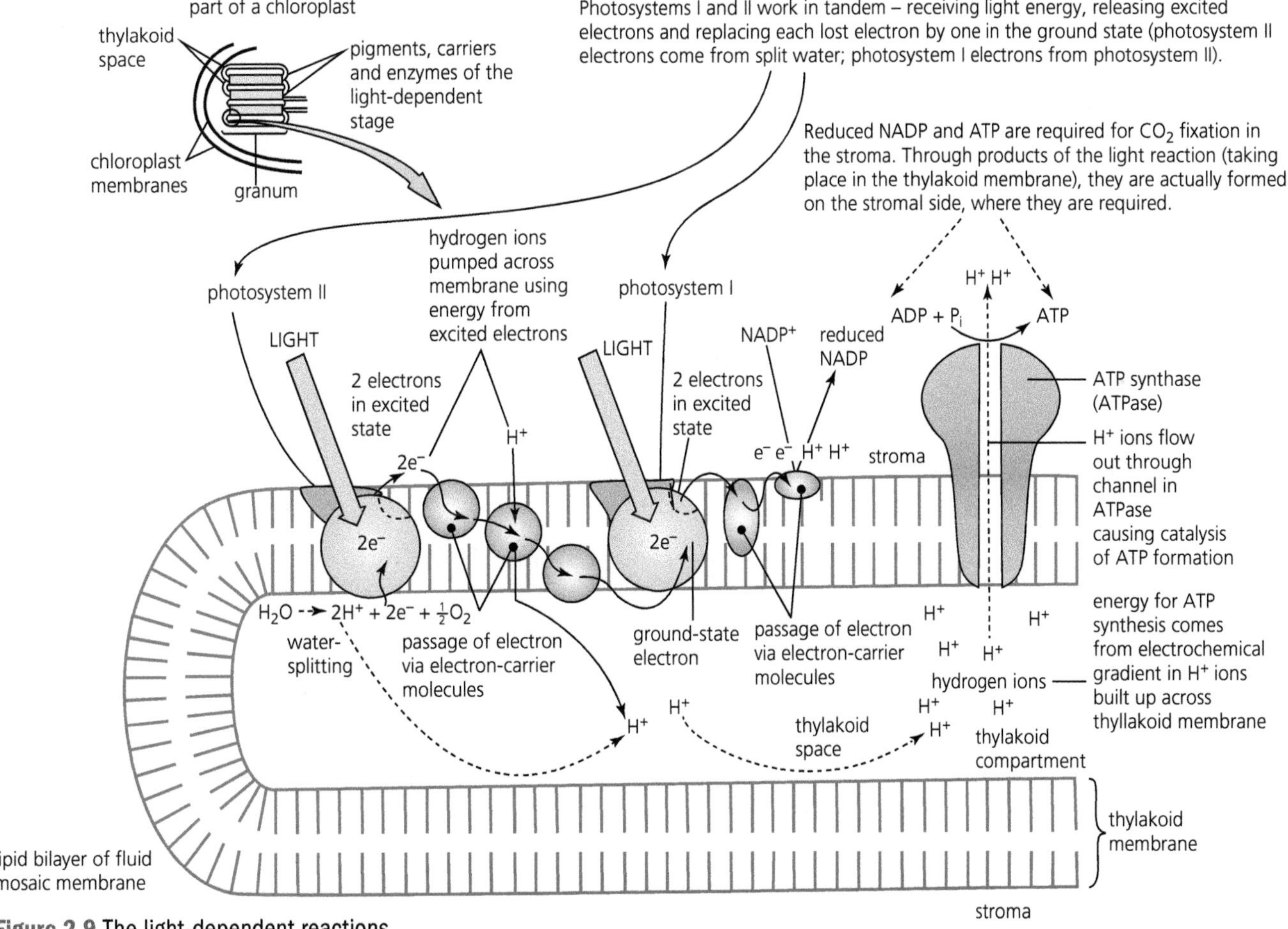

Figure 2.9 The light-dependent reactions

As a result of these energy transfers, the excitation level of the electrons falls back to 'ground state' and they come to fill the vacancies in the reaction centre of photosystem I. Thus, electrons have been transferred from photosystem II to photosystem I.

Meanwhile the 'holes' in the reaction centre of photosystem II are filled by electrons (in their 'ground state') from water molecules. In fact, the positively charged 'vacancies' in photosystem II are powerful enough to cause the splitting of water (photolysis) in the presence of a specific enzyme. The reaction this enzyme catalyses then triggers the release of hydrogen ions and oxygen atoms, as well as 'ground-state' electrons.

The oxygen atoms combine to form molecular oxygen, the waste product of photosynthesis. The hydrogen ions are used in the reduction of $NADP^+$ (see below).

In the grana of the chloroplasts, the synthesis of ATP is coupled to electron transport via the movement of protons by chemiosmosis. Here, the hydrogen ions trapped within the thylakoid space flow out via ATP synthetase enzymes, down their electrochemical gradient. At the same time, ATP is synthesised from ADP and P_i. This is called photophosphorylation.

Key term

Non-cyclic photophosphorylation The production of ATP from ADP using light energy in a linear series of reactions.

You have seen that the 'excited' electrons that eventually provide the energy for ATP synthesis, originate from water. They fill the vacancies in the reaction centre of photosystem II and are subsequently moved on to the reaction centre in photosystem I. Finally, they are used to reduce $NADP^+$. The photophosphorylation reaction in which they are involved is described as **non-cyclic photophosphorylation**, because the pathway of electrons is linear.

- Secondly, the excited electrons from photosystem I are picked up by a different electron acceptor. Two at a time, they are passed to $NADP^+$, which – with the addition of hydrogen ions from photolysis – is reduced to form $NADPH + H^+$.

By this sequence of reactions, repeated again and again at very great speed throughout every second of daylight, the products of the light-dependent reactions (ATP and $NADPH + H^+$) are formed.

ATP and reduced NADP do not normally accumulate, however, as they are immediately used in the fixation of carbon dioxide in the surrounding stroma (in the light-independent reactions). Then the ADP and $NADP^+$ diffuse back into the grana for re-use in the light-dependent reactions.

Test yourself

10 Explain what happens to the energy from photons immediately after they are 'captured' by a chlorophyll molecule.

11 Describe what happens to the H^+ ions released from the splitting of water in photosystem II.

12 State the similarities between ATP production in oxidative phosphorylation in respiration and here in the light-dependent stage.

13 Suggest why the ATP produced in the light-dependent stage is not available for general metabolic purposes in the cytoplasm.

The light-independent reactions

In the light-independent reactions, carbon dioxide is converted to carbohydrate. These reactions occur in the stroma of the chloroplasts, surrounding the grana. Carbon dioxide readily diffuses into the chloroplast where it is built up into sugars in a cyclic process called the Calvin cycle.

In the Calvin cycle, carbon dioxide is combined with an acceptor molecule in the presence of a special enzyme, **ribulose bisphosphate carboxylase** (rubisco for short). The stroma is packed full of rubisco, which easily makes up the bulk of all the protein in a green plant. In fact, it is the most abundant enzyme present in the living world.

The acceptor molecule is a five-carbon sugar, ribulose bisphosphate (referred to as RuBP) and carbon dioxide is added in a process known as fixation (Figure 2.11). The product is not a six-carbon sugar, but rather two molecules of a three-carbon compound, **glycerate-3-phosphate (GP)**. GP is then reduced to form another three-carbon compound called **glyceraldehyde 3-phosphate (GALP)**. Some of the GALP is converted into the products of photosynthesis, such as glucose, or amino acids and fatty acids. The glucose may be immediately respired, or stored as starch until required. But the bulk of GALP is converted to more acceptor molecule, enabling fixation of carbon dioxide to continue.

Key terms

Calvin cycle A sequence of reactions in the light-independent stage of photosynthesis where carbon dioxide is taken in and sugars are produced.

Rubisco The abbreviation for the enzyme ribulose bisphosphate carboxylase, which is important in fixing atmospheric carbon dioxide during photosynthesis.

Ribulose bisphosphate Important 5-carbon sugar in the light-independent stage of photosynthesis, which is used to fix carbon dioxide from the atmosphere.

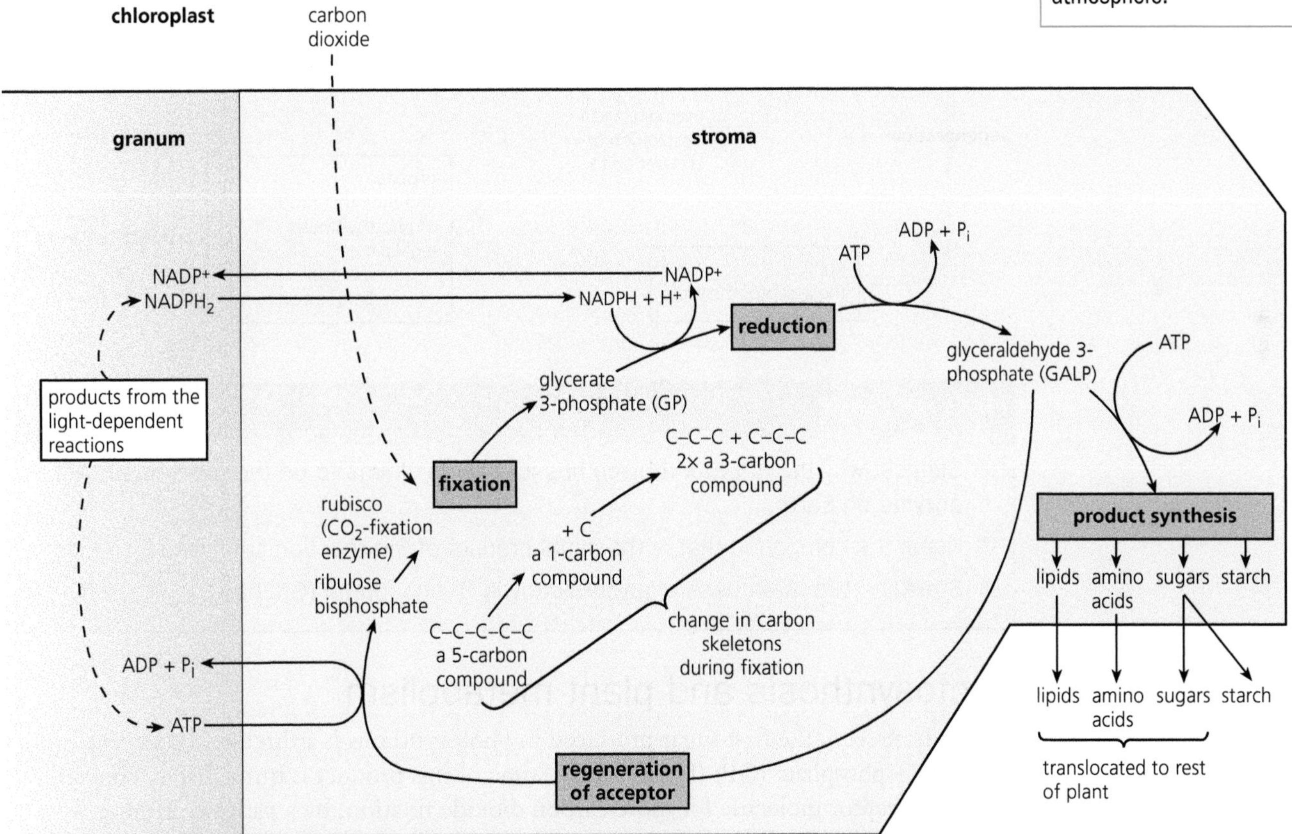

Figure 2.10 The light-independent reactions *in situ*

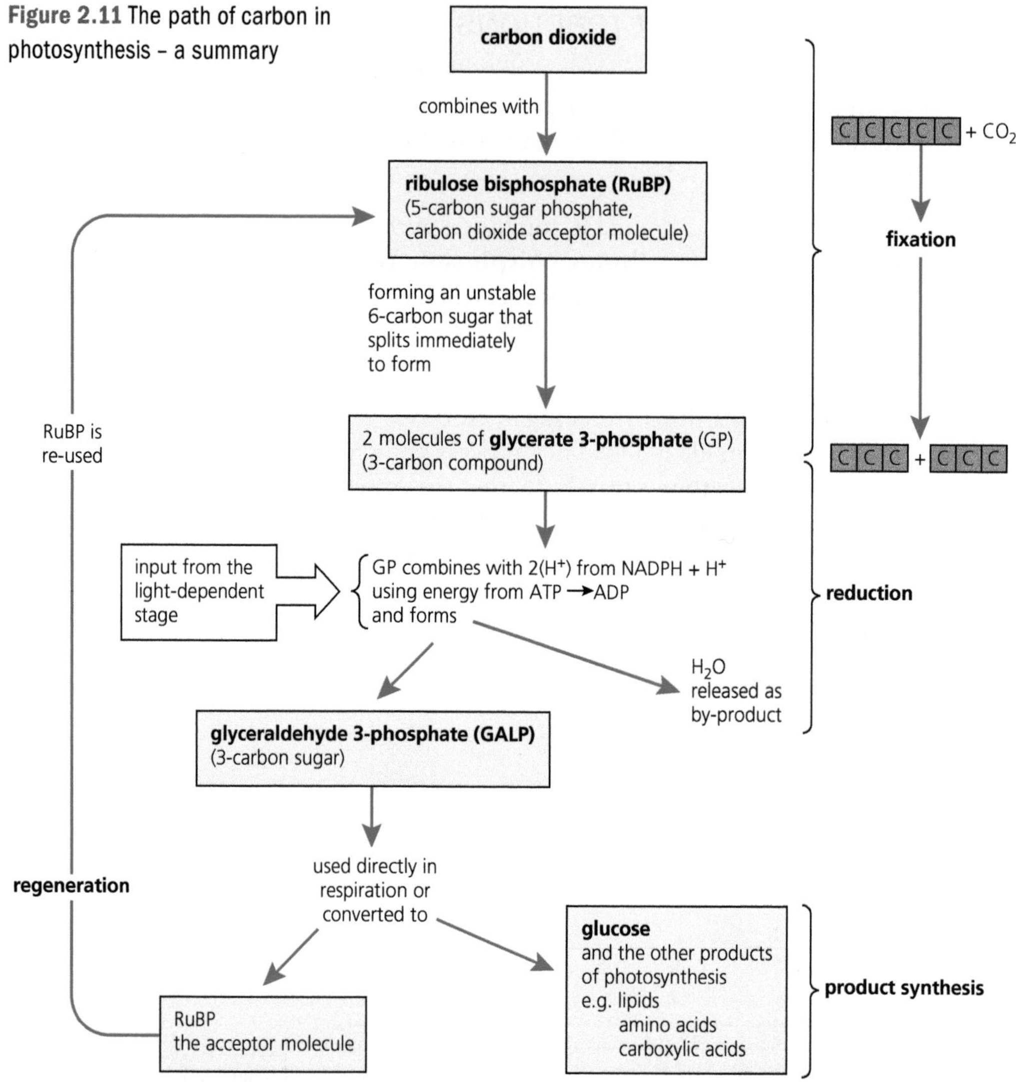

Figure 2.11 The path of carbon in photosynthesis – a summary

Test yourself

14 Suggest why the enzyme rubisco has justifiable claims to be the most important enzyme on Earth.

15 Name the compound that is the initial product of the reaction catalysed by rubisco.

16 State the two main uses of glyceraldehyde 3-phosphate (GALP).

Photosynthesis and plant metabolism

As you have seen, the first sugar produced in photosynthesis is a three-carbon compound, glycerate 3-phosphate (GP) (Figure 2.11). Some of this product is immediately converted into the acceptor molecule for more carbon dioxide fixation, by a pathway known as the Calvin cycle. The remainder is converted into the carbohydrate products of photosynthesis, mainly glucose and starch, or serves as intermediates that are the starting points for all the other metabolites the plant requires. By *intermediates*, we mean all the substances of a metabolic pathway from which the end product is assembled.

Glucose is also the substrate for respiration. By *substrate*, we mean a molecule that is the starting point for a biochemical pathway, and a substance that forms a complex with an enzyme (thereby getting the pathway 'up and running'). The intermediates of respiration are also starting points for the synthesis of other metabolites. In other words, the biochemical pathways of both photosynthesis and respiration interact to supply metabolism with the intermediates required. These include:

- specialist carbohydrates, such as sucrose for transport and cellulose for cell walls
- lipids, including those in membranes
- amino acids and proteins, including those in membranes and those that function as enzymes
- nucleic acids, growth factors, vitamins, hormones and pigments.

The fates of the products of photosynthesis are summarised in Figure 2.12.

Test yourself

17 Name the ion required for the manufacture of all amino acids from the Calvin cycle intermediates.

18 Name the ion required for the synthesis of nucleic acids.

19 Describe exactly where ions enter the plant and how they reach organs such as the leaf.

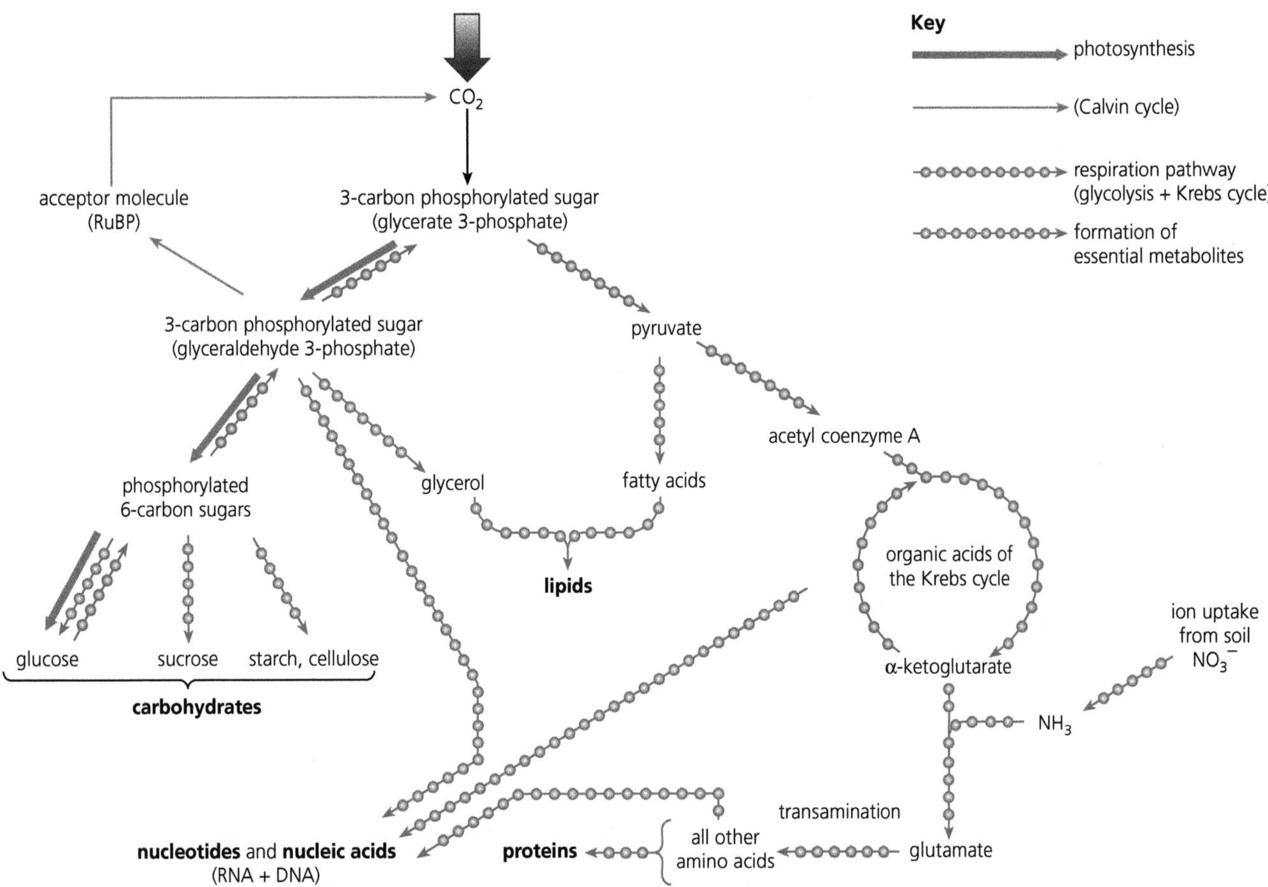

Figure 2.12 How the products of photosynthesis are linked with other plant cell metabolism

Core practical 10

Investigating the effect of different wavelengths of light on the rate of photosynthesis

Background information

You can follow the rate of photosynthesis in several ways but the most common way is to measure the rate of oxygen evolution by aquatic plants. Using aquatic plants means that it is much easier to collect the gas given off with simple apparatus. As you have seen in Chapter 1, this is not always straightforward as plants also respire and therefore use up some of the oxygen before you can collect it. Fortunately, most plants give off far more oxygen than they consume and provided you can assume that the rate of respiration is constant then the rate of oxygen evolution will be directly proportional to the rate of photosynthesis.

The most common plant to use for this investigation is the Canadian pondweed *Elodea canadensis*. This can easily be obtained from ponds and streams in the wild, as well as many aquarium suppliers as it is often used to oxygenate fish tanks. Unfortunately, very few aquatic plants give off oxygen in a controlled way that will allow you to collect it. *Elodea* can be rather unreliable in this respect and the tropical pondweed *Cabomba* sp. is often recommended as a more reliable alternative.

You can collect the gas given off in a given time and draw it into a capillary tube, where its volume can be calculated in a similar manner to the respirometer described in Core practical 9 (Chapter 1).

Lamps to produce different wavelengths of light are very expensive. Coloured filters are normally used to produce a range of wavelengths within the visible spectrum but it is important that the light source is of high intensity. It is useful to have filters of known wavelengths to produce more accurate data.

Carrying out the investigation

Aim: To investigate the effect of different wavelengths of light on the rate of photosynthesis.

Risk assessment: There are no significant risks associated with this investigation. Reasonable care and attention will be needed when manipulating light sources which may be hot. Care should be taken when using large volumes of water close to electricity.

Care should also be taken when using low energy light bulbs (compact fluorescent tubes) in bench lamps to light the pond weed. The bulbs contain small amounts of mercury and are coated in a chemical which fluoresces. If breakages occur, the fragments should be swept up carefully, and the light disposed of in the same way as fluorescent tubes. The room should be well ventilated.

1. Before setting up the apparatus as shown in Figure 2.13, it is important to check that you have a section of pondweed that will produce a good stream of oxygen bubbles consistently. Measure out a length of pondweed (*Cabomba* is recommended) that will fit comfortably into a large boiling tube.
2. Cut off about 1 cm from the end of the stem to ensure that there is a free passage for any gas given off.
3. Cover this with a dilute solution of sodium hydrogencarbonate to provide an excess supply of carbon dioxide (hydrogencarbonate ions) in the water.
4. Place a bench lamp 15 cm from the edge of the boiling tube and observe the end of the stem for a few minutes. It is useful to include a beaker of clean water between the lamp and the tube to act as a heat shield, helping to keep the temperature of the tube constant.
5. Set up the apparatus shown in Figure 2.13, making sure that the whole of the capillary tube is full of water by using the syringe. Move the pondweed until the stream of bubbles is directly under the end of the capillary tube as shown in the diagram and no bubbles are escaping. Check that the lamp is exactly 15 cm from the tube.
6. Start the stopclock and collect the gas given off for at least 5 minutes. You will need to adjust the time so there is sufficient volume of gas to measure, as all plant samples will vary.
7. At the end of your chosen period, use the syringe to draw up the bubble into the capillary tube so that its length can be measured on the scale. Repeat this at least three times with the lamp having no filter.
8. Cover the tube with the first coloured filter, without moving the lamp. Leave the apparatus for at least 5 minutes to settle down to a constant rate with the new filter.
9. Remove any gas bubble formed during this time by drawing it to the far side of the scale using the syringe, then take three more measurements.
10. Repeat the whole process with different coloured filters using the same piece of pondweed.

Questions

1. What effect will an increase in temperature have on the volume of oxygen given off?
2. Will changing the colour of the filter change other light variables?
3. Why do only some aquatic plants give off a predictable stream of bubbles?
4. Some protocols suggest counting bubbles as an alternative to measuring the volume of oxygen. Why would this be very inaccurate?

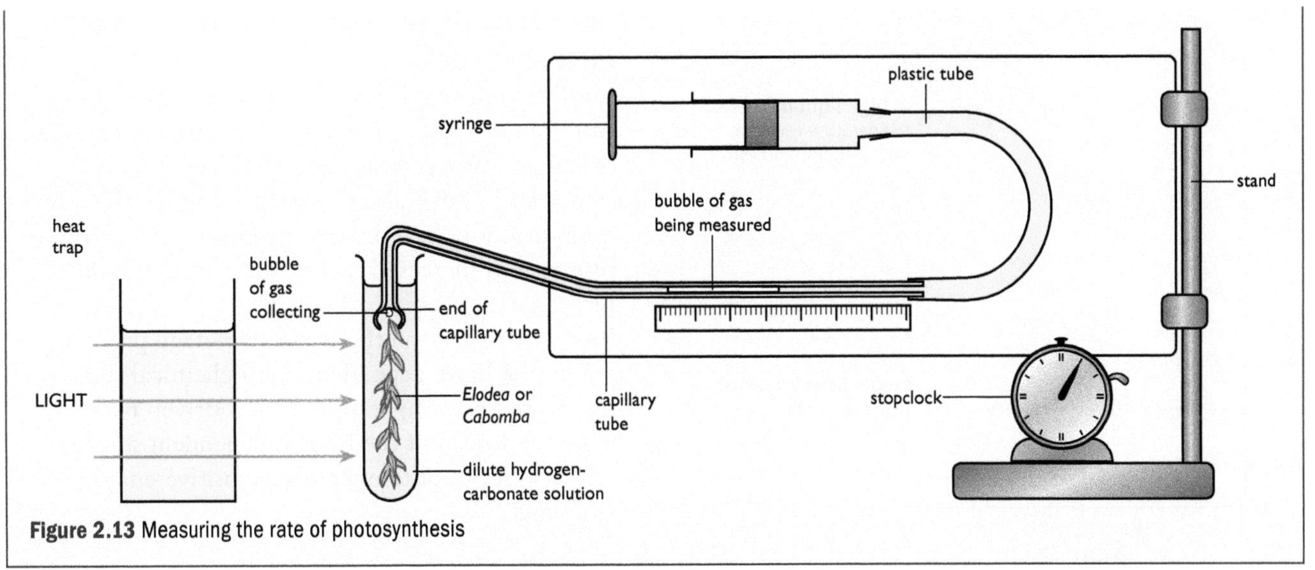

Figure 2.13 Measuring the rate of photosynthesis

Factors affecting the rate of photosynthesis

The rate of photosynthesis is affected by a number of different factors. The main factors are light, carbon dioxide concentration and temperature. In the plant's normal habitat these factors are constantly changing throughout each day, therefore it is not possible to name one overall factor that is most important. What we can say at any one time is that the factor that will have the greatest influence will always be the one that is least favourable. This is known as the **Law of limiting factors.** For example, in open grassland on a sunny day at noon there will be lots of light available and the temperature will be fine, so the availability of carbon dioxide might be the limiting factor for photosynthesis. However, later in the day, as the sun goes down, then it is most likely that light intensity will become the limiting factor. Some of these effects are summarised in Figure 2.14.

> **Key term**
>
> **Limiting factor** In any process controlled by more than one factor then the rate of the overall process will be determined by the availability of the least favourable factor.

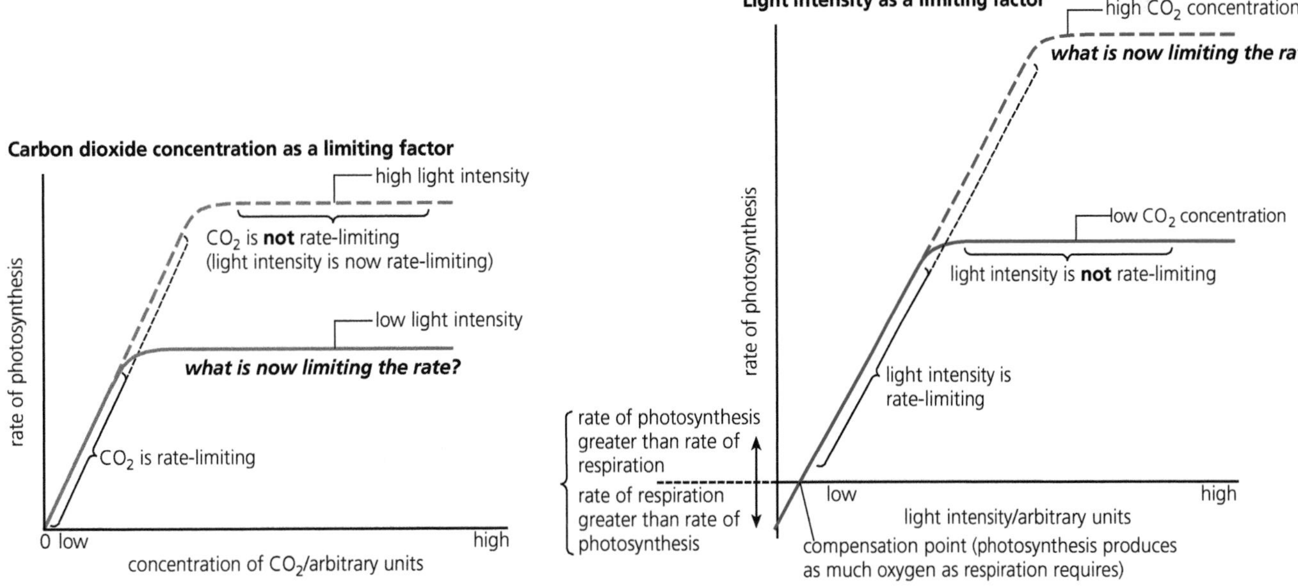

Figure 2.14 Carbon dioxide concentration and light intensity as limiting factors

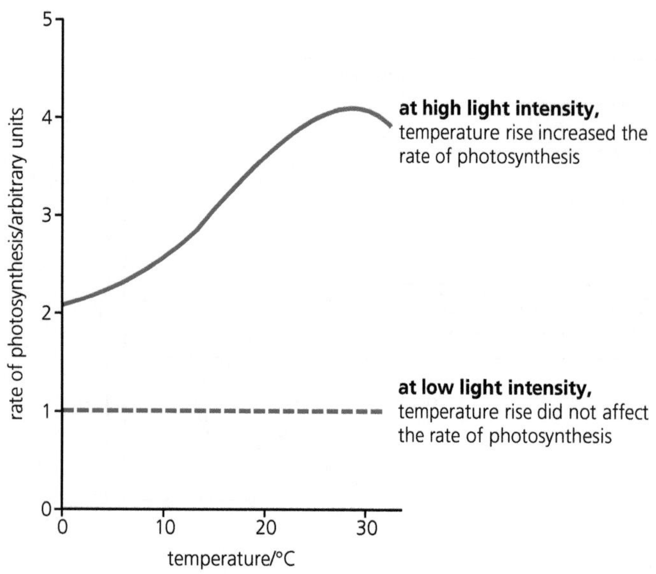

Figure 2.15 The effect of temperature on the rate of photosynthesis

The effect of temperature on the rate of photosynthesis

Unlike a normal series of enzyme-controlled reactions, photosynthesis shows a different response to temperature according to the level of light intensity. In general, temperature has no effect on the rate at low light intensities but shows a familiar response to increased temperature at high light intensities as shown in Figure 2.15. One reason for this is that photosynthesis is a two-stage process where the light-dependent photochemical stage is unaffected by temperature (as it is driven by energy from photons) but the light-independent stage is a typical series of temperature-sensitive enzyme reactions.

Exam practice questions

1 Carotenoid pigments that pass on their electrons to chlorophylls are known as:

A auxiliary pigments

B augmenting pigments

C accessory pigments

D additional pigments *(1)*

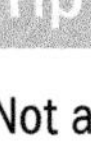

Tip

Not all multiple-choice questions will be simple recall like Question 1. Some, such as Question 2, will need understanding of a principle and some careful thought, so this would be AO2.

2 A mixture of substances, M, contains a compound, X. Chromatography was used to separate compound X from the mixture but two different solvents were needed. The mixture was first separated in solvent A and then the paper was turned through 90 degrees (as shown in the diagram) to be separated again with a different solvent, B. In each case the solvent front was allowed to reach the top of the paper before the paper was removed.

The R_f value of compound X in solvent A was 0.5 and its R_f value in solvent B was 0.75.

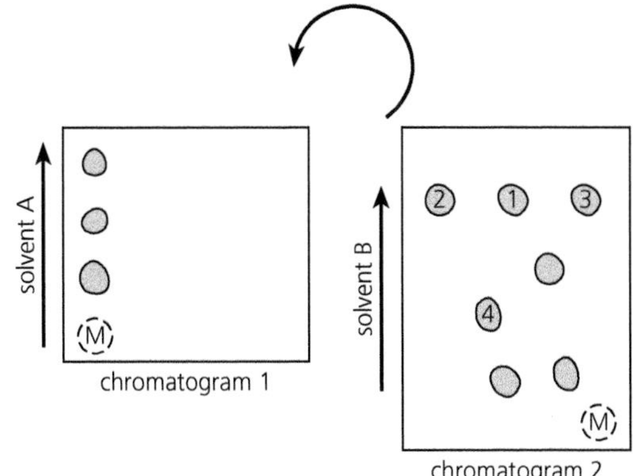

Compound X will be found in chromatogram 2 at the spot labelled:

A 1

B 2

C 3

D 4 *(1)*

Tip

Question 3 is a typical mixture of AO1 – asking you to show you understand basic specification material and AO2 – asking you to apply that understanding to data in a new context.

3 **a)** The graph shows the absorption spectrum of chlorophyll *a* and chlorophyll *b* taken from a marine alga (seaweed).

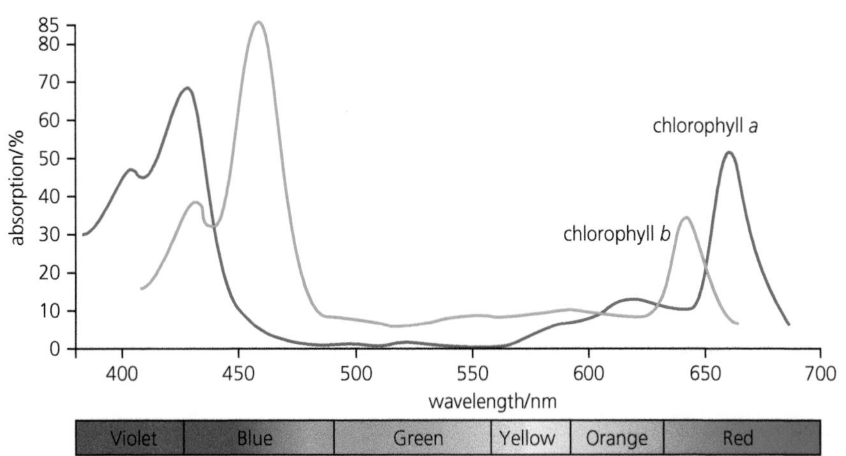

i) Analyse the data and explain how it illustrates the need for more than one pigment in photosynthesis. *(3)*

ii) Explain how chlorophyll molecules are able to trap light energy in the first stage of the light-dependent reaction of photosynthesis. *(3)*

b) Photosynthetic marine algae (seaweeds) are often found growing in oceans at depths of 10 m or more.

The table below shows how the different wavelengths of light are absorbed by clear seawater.

Wavelength of light (nm)	Depth of seawater at which 90% of the light is absorbed (m)
450	40
525	24
575	9
610	5
680	2

i) Describe the trend shown by the data in this table. *(2)*

ii) In addition to chlorophyll *a* and *b*, many marine algae found at this depth also contain the pigment fucoxanthyn. The absorption spectrum of fucoxanthyn shows a strong peak between 510–525 nm. Explain how the presence of this pigment will assist the growth of marine algae at depths below 15 m. *(3)*

4 The diagram represents a simplified scheme for the reactions of photosynthesis.

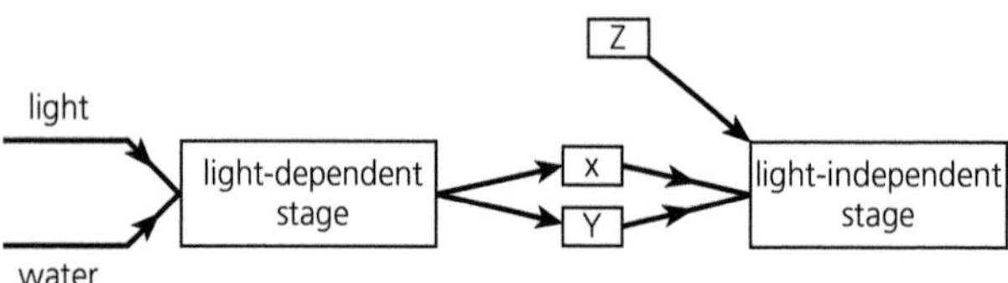

a) Name the intermediates labelled X, Y and Z. *(3)*

b) Explain how oxygen molecules are formed during the light-dependent stage. *(3)*

c) During the early stages of research into the biochemistry of photosynthesis, Otto Warburg used flashes of light to investigate the nature of the whole process. He used a culture of algae and a high concentration of carbon dioxide with a high-intensity lamp. The algal cultures were illuminated by spinning a disc in front of the lamp. The disc had segments cut out so that the speed of rotation would change the frequency of the flashes. The rate of photosynthesis during the period of illumination was measured at different frequency of flashes. Some of his results are summarised in the table on the next page.

Tip

Question 4 is also a mixture of AO1 and AO2 but is a more difficult question as you need to think carefully about the overall process of photosynthesis and only part (a) is simple recall. You are unlikely to have met the data in this form so the question requires you to understand the links between the two stages in some depth.

Frequency of flashes/min^{-1}	Relative rate of photosynthesis
continuous illumination	100
4	110
8000	200

i) Why did Warburg use a high concentration of carbon dioxide? *(1)*

ii) We now believe that the reactions of the light-dependent stage occur very rapidly and that there are two stages in the overall process of photosynthesis. Explain how the results of this experiment support this model of photosynthesis. *(3)*

5 It has been suggested that because of the similarities between mitochondria and chloroplasts, they might have had some common origin in evolutionary time. Is the evidence of their structure and function sufficient to support this conclusion? *(7)*

Stretch and challenge

6 All plants fix carbon dioxide using the enzyme rubisco and the compound RuBP, and are known as C3 plants because they form the 3-C molecule GP. However, some plants such as maize and sugar cane use an additional pathway, called the C4 pathway, to fix carbon dioxide. These plants are able to photosynthesise more rapidly and grow more quickly than C3 plants. You might think that C4 plants would out-compete C3 plants and be the dominant form of photosynthesis, but this is not the case.

a) What C4 compounds are used to fix carbon dioxide in C4 plants?

b) What is the function of bundle sheath cells in the C4 process?

c) Maize and sugar cane are C4 plants and dominate food crops in tropical countries. In temperate zones C3 plants, such as wheat, are the major crops. Why are C4 plants at an advantage in tropical regions but not in cooler climates?

Tip

Question 5 is first of all a synoptic question where you will need to use your knowledge of the functions of organelles from the first parts of the course in addition to respiration and photosynthesis. It also requires that you compose your answer carefully in continuous prose and make sure that you make comparisons (similarities and differences). Simply describing mitochondria and chloroplasts separately will gain you few marks. As an AO3 type of question you will be expected to come to some form of conclusion, for example do you think that the similarities indicate that they may have a common ancestor or are the differences too great?

Microbial techniques

Prior knowledge

In this chapter you will need to recall that:

- → microorganisms include bacteria, fungi and single-celled protoctists
- → many microorganisms are pathogenic
- → aseptic techniques allow microorganisms to be studied safely
- → bacteria have prokaryotic cells, whereas fungi and protoctists have eukaryotic cells.

Test yourself on prior knowledge

1 What is a pathogen?

2 Name two differences between a prokaryotic cell and a eukaryotic cell.

3 Give **two** features that a bacterial cell and a fungal cell have in common.

4 Figure 3.1 shows a student pouring a growth medium into a dish.

a) Name the type of dish into which she is pouring the growth medium.

b) Explain **one** aseptic technique you can see she is using in the drawing.

Types of media used to culture microorganisms

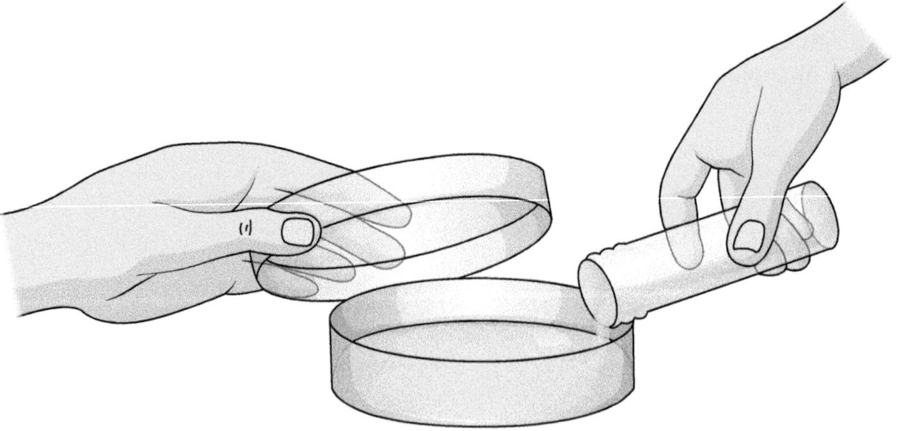

Figure 3.1 A student pouring a liquid growth medium into a dish

It is remarkably easy to grow microorganisms. All you need do is leave food uncovered, especially if the environment is warm. You will soon notice microorganisms growing on the food. Look at the tomato in Figure 3.2 on the next page. After a few days' exposure in a warm room it had been colonised by many microorganisms. The blue-grey patches are colonies of *Penicillium* and the tiny white spots are colonies of yeast. This room was not an unusual environment; bacterial and fungal spores are ever present in the air around us, including the air in your college or school laboratory.

A culture medium provides the essential nutrients that a population of microorganisms needs for its growth. All microorganisms share basic nutritional needs – a source of the elements carbon and nitrogen, for example. Consequently, all synthetic culture media are based on a buffered solution of inorganic ions. To such a solution, growth factors specific to the needs of a particular microorganism are added, for example an energy source or vitamins.

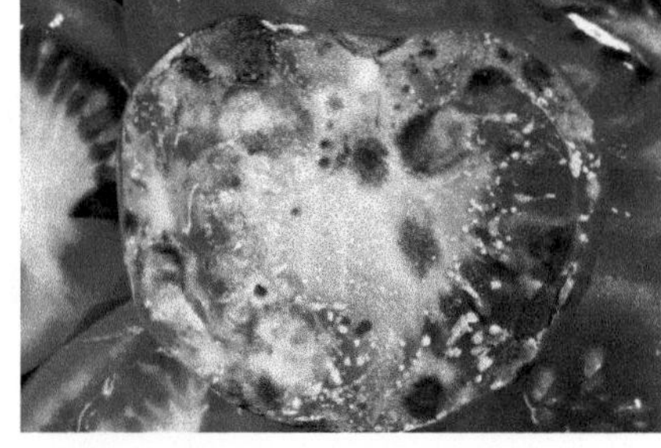

Figure 3.2 This cut tomato has been colonised by microorganisms. They grow by secreting hydrolytic enzymes onto the food and absorbing the products of digestion

Broth and solidified culture media

Culture media can be either liquid or solidified. Table 3.1 outlines the relative advantages of these two types of media.

Table 3.1 A comparison of liquid and solidified culture media

Type of culture medium	Usual method of cultivation in a laboratory	Advantage of this method
Liquid (broth culture)	In a partially filled conical flask, or similar flask that enables maintenance of a large surface area in contact with the air. The broth is usually agitated or stirred and, when culturing aerobic microorganisms, provided with sterile air.	Ensures that the culture does not die, so active cells are always available. Allows harvesting of any useful metabolic products from the microorganisms.
Solidified	The addition of a gelling agent, such as agar (an extract from seaweed), to a liquid medium makes it solidify.	Being solid, there is little risk of spillage, so these cultures are useful for storing microorganisms.
	In a Petri dish	Provides a large surface area for growth and for gas exchange with the air in the dish. Individual cells inoculated onto the surface of the agar develop into a visible colony, allowing isolation and identification of the microorganisms from a mixed inoculum.
	In a glass flat-sided bottle (often called a 'medical flat') or test tube.	Provides a greater depth of agar than a Petri dish, reducing the risk of dehydration and salt crystallisation.

Key terms

Agar An extract from seaweed that is used as a gelling agent to solidify culture media. Since few microorganisms can hydrolyse agar, the medium stays solid as the microorganisms grow.

Petri dish A shallow glass or plastic dish with a slightly larger dish that fits over it as a lid. Invented in 1887 by a German scientist, and used correctly, it enables scientists to transfer cultures to and from a solidified culture medium with minimal exposure to air.

Inoculum The small sample of a microbial culture that, using aseptic techniques, is transferred to a new medium.

In addition to the advantages shown in Table 3.1, a liquid culture allows us to carry out two types of culture:

- **Batch culture** – **inoculation** of microorganisms into a sterile container with a fixed volume of growth medium.
- **Continuous culture** – inoculation of microorganisms into a sterile container containing liquid growth medium. From time to time, some of the culture is removed and replaced by fresh sterile medium. Figure 3.3 shows a typical set up of equipment for continuous culture.

Only batch culture is possible using solidified culture media since it contains a limited mass of agar.

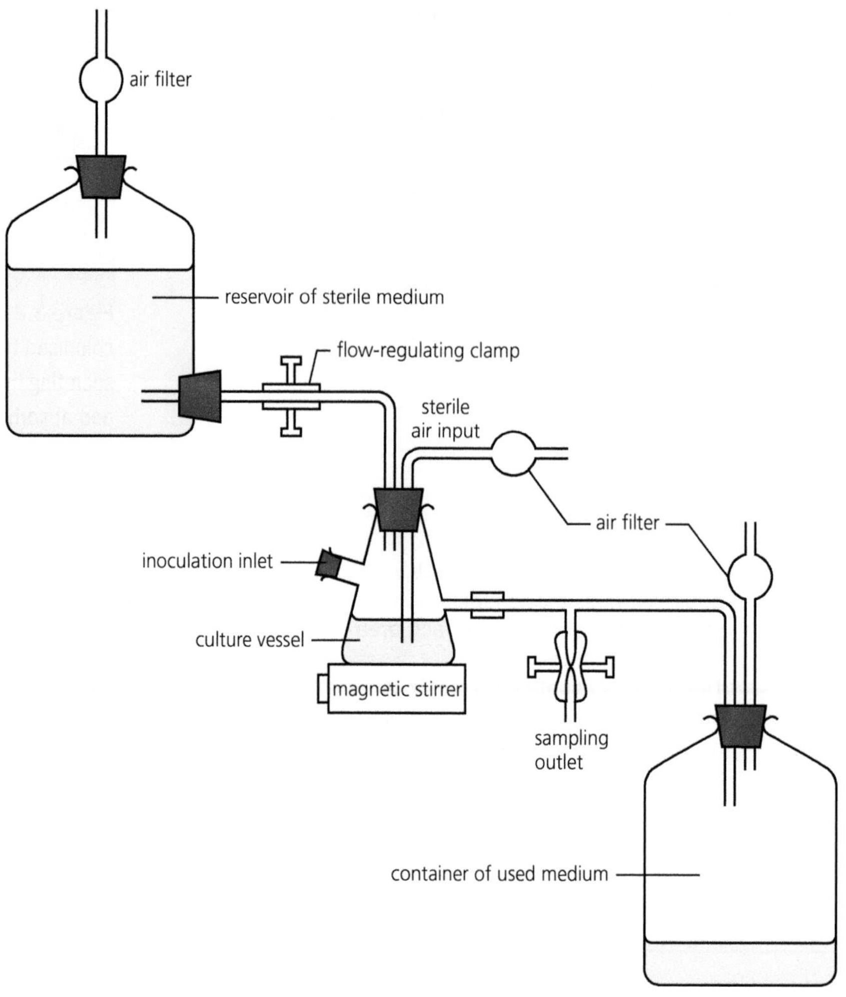

Figure 3.3 A typical arrangement of laboratory equipment needed to maintain a continuous culture of microorganisms

Broad spectrum and narrow spectrum culture media

Many culture media, such as nutrient agar, contain the basic nutrients that most microorganisms need for growth. Consequently, they can be used to grow a wide variety of microorganisms. They are known as general purpose or **broad spectrum media**. They often contain yeast extracts – a mixture of soluble amino acids, peptides, sugars, vitamins, bases and inorganic ions; or peptones – partly hydrolysed protein.

A broad spectrum culture medium is the type you are most likely to use in your college or school laboratory. In addition to being general purpose, you would also use them if you did not know the nutritional requirements of a particular microorganism.

Some media, however, will allow the growth of only a few, or even one, species of microorganism. These are called narrow spectrum media, or **selective media**. If you were to inoculate a mixed culture onto a selective medium, only the specific organism, for which the medium had been designed, will grow. The growth of any others will be suppressed.

You might wonder why selective media are used. Their main role is in diagnostic work in pathology laboratories and veterinary laboratories. MacConkey agar is a selective medium, allowing only the growth of Gram negative bacteria (see *Edexcel A level Biology 1*, Chapter 4). If a medical laboratory technician inoculated a sample of human faeces onto MacConkey agar containing bile salts, any bacteria that grew would belong to the genus *Salmonella*. The technician would then be able to alert surgeons that this patient was harbouring a pathogenic bacterium, so that a suitable drug could be prescribed. Later in this chapter (Figure 3.9), you will see bacteria growing on agar containing bovine blood. This selective medium allows the identification of bacteria belonging to the genera *Staphylococcus* and *Streptococcus*.

In a different context, the bacterium *Acidithiobacillus thiooxidans* is used in a process called bioleaching – the removal of metals from their ores by bacterial action. This bacterium uses sulfates as an energy source. The selective broth medium for *Acidithiobacillus thiooxidans* contains $10\,g\,dm^{-3}$ of powdered sulfur. This discourages the growth of other bacteria so, if inoculated as part of a mixed culture, the growth of any bacteria allows identification of *Acidithiobacillus thiooxidans* in the mixed culture.

Incubation

Following inoculation, a culture medium is incubated. This means that it is placed in a temperature-controlled cabinet for a suitable period of time.

Environmental temperature is a variable that is important in determining the ecological niche of each species of microorganism. You must, therefore, incubate media at a temperature that is appropriate for the microorganism being cultured. Although pathogenic bacteria grow best at your own body temperature, some bacteria grow well in the cold of the Antarctic and others in hot-water springs and deep thermal vents. On the basis of the temperature at which they grow best, bacteria are commonly classified into three groups.

- **Psychrophiles** grow best at low temperatures, in the range −10 °C to 20 °C.
- **Mesophiles** grow best at ambient temperatures, in the range 20 °C to 45 °C.
- **Thermophiles** grow best at high temperatures, in the range 55 °C to 85 °C.

In your college or school laboratory, you will probably incubate at a temperature at, or just above, room temperature. You will certainly not incubate cultures at 37 °C, in other words, your body temperature.

Can you think why this is an important safety rule?

Aseptic techniques

When you grow microorganisms in a laboratory, you need to be aware that all surfaces are contaminated by microorganisms. This applies to the surface of your skin, of the laboratory bench and of every item of laboratory equipment. Normally this does not matter, but when you grow microorganisms, it does. You need to avoid:

- contamination of your **microbial culture**. You normally grow a pure culture, in other words a culture containing only one type of microorganism. Since the air, the laboratory equipment and your skin and clothing are contaminated by microorganisms, they could easily enter your culture
- contamination of yourselves or other laboratory workers. You will be working with microorganisms that are considered safe, that is they will not cause disease. Remember, though, that:
  - you might accidentally grow a harmful microorganism that has contaminated your culture
  - microorganisms can change their nature by, for example, mutation or by passing nucleic acid from one cell to another (Figure 3.4)
  - some people are more susceptible to infection than others.

Test yourself

1. Explain why media for growing bacteria must contain a source of nitrogen.
2. 'Since few microorganisms can hydrolyse agar, the medium stays solid as the microorganisms grow.' Explain why few microorganisms can hydrolyse agar.
3. Give **one** advantage of using a broth culture over a solid culture.
4. Distinguish between a broad spectrum medium and a selective medium.
5. 'You will certainly not incubate cultures at 37 °C, in other words your body temperature.' Explain why.

Key term

Microbial culture
A population of microorganisms growing in a liquid growth medium or on a solid growth medium.

Key terms

Aseptic technique A way of working with microorganisms that ensures that only one type of microorganism, that is one population, grows in each culture and that no microorganisms escape the culture.

Sterilisation methods Steps you can take to remove, or destroy, any microorganisms that might contaminate your cultures.

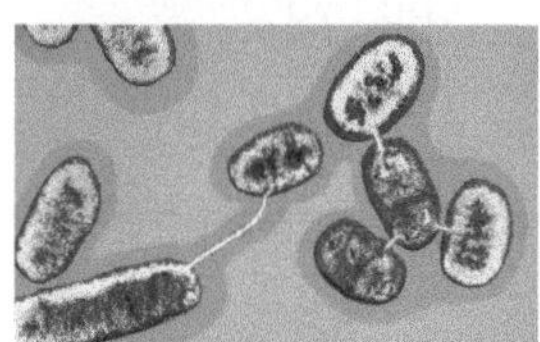

Figure 3.4 A false-colour transmission electron micrograph showing cells of *Escherichia coli* conjugating. The white links between cells are conjugation tubes that allow DNA (coloured orange) to pass from one cell to another (×13 000)

Key term

Autoclave A machine that works in the same way as a domestic pressure cooker. The latent heat of condensation of pressurised steam rapidly kills microorganisms, including spores.

When culturing microorganisms, you must always work as though the microorganisms in your culture are potentially harmful. The microbial techniques you will learn are the same as those that technicians working with potentially lethal microorganisms would use. They are called **aseptic techniques**.

Sterilisation methods

The only way you can be sure that bench surfaces and items of laboratory equipment are free of microorganisms is to kill them or remove them. In other words, you need to *sterilise* any equipment you will use. There are several **sterilisation methods** you can use. Which one you use depends on the nature of the object you wish to sterilise. Table 3.2 summarises commonly used sterilisation methods and shows when you use them.

Table 3.2 A summary of sterilisation methods commonly used when working with microorganisms

Method	Description	Use in aseptic techniques
Chemical agents	Disinfectants are chemicals that stop, or slow, the growth of bacteria. They are ineffective against bacterial spores, though.	Clean laboratory bench before and after working with microorganisms. Dispose of wet laboratory equipment immediately after use, e.g. glass pipettes. Treat any spillages that occur. (Note that disinfectants take time to become effective, so any spillages should be covered with disinfectant and left for at least 15 minutes before being mopped up.)
Heat treatment	Naked flame – hold an object in, or pass an object through, a Bunsen flame. This method is simple and effective as no microorganism can survive exposure to a naked flame.	An inoculating loop is 'flamed' by holding it in the hottest part of a Bunsen flame until it glows red (Figure 3.5, next page). Needles and forceps can also be flamed during manipulation of cultures. Flame sterilisation is often used on glass rods and glass spreaders after they have been dipped in 70% alcohol. The neck of a glass bottle, flask or tube containing a culture of microorganisms is sterilised by passing it through a Bunsen flame without allowing it to become red hot (Figure 3.6, page 40).
	Dry heat – place an object in a hot-air oven at 160 °C for at least 1 hour.	A routine method for the sterilisation of laboratory glassware prior to its use.
	Moist heat – place objects in an **autoclave** at 121 °C for at least 15 minutes.	This is the preferred method for many items of laboratory equipment and for culture media that are not heat-sensitive. It is also used to sterilise old cultures and spent media before disposing of them.
Filtration	Pass a liquid culture through a filtration device that has itself been sterilised by, e.g., dry heat. Using a filter of pore size 0.2 μm will remove bacteria (but not viruses).	The sheer size of the pores involved (typically 0.20 to 0.45 μm) makes this unsuitable for all but the smallest volumes of liquid.
Radiation	Expose items to UV or ionising radiation.	UV radiation with a wavelength less than 330 nm is most effective but, as this can damage the retina, is not used in college or school laboratories. Ionising radiation, such as γ-rays, cannot be used in a college or school laboratory since industrial facilities are needed. Many sterile plastic items, however, are supplied in packages that have been treated using UV or γ-radiation.

Test yourself

6 How would you sterilise a nutrient broth for use in the laboratory in your school or college? Explain your answer.

7 The apparatus in Figure 3.3 contains three air filters. Give **two** functions of these air filters.

8 When working with microorganisms, a scientist works with a Bunsen burner permanently burning. Other than for flame-sterilising equipment, give **one** advantage of having a lit Bunsen on the bench where microorganisms are being transferred.

9 Some laboratories contain a chamber, rather like a fume cupboard, in which scientists perform transfer of microorganisms. Explain why a UV lamp would be useful in such a cabinet.

10 Figure 3.5 shows an inoculating loop being sterilised in a Bunsen flame. Suggest **one** advantage of using a loop for inoculating bacteria rather than a simple wire.

Safety

- It is important to read this safety information before performing any microbial techniques in the classroom.
- **Sterilisation:** It is important to use steam at 121 °C for 15 minutes in an autoclave/sterilising pressure cooker.
- Wash hands thoroughly with bactericidal handwash before and after microbiological work.
- Always work on a surface that has been disinfected properly (for example, with 1% VirKon for 10 minutes), and the surface should be re-disinfected after the microbiology. BIOCIDES such as VirKon are used commercially to keep any work with microbes safe and have guarantees of effectiveness that household disinfectants do not.
- CLEAPSS and SSERC advise that all microbiology work takes place close (within 10 cm) to a roaring Bunsen flame, as this measure controls air movement.

Individual steps used in aseptic techniques

You will perform at least two experiments during your A level course that involve cultures of microorganisms. Once you have started, you cannot contaminate any sterile item or put down any item of equipment until you have sterilised it. Whilst not difficult, some of the steps involved in doing this are fiddly. It is worth looking at them individually and, if you have time, practising each without using any media or microorganisms.

Flame sterilising an inoculating loop

An inoculating loop is a wire, often made of tungsten, embedded in a handle. The end of the wire is made into a loop with a diameter of about 5 mm. Rather like the loop you might have used to blow bubbles as a child, this loop picks up a relatively constant volume of fluid as a film across the loop, in this case a liquid culture of microorganisms. The heated loop can also be used to remove a sample from a bacterial colony on a solid culture medium.

Figure 3.5 An inoculating loop being sterilised in the hottest part of a Bunsen flame

Figure 3.5 shows the end of a wire loop being sterilised. As you can see from the flame, the air inlet of the Bunsen burner is fully open and the loop is held at an angle in the hottest part of the Bunsen flame – above the blue region of unburnt gas. The loop in Figure 3.5 is white hot: no bacteria can survive this.

Flame sterilising the top of a container of microorganisms

You will be given a liquid culture of bacteria from which to take a sample. It will be in a glass container – a bottle or tube – with a 'lid'. The lid might be non-absorbent cotton wool or it might be a screw-cap. When you remove the lid from the container, you must ensure that no contaminating microorganisms get into the container and none escape from it into the air. Remember that once you have started you cannot put down the container or the lid. Figure 3.6 shows you how you should do this. The diagram shows how someone who is more comfortable using their right hand would do this; if you are more comfortable using your left hand, you should do so. Notice in Figure 3.6 that the person has removed the lid from the glass tube but is holding it in the crook of their little finger; they have not put it down. When removing a cotton wool bung, this is easy. When removing the screw cap from a bottle it is a little more difficult.

Removing a screw top from a container is something you are likely to do often, in daily life, for example when opening a bottle of cola or mineral water. You would normally use

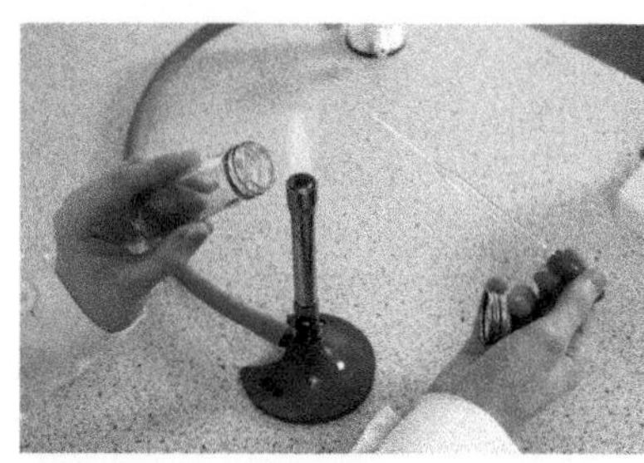

Figure 3.6 The neck of a glass container of a microbial culture is sterilised by passing it through the flame of a Bunsen burner without letting it get red hot. Notice that the scientist has the cap of the container in the crook of her little finger. This is a skill you must practise

one hand to hold the bottle still and the other hand to unscrew the cap. When using aseptic techniques, you do this the other way around – you use the crook of the little finger on one hand to hold the cap still and use the other hand to unscrew the bottle.

Having removed the lid, the person in Figure 3.6 can pass the top of the container through a hot Bunsen flame.

Using a wire loop to inoculate a culture onto a solid growth medium

As described in Table 3.1, solid growth media are usually contained within a Petri dish. This dish enables a small volume of growth medium to have a relatively large surface area.

You would use a sterile inoculating loop to transfer microorganisms to the surface of a sterile growth medium in a Petri dish. The transferred sample is referred to as an inoculum and the process of spreading the sample on an agar surface is called **plating**. Remember, your aseptic techniques prevent unwanted contamination of this sterile medium. If you fully removed the lid from the Petri dish, you would expose the entire agar surface to the air. This would enable airborne bacteria or fungi to contaminate the growth medium. The way you avoid this is simple but, again, is a technique you need to practise. Rather than fully remove the lid, you open the lid of the Petri dish at an angle with just enough room to be able to manipulate the inoculating loop inside. Figure 3.7 shows how to do this; again, showing someone who is more comfortable using their right hand.

Key term

Plating The transfer of an inoculum to the surface of a sterile agar medium.

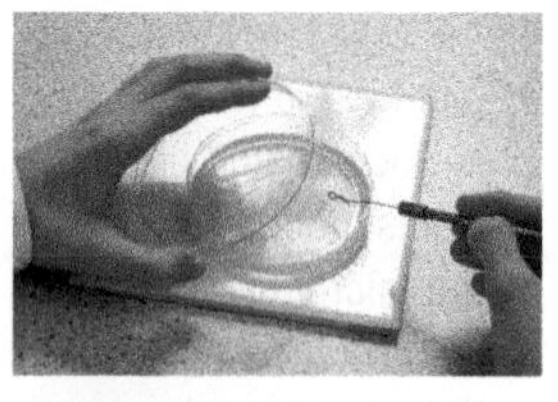

Figure 3.7 During inoculation, contamination of a sterile solid growth medium is avoided by lifting the lid of a Petri dish at an angle and as little as possible

Activity

Transferring an inoculum from a broth culture to an agar plate

Before starting one of the core practicals in this chapter, it is a good idea to practise the three skills described above as a single process. At this stage, you can safely do this with a glass container part-filled with tap water and an empty Petri dish.

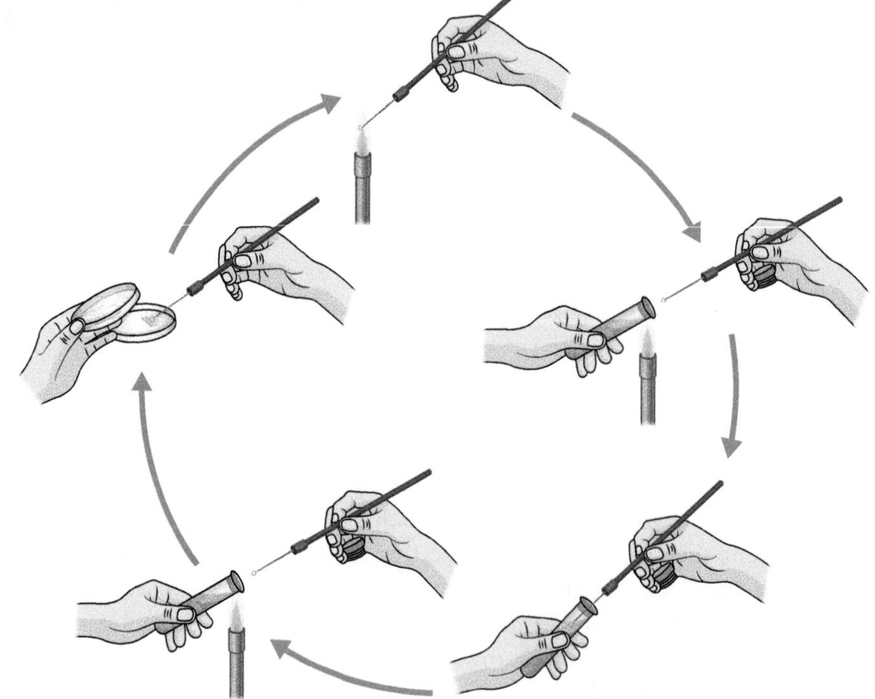

Figure 3.8 Practising the skills involved in transferring an inoculum from a broth culture to a Petri dish

Use Figure 3.8 to help you to practise the following steps.

1 Hold an inoculating loop in the hottest part of a Bunsen flame until it glows red-hot, or white-hot. Do **not** put down this inoculating loop until you have completed step 10.
2 Still holding the inoculating loop, allow it to cool.
3 Using the little finger of the hand in which you are holding the inoculating loop, remove the lid from the top of the glass container.
4 Still holding the lid in the crook of your little finger, pass the top of the glass container through the Bunsen flame. This should be a brief passage, otherwise you will risk the glass becoming so hot that you drop the container.
5 Insert the cooled inoculating loop into the glass container and remove a sample of water.
6 Still holding the lid in the crook of your little finger, pass the top of the glass container through the Bunsen flame.
7 Replace the lid on the glass container and put the container on the bench or into a test-tube rack.
8 Use your free hand to slightly raise the lid of a Petri dish. Put the tip of the inoculating loop into the Petri dish and gently slide it across the dish.
9 Replace the lid of the Petri dish.
10 Hold the inoculating loop in the hottest part of a Bunsen flame until it glows red, or white-hot. Allow it to cool and put down the inoculating loop.

You can repeat the above procedure as often as time and the availability of equipment allow, until you feel fully confident performing it.

Test yourself

11 How can you tell that the air inlet of the Bunsen burner in Figure 3.5 is fully open?
12 Suggest why you should hold an inoculating loop at an angle rather than horizontally when flame sterilising it.
13 Explain why the cotton wool used to close a tube of broth agar would be non-absorbent.
14 Explain why a microbiologist must hold the lid of a container in the crook of their little finger.
15 Other than to avoid dropping a tube that has become too hot to hold, explain why you should only briefly pass the top of a tube through a Bunsen flame when transferring an inoculum from a broth culture.

Streak plating

If you leave an agar plate open to the air, a number of microorganisms will land and start to grow and you end up with a mixed culture. You might want to isolate one type of microorganism from this mixed culture. You can do this using a technique called steak plating.

Using aseptic techniques, you use an inoculating loop to remove a sample from the mixed culture and transfer it to an agar plate. Having made the transfer, you then dilute the inoculum by spreading it time and again. Table 3.3 shows how this is done.

Table 3.3 Steps in the technique of streak plating

Description	Appearance of plate, viewed from above
Add your initials and the date to the base of the Petri dish into which you will transfer the inoculum. Use aseptic techniques to remove a sample from the mixed culture, using a sterile wire loop.	
Opening the lid of the Petri dish as little as possible, hold the loop parallel with the surface of the agar. Smear the inoculum backwards and forwards across a small area of the medium (region A in the diagram), taking care not to cut into the agar surface. Replace the lid of the Petri dish.	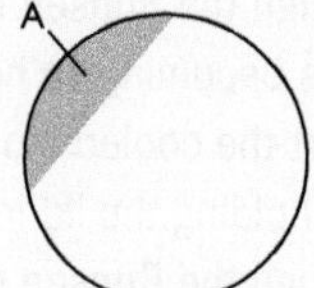
Flame the loop and allow it to cool.	
Turn the Petri dish through about 90°. Again, opening the lid of the Petri dish as little as possible, use the loop to streak the inoculum from A across the surface of the agar in three parallel lines (B). As before, take care not to cut into the agar surface. Replace the lid of the Petri dish.	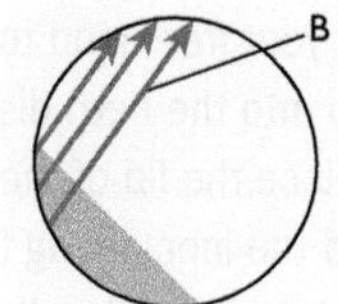
Flame the loop and allow it to cool.	
Turn the Petri dish through another 90°. Again, opening the lid of the Petri dish as little as possible, use the loop to streak the inoculum from B across the surface of the agar in three parallel lines (C), taking care not to cut into the agar. Replace the lid of the Petri dish.	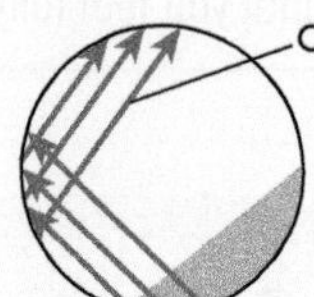
Flame the loop and allow it to cool.	
Turn the Petri dish through a final 90°. Again, opening the lid of the Petri dish as little as possible, use the loop to streak the inoculum from C across the surface of the agar as shown (D). Be careful not to cut into the agar and not to touch area A. Replace the lid of the Petri dish. Flame the loop and allow it to cool.	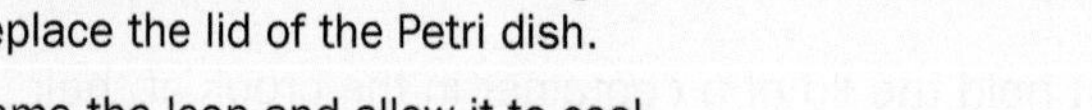 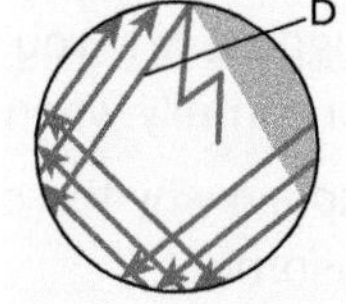

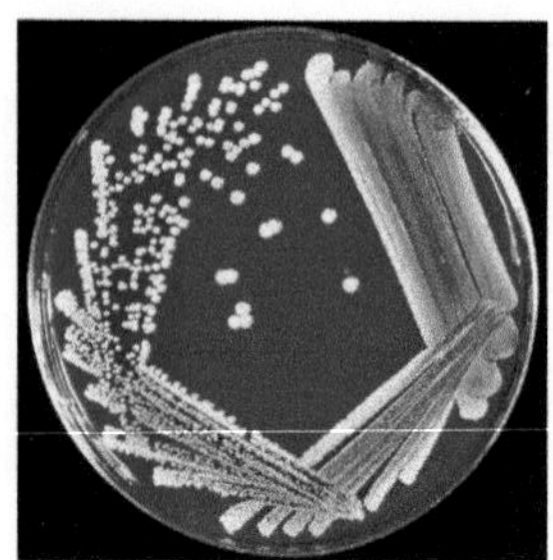

Figure 3.9 The result of streak plating. Here, the technique has been used on a culture of the skin bacterium *Staphylococcus aureus* on an agar plate containing bovine blood. This combination of agar and pathogen has been used to produce a particularly vivid photograph - it should not replicated in school laboratories

After a suitable period of incubation, colonies of microorganisms will have grown on the agar. In the areas A, B and C in Table 3.3, they will probably form a continuous mat. In area D, however, you should find small colonies formed by the division of a single microbial cell. Figure 3.9 shows the result of streak plating after incubation of the agar plate.

Safety note

You are now going to cover one of the core practicals in the Edexcel specification. Before you carry out this practical, or the one later in this chapter, you need a reminder of the safety issues involved in manipulating microorganisms in the laboratory. They include the following general rules.

- Wipe down the bench with disinfectant both before and after working with microorganisms.
- Have a container of disinfectant readily accessible. You might need it for a sudden emergency.
- Organise your bench, making sure you have everything you need to use within easy reach. Once you have started, you cannot put things down without first sterilising them.

- Work in a controlled way. On a crowded bench, any sudden movements could put one of your fellow students in peril.
- Do not make any hand-to-mouth movements. Putting anything to your mouth, including food, the end of your pen or even your finger, could be dangerous to you as it might result in infection.
- Cover any cuts or abrasions with a plaster or wear disposable plastic gloves. This prevents infection via your broken skin surface.
- Take care when using sharp instruments, such as mounted needles or glass pipettes.
- Do not put any waste culture medium down the sink. It must be autoclaved before being disposed of.
- Put contaminated items into disinfectant immediately after use. This applies to items such as pipettes, slides and glass spreaders.
- Keep flammable objects well away from the Bunsen flame. You will be working with a Bunsen burner constantly burning with its hottest flame. You must not let anything catch fire.

Core practical 13

Isolate individual species from a mixed culture of bacteria using streak plating

Before starting this investigation, make sure you thoroughly understand the steps shown in Figure 3.8, Table 3.3 and the text accompanying both. If possible, you should practise the skills described before carrying out this practical with microorganisms.

This investigation involves two practical sessions. In the first, you will use the streak plate method to inoculate sterile agar plates with a broth culture with which you have been provided. These agar plates will then be incubated for you.

In the second session, you will remove bacteria from one colony and inoculate them onto sterile agar plates. These will then be incubated for you.

Session 1

1. Ensure the bench is clear of anything you will not need to use.
2. Clean the bench using a disinfectant solution. Ensure that all the equipment you need is arranged on the bench so that it is easily accessible.
3. Light a Bunsen burner and turn the air inlet valve to produce the hottest flame.
4. Use a Chinagraph pencil or permanent marker to label the **base** of two Petri dishes containing sterile nutrient agar. Label with your initials and the date.
5. Sterilise the inoculating loop in the Bunsen flame.
6. Remove the lid from the tube of broth culture by grasping it in the crook of the little finger of the hand in which you are holding the inoculating loop.
7. Pass the neck of the tube through the Bunsen flame.
8. Remove a loopful of broth culture.
9. Pass the neck of the tube through the Bunsen flame and replace its lid.
10. Lift the lid from one of the Petri dishes just high enough to be able to insert the loop.
11. Using free arm movement from your elbow, smear the loop across part of the surface of the agar (see Table 3.3, area A).
12. Replace the lid of the Petri dish and sterilise the loop in the Bunsen flame.

13 As described in Table 3.3, turn the Petri dish through about 90°. Using free arm movement from your elbow, streak the inoculum across the agar in three parallel lines.
14 Repeat steps 12 and 13 until you have replicated the process described in Table 3.3.
15 Repeat steps 5 to 14 using the second Petri dish of sterile agar.
16 Sterilise the inoculating loop by passing it through the Bunsen flame.
17 Seal the lids of both Petri dishes to their bases using strips of sticky tape.
18 Hand your inoculated plates to your teacher to be incubated.
19 Clean the bench with disinfectant.
20 Wash your hands.

Session 2

1 Ensure the bench is clear of anything you will not need to use.
2 Clean the bench using a disinfectant solution. Ensure that all the equipment you need is arranged on the bench so that it is easily accessible.
3 Light a Bunsen burner and turn the air inlet valve to produce the hottest flame.
4 Use a Chinagraph pencil or permanent marker to label the **base** of two Petri dishes containing sterile nutrient agar. Label with your initials and the date.
5 **Without removing their lids**, examine your agar plates from session 1. Choose whichever has produced single colonies of bacteria in the most diluted of the streaks you made. Remove the strips of sticky tape from this Petri dish.
6 Sterilise the inoculating loop in the Bunsen flame.
7 Lift the lid from the chosen Petri dish just high enough to be able to insert the loop.
8 Carefully remove a small sample from a single colony that has grown on this plate. Replace the lid.
9 Take a Petri dish containing sterile agar and inoculate this plate with the sample you have just taken. Smear the inoculum across the surface of the agar plate.
10 Repeat steps 5 to 8 using a second Petri dish containing sterile agar.
11 Sterilise the inoculating loop by passing it through the Bunsen flame.
12 Re-seal the lid of the Petri dish you have used from session 1. Hand both Petri dishes from session 1 to your teacher for safe destruction.
13 For the Petri dishes you have just inoculated, seal the lids to their bases using strips of sticky tape.
14 Hand your newly inoculated plates to your teacher to be incubated.
15 Clean the bench with disinfectant.
16 Wash your hands.
17 After incubation, check your plates from session 2 and record the appearance of the colonies that are growing. Can you conclude that you have successfully isolated a pure culture?

Questions

1 Suggest why you were told to arrange the equipment you needed so that it was easily accessible.
2 Why should you label the base of the Petri dish, rather than its lid?
3 Suggest why you were instructed to use an arm movement from your elbow when smearing the loop across the agar.
4 Why did you seal your Petri dishes with sticky tape?
5 In session 2, you were told not to remove the sticky tape from the Petri dishes. How could you see which Petri dish had produced single colonies of bacteria?

Measuring the growth of bacterial cultures

To measure the growth of a bacterial population, you need to find out how many bacterial cells are in your culture medium at repeated time intervals. You can do this by making a:

- **total count** – count all the cells in the culture
- **viable count** – count only the living cells in the culture.

Although you can easily count the number of large organisms in a population, you cannot easily do so with a bacterial population. Even with the best light microscopes, bacteria are difficult to see and you can't distinguish between a dead bacterium and a living one just by looking at it.

Serial dilutions

Another problem you encounter is the sheer number of bacteria in a single population. For example, a single inoculum into a liquid growth medium, incubated overnight under optimal conditions, is likely to contain in the order of 10^8 bacteria per cm^3 of medium. This is far too many to count! You can overcome this problem by repeatedly diluting a sample from the population until you find a number of cells that you can count accurately and are confident will give you a reliable estimate of the population size. Then, knowing your dilution factor, you can multiply your count by this dilution factor to obtain an estimate of the number of cells in the undiluted culture.

You might ask, "But how do I know how many times to dilute the culture before I can count individual cells?" The simple answer is that you don't. Instead, you make a succession of dilutions, called a **serial dilution**, from which you hope to find one that will enable you to count cells. Figure 3.10 shows how you make a serial dilution; in this case, each dilution factor is 1 in 10.

Key term

Serial dilution A repeated dilution, by a constant dilution factor, of an original solution or microbial culture.

Activity

Making a serial dilution

You need a microbial culture in a liquid medium in order to make a serial dilution. If you follow the steps in Figure 3.10 you will see how you would do this.

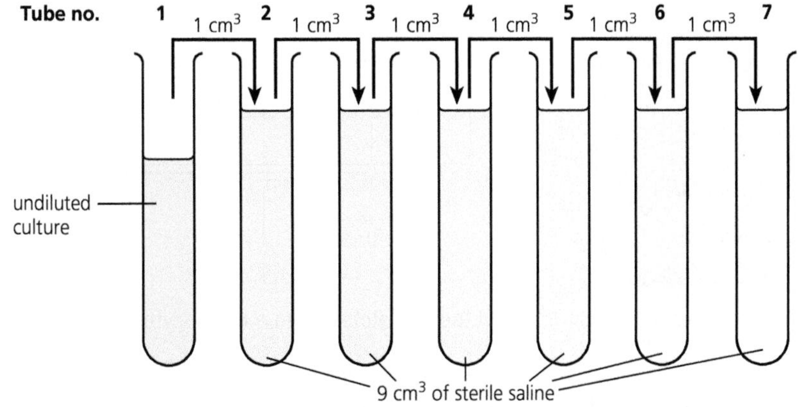

Figure 3.10 The steps involved in making a serial dilution of a broth culture of bacteria

Safety tip

It is extremely dangerous to use hand-to-mouth movements when dealing with microorganisms. When making serial dilutions, you should never pipette by mouth.

Firstly, using aseptic techniques, you transfer 1 cm^3 of your culture solution in tube **1** into 9 cm^3 of sterile diluent in tube **2**. Then stir or agitate tube **2** to ensure complete mixing before the next 1 cm^3 sample is withdrawn and added to tube **3**. Repeat this procedure to follow the serial dilution steps in Figure 3.10.

1 Describe how you would use aseptic techniques to begin this transfer.
2 How would you sterilise a dry glass pipette?
3 How would you draw 1 cm^3 of liquid medium into the pipette?
4 If your dilution series is to give you an estimate of the population size that is close to its true value (in other words, is accurate), what size of pipette would you use to measure a 1 cm^3 sample?
5 What is the dilution factor in tube **2**?
6 You flame-sterilised the pipette before you used it to transfer the bacterial culture. Why can't you flame-sterilise it after you have used it?
7 Why must you use a new pipette for each transfer of medium?
8 Suggest why you would use a 0.9% saline solution as the diluent, rather than water.

Making total counts

The total count estimates the number of cells in a culture, regardless of whether they are alive or dead. We can make these estimates in one of three ways:

- A direct count, using a counting chamber and a light microscope.
- An indirect count, measuring the dry mass of a filtered culture.
- An indirect count, using a colorimeter to measure the turbidity of a culture.

Direct count using a haemocytometer

A haemocytometer is a special microscope slide. As its name suggests, it was originally designed for counting red blood cells. You can see in Figure 3.11(a) that a haemocytometer is thicker than a normal microscope slide. Its central platform, between the two grooves, has a grid etched into it and is slightly lower than the main glass slide. When a cover slip is placed over this central part, it produces a film of known depth over the grid.

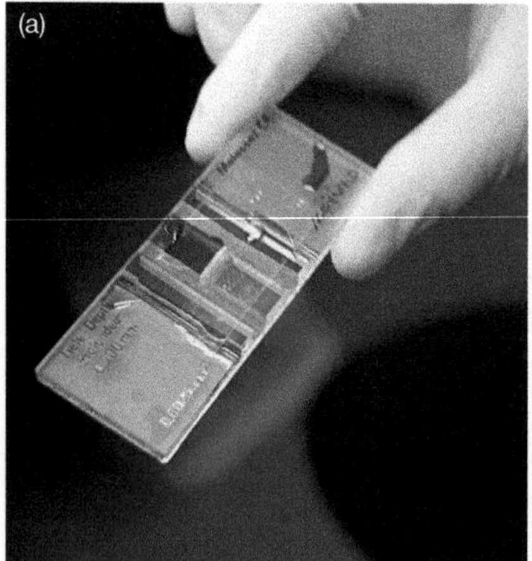

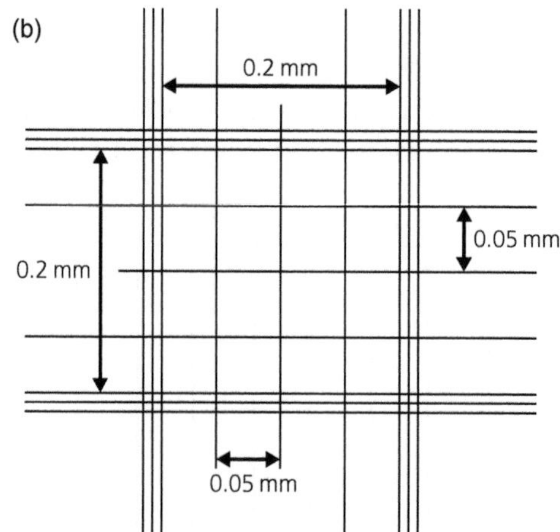

Figure 3.11 (a) A haemocytometer slide and (b) part of the grid etched onto a haemocytometer slide

Figure 3.11(b) shows part of a grid etched onto the platform of a haemocytometer. You can see that the grid is formed of squares of known dimensions. This grid is very accurate, which is why this slide is very expensive. Since you know the depth of the film of fluid and the size of an individual square, you know the volume of liquid covering each square that you view using a light microscope. So, by viewing a sample from each tube in your serial dilution, you quickly find one with a number of cells that enables you to count bacterial cells accurately, yet contains sufficient cells for you to be confident that it is a representative sample. There is, however, a further complication you need to deal with.

Look at Figure 3.12. It shows a number of cells within a 0.2 mm × 0.2 mm section of the haemocytometer grid. How many do you count? In this field of view, there are 11 cells. Some, however, are overlapping the lines delineating the 0.2 mm × 0.2 mm square. How do you deal with those?

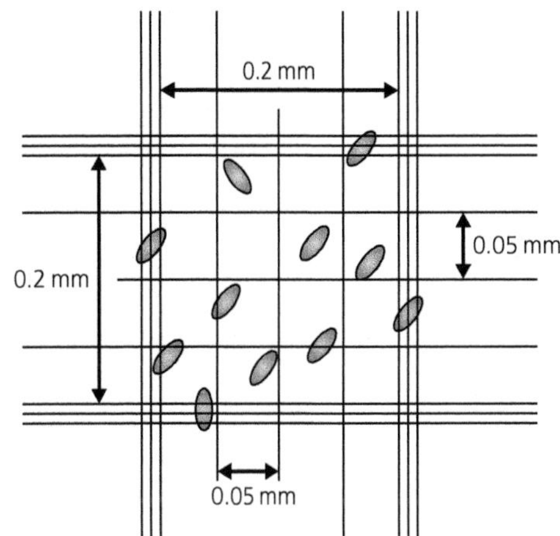

Figure 3.12 How to count cells using a haemocytometer

One important aspect of scientific research is repeatability, that is other scientists must be able to replicate your method and obtain similar results. You cannot, therefore, tolerate an *ad hoc* approach to counting cells that overlap the grid lines of a haemocytometer. By international convention, you:

- **do** include in your count any cell that touches or overlaps the middle of the three lines at the top and right-hand side of a 0.2 mm × 0.2 mm square
- do **not** include any cell that touches or overlaps the middle of the three lines at the bottom and left-hand side of a 0.2 mm × 0.2 mm square.

Thus, in Figure 3.12, your cell count would be eight cells.

Indirect count measuring dry mass of cells

You are unlikely to use this technique since it is slow and involves the use of expensive filtration equipment that must be sterilised before use.

Since the technique involves filtering a culture to remove the bacterial cells, it is only useful with small volumes of liquid media. After finding the dry mass of the sterile filtration membrane, you would filter a known volume of liquid culture. You would then heat the filter membrane, with the bacteria on its surface, in an oven at 100 °C until its mass remained constant. By subtracting the mass of the sterile filter membrane from the final mass, you find the dry mass of the bacteria. Finally, knowing the volume of medium you filtered, you can calculate the mass of cells per unit volume of culture.

> **Safety**
>
> Cuvettes should be sterilised immediately after using them for microbial work.

Indirect count measuring turbidity

As a population of cells increases, it will make the culture medium in which it is growing, cloudier. You can measure the degree of this 'cloudiness', or **turbidity**, using a colorimeter.

Using aseptic techniques, you would place a sample from one of the tubes in your serial dilution (Figure 3.10) in a special flat-sided tube, called a **cuvette**. You then place this cuvette into a colorimeter (Figure 3.13) and pass light through it. You can set the colorimeter to measure the amount of light absorbed by the contents of the cuvette (the **absorbance**) or the amount of light that passes through the contents of the cuvette (the **transmission**). The more bacteria present, the greater the absorbance or the less the transmission. You use these measures as an indicator of the size of the microbial population in your sample.

Figure 3.13 A cuvette being placed in a colorimeter

This technique has its drawbacks. You can only use it with liquid cultures and you cannot distinguish between living and dead cells. Additionally, you need to calibrate your measurements so that you can relate your readings of absorbance or transmission to the actual density of cells. This means that you must first produce a calibration curve, plotting the density of cells found, for example using a haemocytometer, against the measured turbidity found using a colorimeter.

Extension

Use of a Coulter counter to measure changes in electrical resistance

A method similar to measuring the turbidity of a bacterial mixture involves the use of a Coulter counter. Like the haemocytometer, the Coulter counter was originally designed for use with red blood cells. It has a probe, with two electrodes, that is put into the liquid culture. As you can see in Figure 3.14, one of its electrodes is inside a glass tube that has a tiny hole in it. As bacteria pass through this hole, the electrodes detect the changes the bacteria cause in the electrical resistance of the medium. These measurements can be relayed to a computer for data processing.

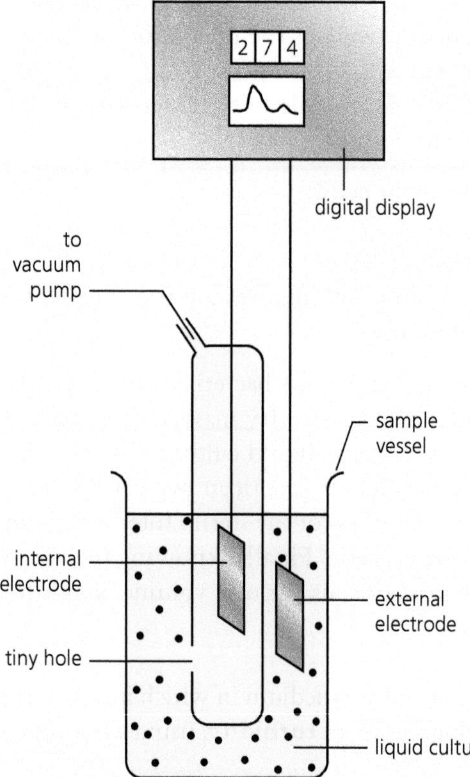

Figure 3.14 The electrodes of a Coulter counter detect changes in the electrical resistance of the culture medium caused by bacterial cells

Making viable counts using spread plates

When making a viable count, you are only interested in those bacterial cells that are capable of growing. Once plated onto a solid culture medium, the growth of each viable cell will produce a visible colony. If you look back to the streak plate in Figure 3.9, you can remind yourself of the different appearance of a single colony and the mat formed by the merger of many colonies.

As with the total count, you need to use a serial dilution so that you can obtain a growth of bacteria in which you can count a reliable number of individual colonies. You would, therefore, produce a serial dilution, as shown in Figure 3.10. You would then use aseptic techniques to pipette a small, known volume (usually $\leq 0.5\,cm^3$) of each dilution onto the surface of sterile, solidified medium in a separate Petri dish. You can see this being done in Figure 3.15(a). Having done this, you would use an L-shaped glass rod as a spreader to gently spread the pipetted suspension of bacterial cells over the whole surface of the culture medium. This is shown in Figure 3.15(b).

Of course, the glass spreader must be sterilised before (and after) use. You would traditionally do this by dipping the end of the spreader into a beaker of 70 per cent alcohol for at least five minute, allowing the excess alcohol to drain off the spreader and then igniting the remainder in a Bunsen flame. After cooling, you can then use the spreader, as shown in Figure 3.15(b) to distribute the cell suspension over the culture medium. Following incubation, each viable cell in the dilution will produce a colony on the agar plate, as you can see in Figure 3.15(d).

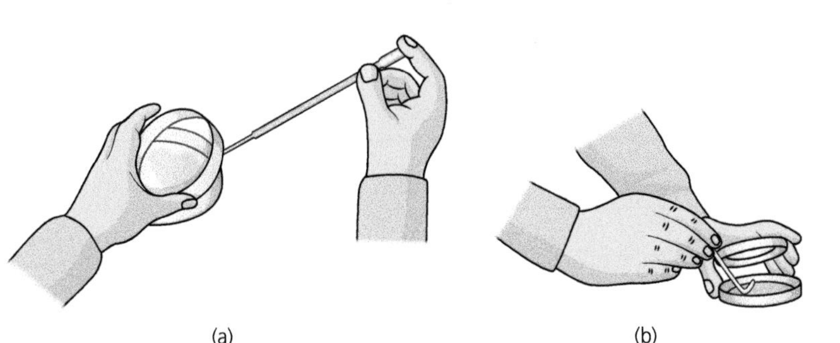

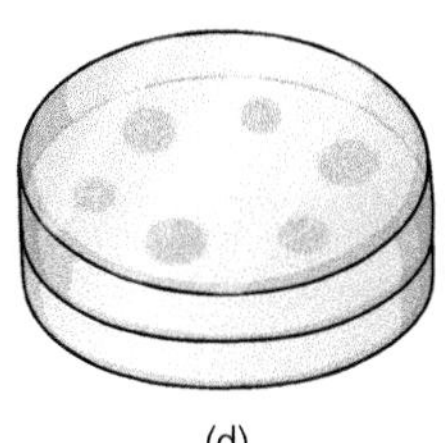

Figure 3.15 Preparing a spread plate

Safety

Sterilising the L-shaped spreader involves a significant fire risk. It is important to drain excess alcohol from the spreader before igniting, to prevent the formation of flaming droplets of alcohol. It is also important to keep the Bunsen burner well away from the alcohol and to ensure the spreader is no longer flaming and has cooled before putting it back into the beaker of alcohol.

Alternatively, for sterilising spreaders, schools can choose to wrap glass/metal spreaders in greaseproof paper/aluminium foil, and then sterilise the spreader by heating at 160 °C for two hours in an oven. The spreaders will stay sterile until unwrapped just before point of use.

Alternatively, schools may choose to use sterile plastic spreaders, which can be obtained from many suppliers.

Table 3.4 provides a summary of the relative merits of each of the methods of estimating cell numbers you have examined in this chapter.

Table 3.4 A summary of the methods for estimating cell numbers

Method of estimating cell number	Relative advantage/disadvantage
Direct microscopic count	• Can only be used for total counts, since you cannot distinguish between living and dead cells • A relatively slow method • Kills the cells examined • Useful for obtaining data to produce a calibration curve for turbidity measurements
Measuring dry mass of cells	• Can only be used for total counts • A slow method, so can be used only with small volumes of culture, making results unreliable • Kills cells being weighed
Turbidity measurement	• Can only be used for total counts • A fast method • Does not kill cells examined • Unreliable for cell densities less than 10^7 cells cm^{-3}
Spread plate (colony count)	• Can be used for viable counts • A two-step process, so speed restricted by incubation time • Does not kill cells being examined

Test yourself

16 Distinguish between a total count and a viable count.

17 A student used aseptic techniques to pipette 0.1 cm^3 of broth culture into a tube containing 9.9 cm^3 of sterile saline. Calculate the dilution factor that she was using.

18 State restricts your ability to use the dry mass method to measure the growth of a bacterial culture.

19 Using a haemocytometer, you find a cell touches the bottom line of the square you are viewing. Should you include this cell in your count? Explain why.

20 Some bacteria clump together as they grow in a liquid growth medium.

a) Explain how this might affect a cell count made using a colorimeter.

b) Suggest how could you avoid this.

Core practical 12

Investigate the growth of bacteria in liquid culture

Before starting this investigation, make sure you understand the steps involved in making a serial dilution (Figure 3.10 and the accompanying text).

To enable this core practical to be completed in a single practical session, it is assumed that:

- it will be carried out as a class exercise
- the class has been provided with samples of a broth culture taken at known times after its inoculation
- a few drops of 40% methanal have been added to each sample. Methanal (also known as formaldehyde) will kill the bacteria in each sample, preventing any further growth. It also means that you no longer need to use aseptic techniques
- each person, or pair, will count the cells in one of the samples.

The investigation involves two stages. In the first, you will be given one sample and use it to produce a serial dilution. You will then use a light microscope and haemocytometer slide to find a suitable dilution with which to estimate the density of cells in your sample.

Preparing the serial dilution of broth culture

1 Label six test tubes **−1** to **−6** and place them in a test tube rack.

2 Pipette 9 cm^3 of water into each test tube.

3 Pipette 1 cm^3 of the broth culture you have been given into the tube labelled **−1**. Thoroughly mix the contents of this tube.

4 Use a clean pipette to transfer 1 cm^3 of cell suspension from the tube labelled **−1** to the tube labelled **−2**. Thoroughly mix the contents of this tube.

5 Repeat step 4, transferring cell suspension from tube **−2** to **−3**, then from tube **−3** to **−4** and so on until you add 1 cm^3 of suspension to tube **−6**.

Counting the cells in the diluted broth cultures

6 Set up a light microscope.

7 Place the special cover slip over the platform of the haemocytometer and **very gently** press it down to ensure it has made contact with the slide. When positioned correctly, you should be able to see interference rings (Newton's rings) at the edge of the cover slip.

8 Using a dropper pipette from the side of the slide, add a small amount of broth culture from tube **−1** to fill the central space above the grid (Figure 3.16). Allow it to settle for a couple of minutes.

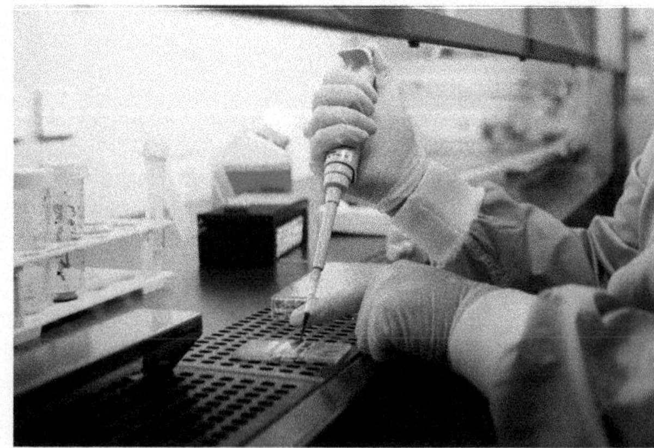

Figure 3.16 Loading a sample onto the haemocytometer slide using a pipette

9 Put the haemocytometer on the stage of the microscope and use the ×10 objective lens to examine the grid.

10 Examine the slide to see whether you can use it for a cell count. You need to find a dilution in which there are sufficiently few cells for you to be able to count them accurately but not so few that the sample will be a poor representation of the broth culture.

11 If you cannot see any cells, you might need to stain them. If so, wash the haemocytometer slide and repeat steps 7 to 9 with a fresh sample to which you have added a drop of 0.1% methylene blue stain.

12 If the dilution is not suitable because it contains too many cells, wash the haemocytometer slide.

13 Repeat steps 7 to 11 until you find a dilution that is suitable for counting cells.

14 Once you have found a dilution that is suitable for counting, switch from the ×10 objective lens to the ×40 objective lens. In doing so, take care not to scratch the objective lens on the cover slip, which is much thicker than a normal cover slip.

15 Count and record the number of cells in this dilution.

a) Select which 0.2 mm squares to count in a pre-arranged way, for example every 4th square from left to right.

b) Use the fine focusing screw to focus the microscope at different levels so that you include all the cells in each square.

c) Include in your count any cell that touches or overlaps the middle of the three lines at the top and right-hand sides of each 0.2 mm square.

d) Keep counting squares until you have included several hundred cells.

e) Use a hand-held tally counter, if one is available.

16 Divide the total number of cells by the number of 0.2 mm squares counted to find the mean number of cells in each square.

17 Use this mean value to calculate the number of cells in the original broth culture. Your teacher will tell you the depth of the platform in the haemocytometer you used.

18 If you have time, repeat the entire process and find the mean value of your repeated results.

19 Hand in your results to your teacher so that the whole class results can be collated.

20 Use the collated class results to plot a growth curve of the broth culture. Use your graph to determine the growth rate of the bacteria.

The bacterial growth curve

Under continuous culture, using equipment such as that shown in Figure 3.3, a bacterial population could continue to grow indefinitely. If you plotted the number of bacteria against time, it would result in a curve showing exponential growth.

In a batch culture, however, a bacterial population cannot grow indefinitely. As the bacteria in a batch culture grow, they use the nutrients in the culture medium. They also excrete the waste products of their own metabolism into the culture medium. As a result, the culture medium provides an increasingly less favourable environment for the bacteria. For this reason, the growth curve of bacteria in batch culture is not exponential. Instead, it shows the four stages shown in Figure 3.17.

The lag phase

This part of Figure 3.17 shows a period after inoculation to a new culture medium during which there is no increase in the number of bacteria. The bacteria *are* active, however. They are absorbing water from the medium, synthesising ribosomes and, often under the stimulation of substances in the medium, switching on genes and beginning to make new mRNA. The length of this phase depends on the culture medium used and the activity of the bacteria before they were inoculated to the new medium.

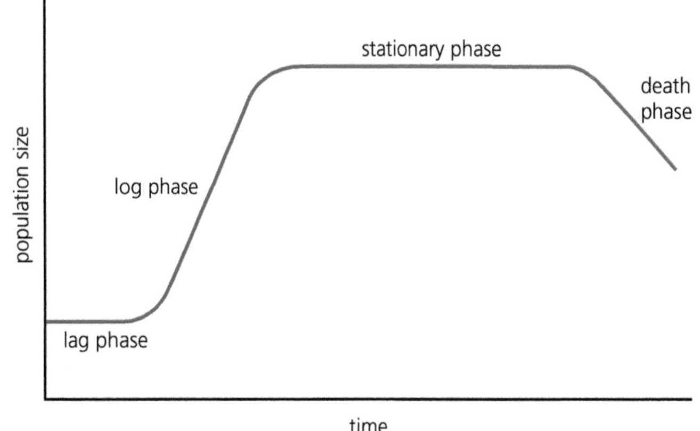

Figure 3.17 A typical growth curve of bacteria grown in batch culture

The log phase

Sometimes called the exponential phase, during this part of Figure 3.17 bacterial cells are dividing by binary fission at their maximum rate. The generation time differs from species to species, but is usually very short. Table 3.5 shows the generation time of some common bacteria, grown under optimum conditions in the laboratory.

Table 3.5 The generation time of some common bacteria grown under optimal conditions

Species of bacterium	Growth medium	Generation time/minutes
Escherichia coli	Glucose and salts	17
Staphylococcus aureus	Heart infusion broth	27–30
Lactobacillus acidophilus	Milk	66–87
Mycobacterium tuberculosis	Selective medium	792–932

The stationary phase

The growth of bacteria during the log phase changes the nature of the culture medium. For example, it removes nutrients from it, adds waste products to it and changes its pH. As a result, the conditions become no longer optimal for bacterial growth and cells begin to die. During the stationary phase, the rate at which new cells are formed by binary fission is the same as the rate of cell death, in other words:

'birth rate' = 'death rate'

The death phase

As the conditions in the culture medium become less and less suitable for growth, an increasing number of cells die. The viable count will obviously fall. As many of the dead cells undergo autolysis, the total cell count might also fall.

Calculations involving the log phase of the growth curve

You can analyse the log phase quantitatively in three different ways.

Finding the number of cells uses your knowledge that bacteria divide by binary fission:

- 1 cell becomes 2; 2 cells become 2 × 2 = 4; 4 cells become 2 × 2 × 2 = 8, and so on.

Instead of writing 2 × 2, you could write 2^2 and instead of writing 2 × 2 × 2, you could write 2^3. This is the basis of exponential (or logarithmic) growth. At each generation, the number of cells in the starting inoculum is increasing by a factor of two, in other words:

- 1 cell becomes 2, 2 cells become 2^2, 2^2 cells become 2^3, and so on.

You could, therefore, calculate that after n generations, the original number of cells in the inoculum (N_0) will have grown to a number (N) given by:

$$N = N_0 \times 2^n$$ **Equation 1**

The exponential growth rate constant (μ) is the rate at which bacteria grow during the log phase of the growth curve. If the number of cells at time t_0 is N_0 and the number of cells at the later time t_x is N_x, you can find the exponential growth rate constant using the following formula.

$$\mu = \frac{2.303\ (\log N_x - \log N_0)}{(t_x - t_0)}$$ **Equation 2**

Example

At 10:00 hours you begin a practical class and count 2×10^3 cells in a sample of a bacterial culture. At 12:00 hours you take another sample and count 5.7×10^4 cells in this sample. You would use equation 2 above to calculate the exponential growth rate constant of this culture as follows.

$N_0 = 2 \times 10^3$ so $\log N_0 = 3.30$

$N_x = 5.7 \times 10^4$ so $\log N_x = 4.76$

$t_x - t_0 = 2$ hours

$$\text{so } \mu = \frac{2.303\,(4.76 - 3.30)}{2}$$

$$= \frac{2.303 \times 1.46}{2}$$

$$= 1.681 \text{ hour}^{-1}$$

Tip

Calculations in your A level examination will be at GCSE higher tier level (Level 2). You need to own a good scientific calculator, know how to use it and remember to take it to your examination. You will be able to find one in national chain stores for less than £10.

The generation time (**g**) is the time between two consecutive divisions. Since each division produces two new cells, this is also referred to as the **doubling time**. It can be calculated using the same symbols as Equation 2 above.

$$g = \frac{0.301\,(t_x - t_0)}{\log N_x - \log N_0}$$ **Equation 3**

Test yourself

21 What can you conclude about generation time from the data in Table 3.5?

22 Explain why the growth curve for bacteria grown in batch culture is different from that for bacteria grown in continuous culture.

23 An inoculum contains 2×10^3 bacterial cells. If grown in liquid growth medium under optimal conditions, calculate how many cells will be present after six generations.

24 Figure 3.15(d) was the result after incubating $1\,cm^3$ of a 10^{-5} sample from a serial dilution. Calculate how many cells were present in the undiluted culture.

25 At 10:00 hours you begin a practical class and count 2×10^3 cells in a sample of a bacterial culture. At 12:00 hours you take another sample and count 5.7×10^4 cells in this sample. Use Equation 3 to find the doubling time of the population.

Exam practice questions

1 Which of the following describes the order of stages of growth of a bacterial batch culture?

A lag phase, log phase, stationary phase, decline phase

B lag phase, stationary phase, log phase, decline phase

C stationary phase, log phase, decline phase, lag phase

D stationary phase, log phase, lag phase, decline phase *(1)*

2 The table shows information about the stability of proteins from three different bacteria.

Species of bacterium	Percentage of proteins denatured at 60 °C
Escherichia coli	55
Bacillus subtilis	57
Unknown *Bacillus*	4

To which group of bacteria does the unknown *Bacillus* belong?

A hydrophiles **C** psychrophiles

B mesophiles **D** thermophiles *(1)*

3 Which method of sterilisation would be most appropriate for a heat-sensitive liquid growth medium?

A disinfectant **C** filtration

B dry air oven **D** flaming *(1)*

4 a) Give one way in which the use of aseptic techniques can be described as *ethical.* *(1)*

b) A student was given seven tubes containing a serial dilution of a bacterial culture.

i) Describe how he would transfer 1 cm^3 of the culture from the 10^{-3} dilution to an agar plate. Include descriptions of the aseptic techniques he would use. *(6)*

ii) The diagram shows some of the student's agar plates after incubation. Which should he use to estimate the number of cells in the undiluted culture? Give the reasons for your choice. *(3)*

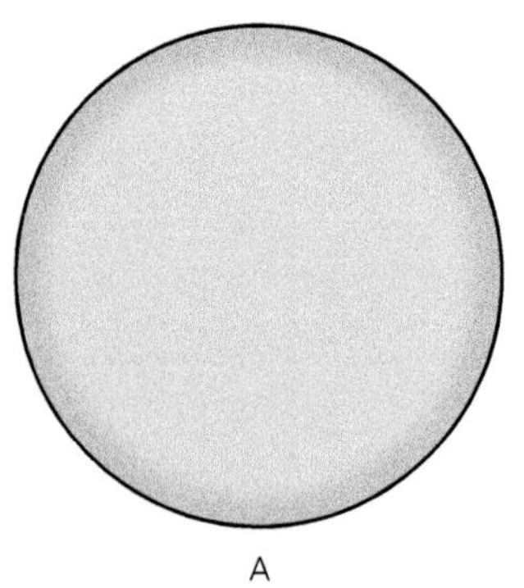

A

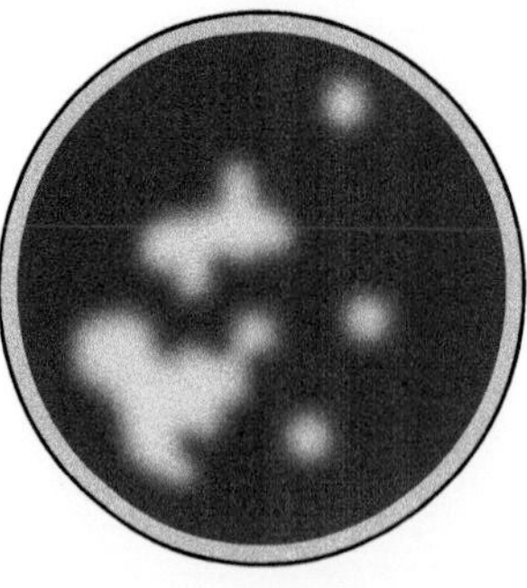

B

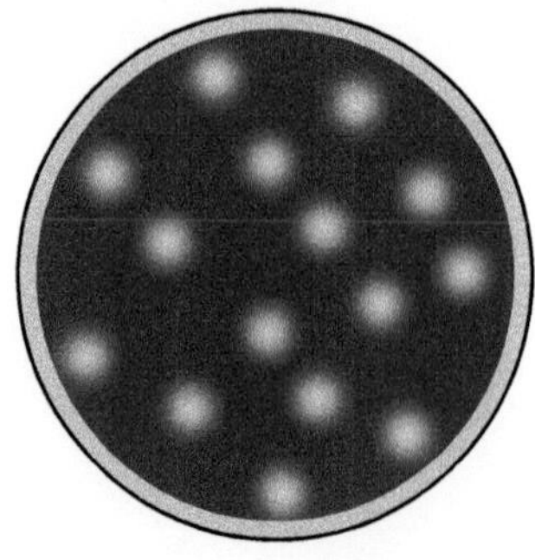

C

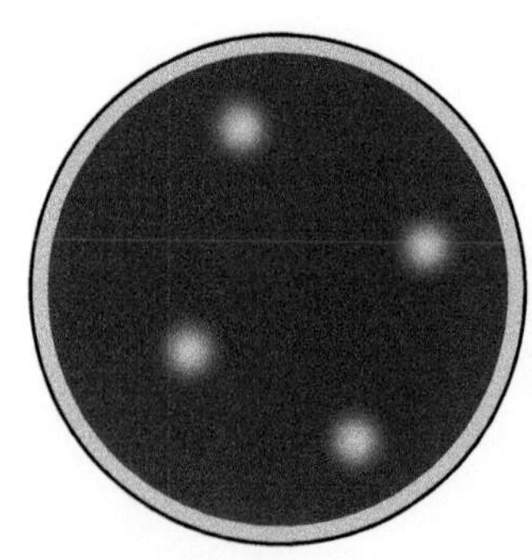

D

5 A student counted yeast cells using a light microscope and haemocytometer. The diagram on the right shows part of one of her fields of view using a dilution of the pure culture of 10^5.

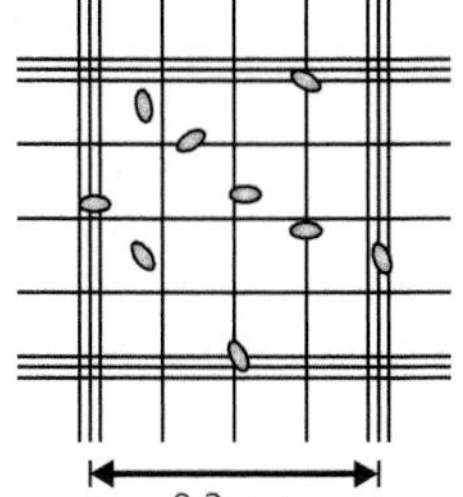

a) How many cells should she count in the square shown? Explain your answer. *(2)*

b) The mean number of cells she counted was 9.3 cells per 0.2 mm × 0.2 mm square. With the haemocytometer she used, the depth of liquid between the cover slip and the haemocytometer platform was 0.1 mm.

Use the information in this question to estimate the density of cells in the pure culture. Give your answer as cells cm^{-3}. Show your working. *(3)*

6 The graph shows a typical growth curve of a bacterial culture.

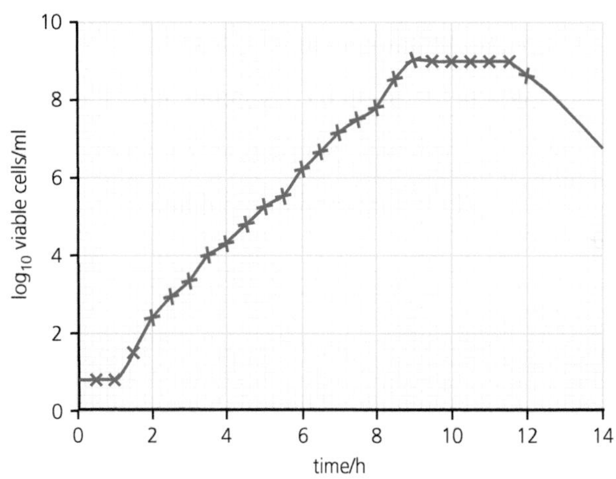

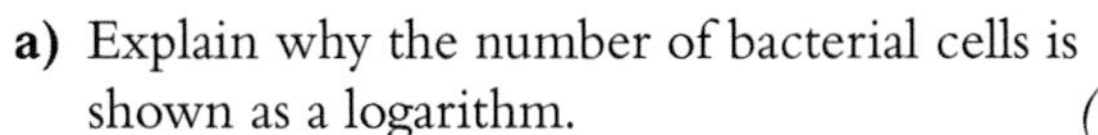

a) Explain why the number of bacterial cells is shown as a logarithm. *(1)*

b) Explain the shape of the curve during the first 30 minutes. *(2)*

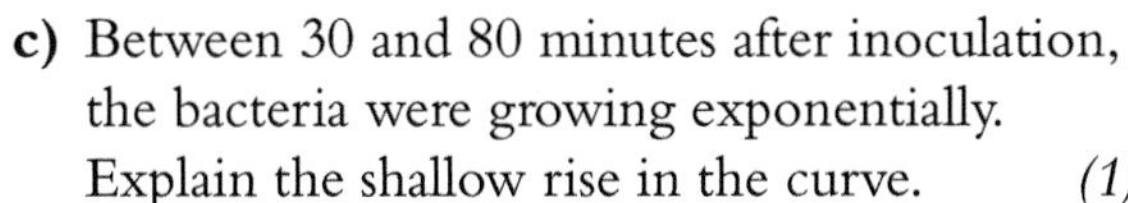

c) Between 30 and 80 minutes after inoculation, the bacteria were growing exponentially. Explain the shallow rise in the curve. *(1)*

d) Use data from the graph to calculate the growth rate constant (μ) between 90 minutes and 510 minutes. Use the formula:

$$\mu = \frac{2.303\ (\log N_x - \log N_0)}{(t_x - t_0)}$$ *(3)*

7 The total count and viable count of a bacterial culture can be investigated.

a) What is the advantage of performing a viable count over a total count? *(2)*

b) A student wishes to investigate the viable count of a broth culture of bacteria.

Devise and outline a method by which she could do this.

Assume she has the equipment and materials she needs in her school laboratory.

Do not include any details of aseptic technique in your outline of the method. *(6)*

Stretch and challenge

8 Use a search engine and/or the resource centre in your college or school to learn more about culture media. What features of selective media can you relate to other topics in your biology studies?

9 The European Union (EU) regulations relating to mineral water for human consumption, specify a maximum plate count of 100 colony-forming units per cm^3 ($CFU\,cm^{-3}$) after incubation at 22 °C.

A serial dilution was produced from a sample taken from a freshly opened bottle of mineral water. A pour-plating technique was used to add 500 mm^3 samples of each dilution to agar plates. After incubation at 22 °C, the colony count of three plates of the 10^{-1} dilution were 28, 32 and 39. What can you conclude about this mineral water? Use the following formula to explain your answer.

$$\text{viable count per cm}^3 = \frac{c}{v} \times D$$

Where c = mean colony count at each dilution

v = volume of liquid transferred to each plate

D = reciprocal of the dilution

4 Pathogens and antibiotics

Prior knowledge

In this chapter you will need to recall that:

- ➔ communicable diseases can be caused by viruses, bacteria, protoctists and fungi
- ➔ animals and plants are susceptible to communicable diseases
- ➔ viruses are non-living particles that depend on their host cell's metabolism to produce more virus particles
- ➔ some viruses undergo a lytic cycle, at the end of which the cell they infect lyses
- ➔ bacteria are prokaryotic organisms; their metabolism differs from that of eukaryotic organisms in many respects
- ➔ protoctists and fungi are eukaryotic organisms
- ➔ the spread of communicable diseases can be reduced or prevented in animals and plants using a variety of techniques.

Test yourself on prior knowledge

1 Name **two** components common to all viruses.
2 List the stages of the lytic cycle of a virus.
3 Give **two** ways in which the structure of a prokaryotic cell differs from that of a eukaryotic cell.
4 Explain **one** way in which the spread of HIV/AIDS can be slowed.
5 How do bacteria and fungi feed?

Bacteria as pathogens

You examined the structure of bacteria in *Edexcel A level Biology 1*, Chapter 4, page 86. To jog your memory, Figure 4.1 on the next page shows a false colour scanning electron micrograph of the intestinal bacterium *Escherichia coli*, together with a drawing to interpret its structure. Remember that bacteria are prokaryotic cells that lack a nucleus or membrane-bound organelles and have 70S ribosomes. This means that their metabolism is different from that of eukaryotic cells, which we will return to later in this chapter.

Like most species of bacteria, *Escherichia coli* is normally harmless. In fact, it is a very common commensal in the guts of mammals, including humans, and is a major component of their faeces. Some species of bacteria, however, are pathogens. This means that they invade the tissues of another organism, their host, and cause harm to it. The damage results from either:

- release of toxins – substances produced by the pathogen
- invasion and destruction of the host's tissues.

Key terms

Commensal A partner in an association between two organisms of different species where one (*E.coli* in our example) benefits and the other (a mammal) neither benefits nor is harmed.

Pathogen An agent of infection that invades another organism, the host, causing harm to it.

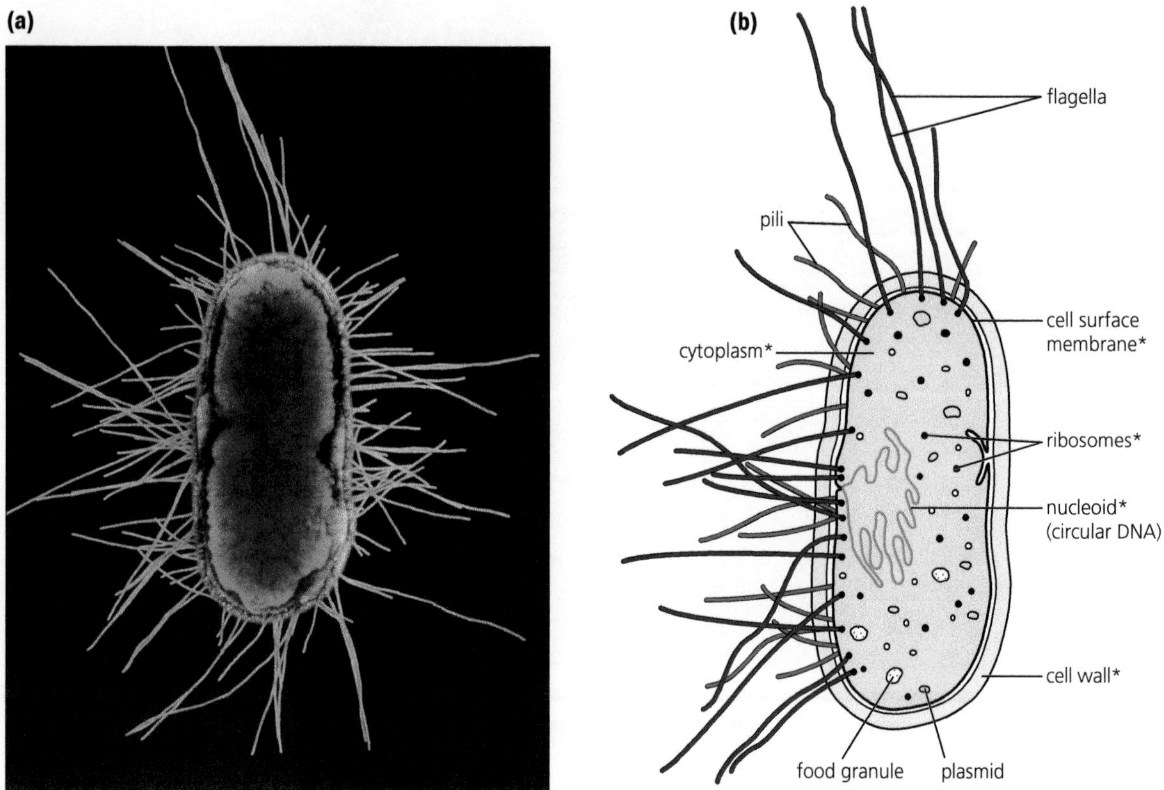

Figure 4.1 The structure of *Escherichia coli.* (a) A false colour scanning electron micrograph; (b) an interpretative drawing of the micrograph in which the asterisks show those structures that occur in all bacteria

Pathogenic effects produced by toxins

A toxin is a poison produced by an organism. Many animals and plants produce toxins, but here we will restrict ourselves to bacterial toxins. Pathogenic bacteria produce toxins that you can classify into two types: **endotoxins** and **exotoxins**. Table 4.1 summarises some of the important properties of these two types of bacterial toxin.

Table 4.1 A comparison of endotoxins and exotoxins

Property	Endotoxin	Exotoxin
Type of bacterium able to produce toxin	Gram negative only	Gram positive and Gram negative
Chemical nature of molecule	Lipopolysaccharide	Polypeptide or protein
Size of molecule/kDa	≈ 10	≈ 1000
Relationship with cell	Part of cell surface membrane	Secreted by cell
Potency/μg needed to cause symptoms	≈ 100	≈ 1
Can be denatured by boiling	No	Yes

Salmonella and endotoxins

Endotoxins are produced by bacteria with Gram-negative cell walls (see *Edexcel A level Biology 1*, page 87). Each endotoxin is a lipopolysaccharide that is embedded in the cell surface membrane of the bacterium. Figure 4.2 shows the general structure of a lipopolysaccharide.

The lipid-A component is the part that is embedded in the outer phospholipid layer of the cell surface membrane. It is this part of the lipopolysaccharide that is toxic.

The O-specific component lies outside the cell surface membrane. This part of the lipopolysaccharide is important in enabling the bacterium to invade its host. It is also this part of the lipopolysaccharide that has antigenic properties, that is, it can cause the production of antibodies against it.

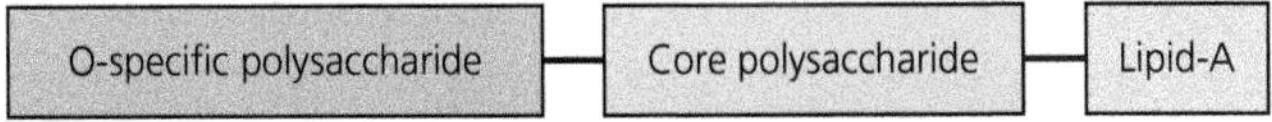

Figure 4.2 Endotoxins are lipopolysaccharide molecules

Figure 4.3 shows a scanning electron micrograph of one species of *Salmonella*, a Gram negative bacillus. Non-typhoidal *Salmonella enterica* produces a localised infection in the human intestines. The bacterium usually invades the body in contaminated food and results in food poisoning. A large number of bacteria must be ingested to cause symptoms in healthy adults. Once in the small intestine, *S. enterica* cells can invade cells lining the intestinal wall and disrupt the junctions between them. On death and lysis, the endotoxins are released from *S. enterica* cells. This causes inflammation and reduces the ability of the cells lining the intestinal wall to stop the movement of water and ions into the lumen of the intestine, resulting in diarrhoea.

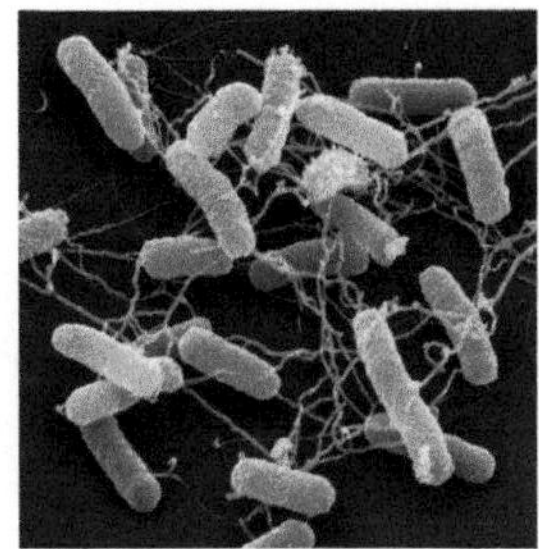

Figure 4.3 Scanning electron micrograph of *Salmonella enterica*

In contrast, typhoidal *Salmonella typhi* causes widespread symptoms. Once ingested in contaminated food, this bacterium invades the body via the lymphatic system, usually via lymph nodes in the tonsils or small intestine. Distributed via the lymph to all parts of the body, *S. typhi* invades body cells and multiplies inside them. Its endotoxins are released when the bacteria die and are lysed. Their release brings about the symptoms of typhoid fever, which include temperatures of up to 40 °C, ulcerations of the gut and, in extreme cases, death.

Staphylococcus and exotoxins

Figure 4.4 shows a scanning electron micrograph of *Staphylococcus aureus*. There are many species of *Staphylococcus* but they all show the same appearance, like bunches of grapes. In fact, that is how they get their generic name, from the Greek words for bunch of grapes (*staphyle*) and granules (*kokkos*).

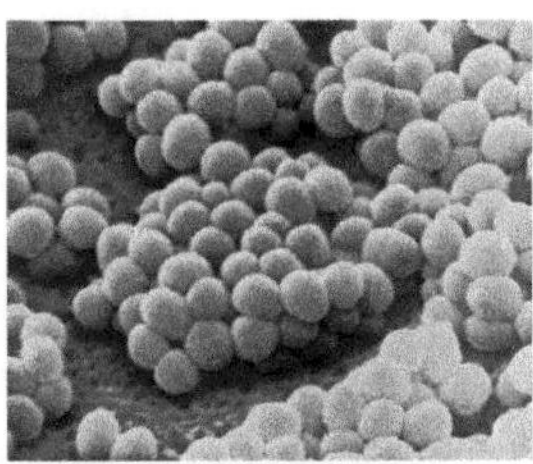

Figure 4.4 A scanning electron micrograph of *Staphylococcus aureus*, a very common bacterium on human skin. This bacterium is the causative agent of a number of human infections, including food poisoning, styes and septicaemia

Cells of *Staphylococcus* secrete a large number of enzymes, including collagenases, lipases, nucleases and proteases, which digest the tissues of their host. The products of this digestion are used by the bacteria as nutrients. *Staphylococcus* cells are also able to secrete two types of exotoxins:

- **Haemolysins** – polypeptides that become integrated in the cell surface membranes of the host's cells, creating pores. These pores cause the host's cells to lose water and ions.
- **Superantigens** – polypeptides that stimulate large numbers of cells of the immune system, resulting in a massive release of cytokines (see Chapter 5, page 87) into the blood. Toxic shock syndrome is one effect of a massive release of cytokines. In toxic shock an otherwise healthy individual develops a high fever and low blood pressure. This may progress to a coma and multiple organ failure.

Figure 4.5 A colony of the Gram positive bacillus *Mycobacterium tuberculosis*

Pathogenic effects produced by invasion of host's tissue

Tuberculosis is a disease of the lungs caused by the bacterium *Mycobacterium tuberculosis*. Figure 4.5 shows a scanning electron micrograph of a colony of this bacterium. Infection occurs when it is inhaled, usually by droplets in coughs and sneezes. However, since its walls are rich in lipid, it can survive for many months on dry surfaces and so can be inhaled in dry dust.

Once inside the lungs, the bacteria are engulfed by macrophages (see Chapter 5, page 81) in the alveoli and bronchioles. If the recipient is in good health, the growth of bacteria is restricted in the lungs. Figure 4.6 shows how this is done. The macrophages that have engulfed the bacteria become surrounded by other cells of the immune system in a structure known as a **granuloma**. The macrophages normally kill the bacteria but a few often survive as a latent infection that could result in infection years later. In fact, between 60–80 per cent of people in the UK are thought to carry these bacteria in a dormant form like this.

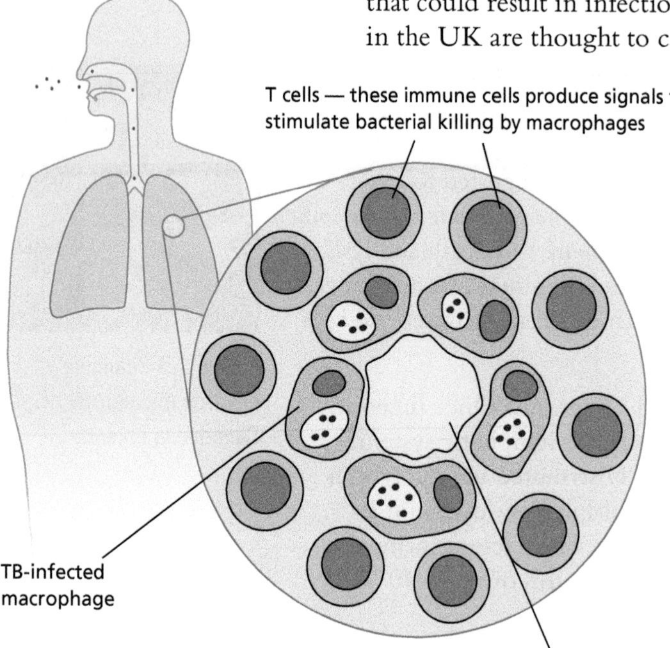

Figure 4.6 A granuloma containing macrophages that have engulfed *M. tuberculosis*

If the recipient has an immune system already weakened by malnourishment or poor health, however, a chronic infection might develop within the lungs. *M. tuberculosis* secretes hydrolytic enzymes into the host's cells and digests them. Unusually, its energy source is cholesterol, which is a component of the surface membranes of mammal cells. As the bacterium digests cells, cavities appear in the lung tissues, blood vessels are broken down and fluid collects. The patient coughs blood in the sputum. The chest X-ray in Figure 4.7 shows the lung damage caused by tuberculosis.

Although initially a lung infection, macrophages can carry *M. tuberculosis* to almost any part of the body, including the central nervous system, the membranes surrounding the brain (the meninges), bone tissue, lymph glands, liver, kidneys and genital organs. Once there, the pathogen can cause cell destruction in its new location.

Test yourself

1 Distinguish between a parasite and a commensal.
2 What is a pathogen?
3 Explain how the properties of an endotoxin enable it to be embedded in the cell surface membrane of a bacterium.
4 Is the toxin produced by *Salmonella* denatured by cooking? Explain your answer.
5 It is estimated that 60 to 80% of the population carry *Mycobacterium tuberculosis*. Explain why the bacterium does not cause disease in these people.

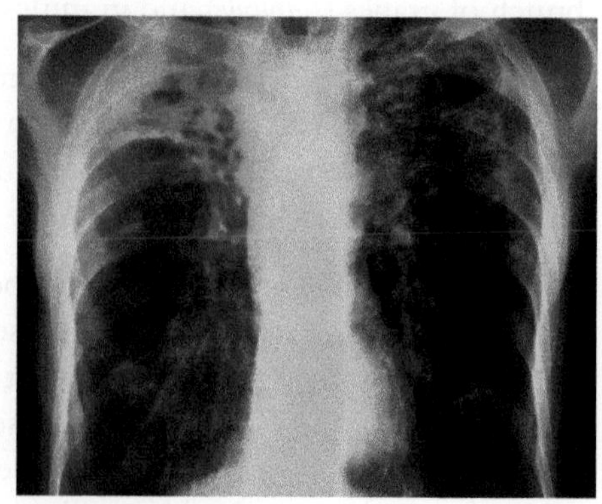

Figure 4.7 A chest X-ray showing the tissue damage caused by tuberculosis

Antibiotics

An antibiotic is a chemical substance that, in low concentrations, kills or inhibits the growth of microorganisms. Most are derived from fungi and bacteria commonly found in the soil, where they provide a competitive advantage to the producers.

Key term

Antibiotic A substance that, in low concentrations, kills or inhibits the growth of microorganisms.

Figure 4.8 shows an agar plate that has been inoculated with bacteria, using the spread plate technique (pages 48–49). A mast ring, with each arm impregnated with a different antibiotic, was placed on this plate before it was incubated. As each antibiotic diffused into the agar, it prevented the growth of the bacteria on the plate, leaving an area of clear agar. You can see from the diameter of the clear areas that some antibiotics were more effective against this bacterium than others.

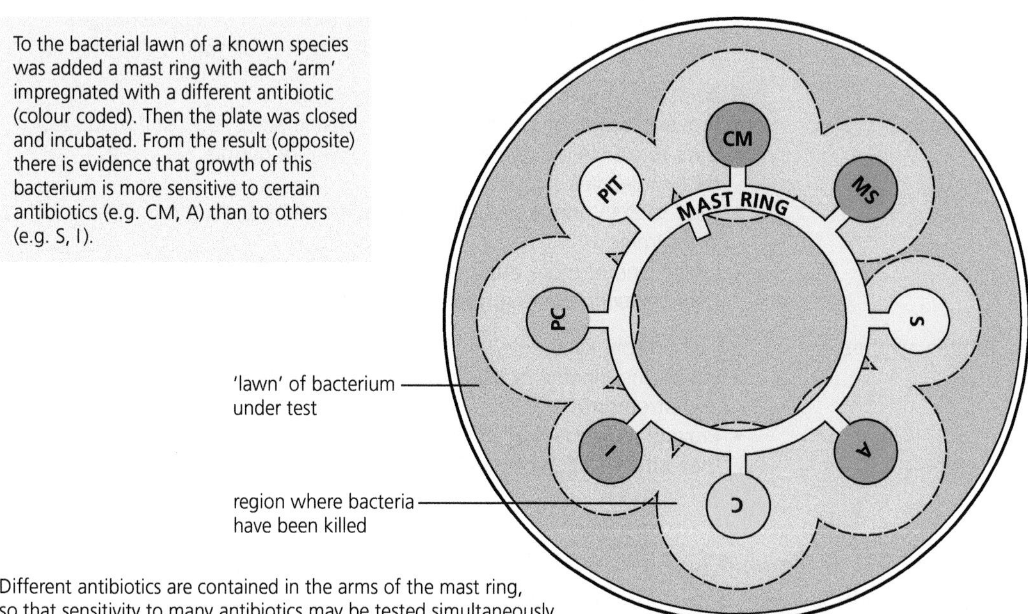

Figure 4.8 Investigating sensitivity to antibiotics. The mast ring shown here is available from suppliers

Some antibiotics are effective against a wide range of pathogenic bacteria; they are called **broad-spectrum antibiotics**. Tetracycline is an example of a broad-spectrum antibiotic. Others, including penicillin, are effective over a limited range of bacteria; they are called **narrow-spectrum antibiotics**. Although these terms remain useful, the development of antibiotic resistance, discussed below, results in what were once broad-spectrum antibiotics becoming restated in the range of pathogens against which they are effective.

The action of antibiotics

There are several ways in which antibiotics work. In general, though:

- **bacteriostatic antibiotics** prevent the multiplication of, but do not kill, bacteria. As a result, they prevent an infection spreading, but it is the host's immune system that finally overcomes the pathogen
- **bactericidal antibiotics** do kill bacteria.

Table 4.2 summarises the three major biochemical mechanisms that are targeted by antibiotics. Ideally, antibiotics should affect the metabolism of bacterial cells without interfering with that of the host. In some cases, however, the particular bacterial components, metabolites or enzymes targeted by the antibiotic are also components of the eukaryotic cells of the mammalian host. The extent to which this occurs influences just how toxic the drug is to the mammalian host tissues. This, in turn, determines whether, and to whom, the antibiotic might be prescribed.

Table 4.2 The biochemical mechanisms of antibiotics

Bacterial mechanism targeted	Effects of antibiotics, with specific examples
Cell wall synthesis	Disrupt the synthesis of bacterial cell walls • Inhibit the enzyme that catalyses the formation of cross-linkages between the peptidoglycan molecules in the cell wall, e.g. penicillin
Nucleic acid synthesis	Disrupt DNA replication or DNA transcription • Prevent formation of precursors of nucleic acids, e.g. sulfonamides • Bind to a DNA molecule and break down its double helix, e.g. nitroimidazoles • Bind to DNA and prevent its replication and transcription, e.g. clofazimine • Inhibit one or more enzymes that catalyse either DNA replication or DNA transcription, e.g. quinolones
Protein synthesis	Inhibit protein synthesis • Inhibit the binding of tRNA to bacterial ribosomes, e.g. tetracycline and streptomycin • Prevent movement of tRNA-peptide complex, e.g. erythromycin • Prevent formation of peptide bonds, e.g. chloramphenicol

The mechanism of action of penicillin – a narrow-spectrum antibiotic

Looking back to Figure 4.1, you will see that possession of a cell wall is a feature common to all bacteria. The wall contains layers of peptidoglycan linked together by cross-linkages. Whilst Gram negative bacteria have only a single layer of peptidoglycan in their walls, Gram positive bacteria might have as many as 40 layers.

As it grows, an individual bacterium constantly remodels its cell wall, breaking parts down and re-building them. Penicillin is an irreversible inhibitor of transpeptidase – the enzyme that catalyses the formation of the cross-linkages between the layers of peptidoglycan. It does not, however, affect the hydrolase enzymes that break the wall down. Consequently, the cell wall of a bacterium becomes thinner and thinner.

Under the influence of penicillin, Gram positive bacteria lose their cell walls entirely and are left as naked **protoplasts**. These cells are much more vulnerable than their counterparts with intact walls and are rapidly killed. Gram negative bacteria are affected in a similar way but do not completely lose their cell walls.

The mechanism of action of tetracycline – a broad-spectrum antibiotic

Bacteria possess 70S ribosomes. Each has two sub-units – a 30S and a 50S sub-unit – made of RNA and proteins. During the translation of mRNA, tRNA molecules, each carrying its appropriate amino acid, bind to the 30S sub-unit of the bacterial ribosome. Tetracycline prevents this binding, so that the affected ribosome is no longer able to manufacture a polypeptide chain. Interestingly, tetracycline does not affect eukaryotic cells because, unlike prokaryotic cells, they do not actively transport tetracycline into their cytoplasm.

Test yourself

6 Antibiotics are derived from soil bacteria and fungi. Suggest how these bacteria and fungi benefit from the antimicrobial substances they secrete.

7 Which of the antibiotics in Figure 4.8 is the most effective against the bacteria? Explain your answer.

8 Explain why the administration of bacteriostatic antibiotics usually lasts longer than that of bactericidal antibiotics.

9 Tetracycline slows bacterial protein synthesis by combining with bacterial ribosomes. It does not stop it, however. Suggest how bacteria are able to continue to produce some protein.

10 Rapidly acting bactericidal antibiotics against non-typhoidal *Salmonella enterica* can sometimes lead to symptoms. Suggest why.

Antibiotic resistance in bacteria

As you have seen, antibiotics can be classed as broad-spectrum or narrow-spectrum. Since narrow-spectrum antibiotics affect only a few groups of bacteria, it follows that others are naturally resistant to these antibiotics. This is called **primary resistance**. In recent years, groups of bacteria that were once susceptible to an antibiotic have become resistant to it. This is called **secondary resistance** and is becoming a major problem in treating disease.

Bacteria might acquire secondary resistance to antibiotics by one of three biochemical mechanisms:

- A decrease in the uptake, or increase in the expulsion, of the antibiotic by the bacterial cell.
- Production by the bacterial cell of enzymes that can modify or inactivate the antibiotic.
- Development by the bacterial cell of a biochemical pathway that bypasses the reaction affected by the antibiotic.

You might wonder how a bacterium can suddenly become resistant to an antibiotic to which it has previously been susceptible. The following facts might lead you to understand how this can happen.

- In many cases, resistance is known to be the result of a single gene. For example, resistance to the sulfonamide antibiotics is controlled by a single gene, encoding a synthase enzyme.
- During DNA replication, errors occur at an estimated rate of 1 in 10^6 base pairs. These errors are **gene mutations** that produce different alleles of a gene.
- Under optimal conditions, some bacteria replicate their DNA and divide by binary fission every 30 minutes or so.

Taking these observations together, it is easy to see how a gene mutation in a single bacterial cell might produce a new allele of a gene that confers resistance to an antibiotic. If this happened 100 years ago, before the development of antibiotics, this cell would not have been at an advantage over cells that did not possess this new allele. In fact, it might have been at a disadvantage. If this happened today, in a mammal being treated with the antibiotic, this cell would be at an advantage. Figure 4.9 shows what would then happen. The antibiotic would kill, or stop the growth of, the susceptible cells but the cell carrying the mutant allele would be unaffected. In time, it would give rise to millions of daughter cells by binary fission. As a result of DNA replication prior to binary fission, each of these daughter cells would carry the new allele, conferring resistance against the antibiotic in question. The entire population of bacteria would be antibiotic resistant.

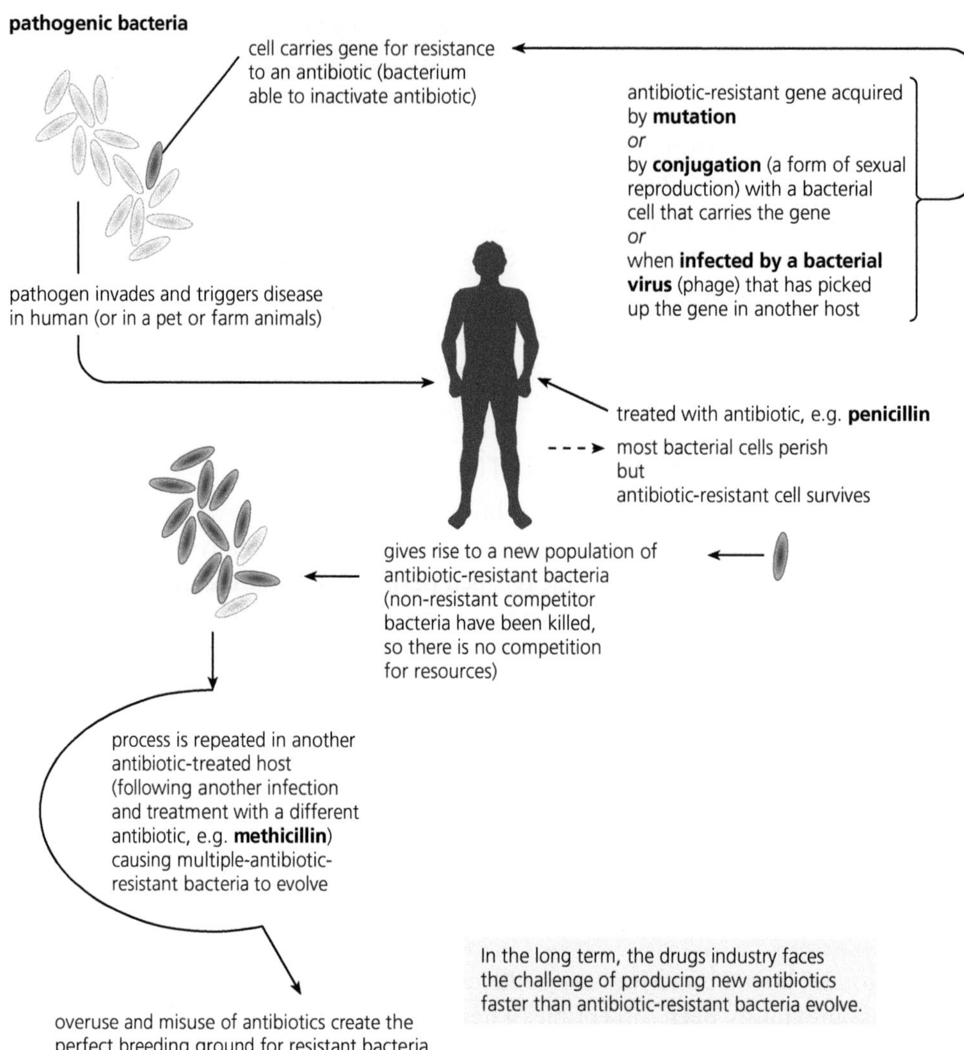

Figure 4.9 The evolution of a multiple-antibiotic-resistant population of bacteria

Above, we have considered a bacterium that replicates its own DNA, including a mutant gene, and passes a copy to each of the daughter cells it produces by binary fission. This is known as **vertical gene transfer**. It is not the only way that bacteria can acquire new genes, however. Figure 4.10 shows a transmission electron micrograph of bacterial cells that are passing genetic material between each other. You see slender white connections between cells. These are called **conjugation tubes**. The diagrams in Figure 4.10 show that the bacteria involved can pass DNA, either one of their plasmids or a part of their own nucleoid, through the tube to a different cell. This is called **horizontal gene transfer**. If, by chance, the DNA that is passed from one cell to another contains a gene conferring antibiotic resistance, the recipient cell immediately becomes resistant to the antibiotic. This exchange can occur between bacteria of the same species, as shown in Figure 4.10, or between bacteria of different species. In the latter case, the gene conferring antibiotic resistance has 'jumped the species barrier'. Medical scientists believe that horizontal gene transfer accounts for the increasing number of bacteria that show resistance to many antibiotics – **multiple-antibiotic-resistant bacteria**.

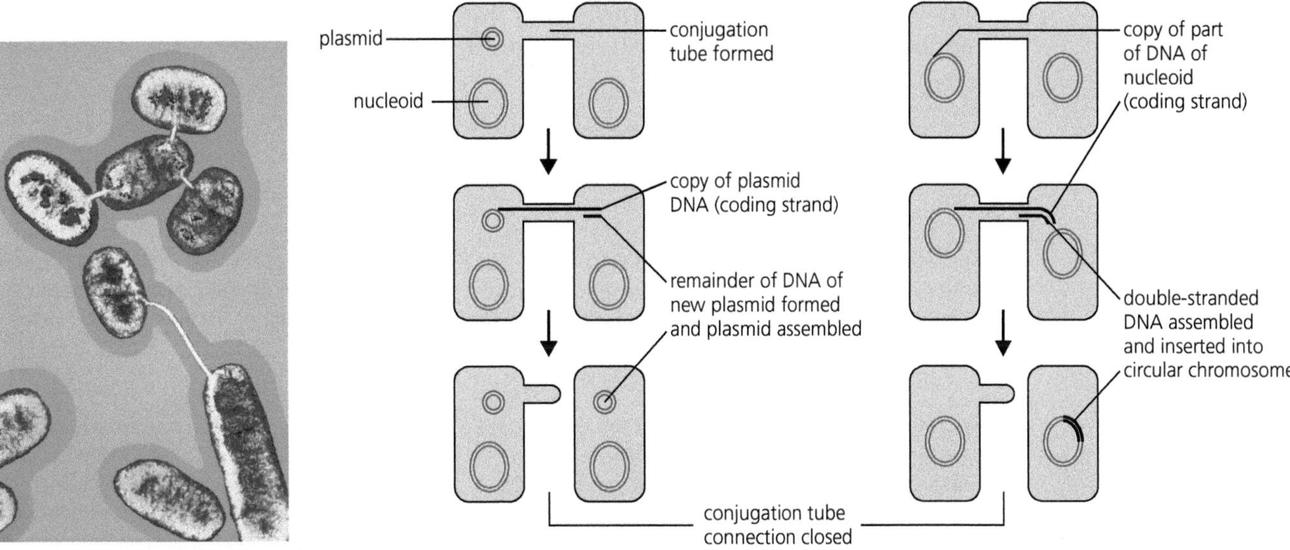

Figure 4.10 Horizontal gene transfer occurs when DNA is passed from one bacterial cell to another via a conjugation tube

Controlling the spread of antibiotic resistance in bacteria

When they were introduced in the 1940s, most bacteria were susceptible to antibiotics. As the bar chart in Figure 4.11 shows, however, even then some were not. The bar chart clearly shows that *Staphylococcus aureus* in about 3% of samples obtained worldwide were resistant to penicillin. It also shows how that percentage increased to the end of the last century. This occurred in many other species of bacteria and with many other antibiotics.

Given their rapid rate of growth, the evolution of bacterial resistance to antibiotics was almost inevitable. What has alarmed people is the rapidity of this evolution. To understand why, you need to look at ways in which antibiotics have been misused. They include the following practices.

- Sub-clinical doses of antibiotics are routinely used in agriculture as growth promoters.
- Doctors, even in the UK, prescribe antibiotics for patients with viral infections, despite the fact that viruses are unaffected by antibiotics.
- In many countries antibiotics are sold over the counter without prescription and, therefore, are often an inappropriate antibiotic.

In all three instances above, animals or humans often take sub-clinical concentrations of antibiotics. Since they are sub-clinical, they do not destroy bacteria but the presence of antibiotics does create a selective pressure for resistance. You might wonder how a doctor prescribing antibiotic for a viral infection results in people taking sub-clinical concentrations of antibiotic. Whilst taking the antibiotic, these people often begin to feel better as their own immune systems overcome the infection and so they do not complete the full treatment of antibiotic.

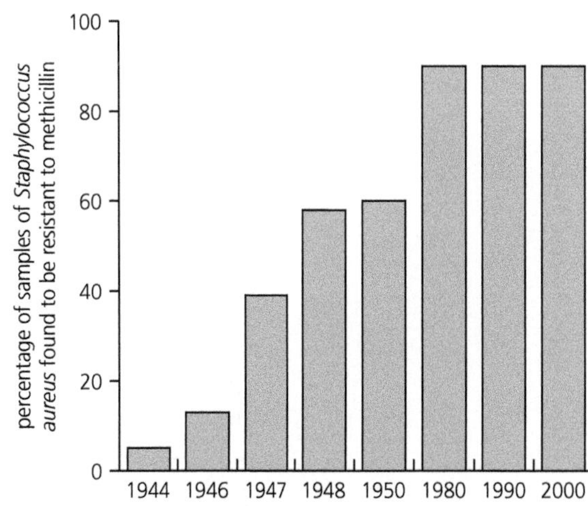

Figure 4.11 The percentage of samples of *Staphylococcus aureus* found to be resistant to penicillin from a large number of studies worldwide

The UK and the European Union parliaments have issued guidelines about the use of antibiotics. Some of the measures proposed are shown in Table 4.3. They are, however, voluntary guidelines. The parliaments feel they cannot legislate because, for example, it is not considered ethical to prevent doctors acting in what they consider to be the best interest of their patients.

Table 4.3 Measures that can reduce the use of antibiotics

Measure	Explanation	Problems that might be encountered
Restrict addition of antibiotics to animal feeds	Animal feeds account for about half all sales of antibiotics. Used at sub-clinical concentrations, they do not destroy bacteria but create a selection pressure that favours bacteria that acquire resistance.	The agricultural industry forms a strong lobby group and would resist any restriction on practices that increase the profitability of farming or the quality of their products.
Doctors to restrict prescriptions for antibiotics	If family doctors stop prescribing antibiotics as a precautionary measure and prescribe them only in cases of established need, the selective pressure favouring bacteria that acquire resistance will become less.	This might mean that doctors advise patients to go home to bed for a few days until their own immune systems overcome the infection. Many patients are unwilling to take this advice or to leave their doctor's surgery without an antibiotic, putting their doctors under severe pressure.
Patients must complete their full course of antibiotics	Many patients begin to feel better as their own immune system overcomes the bacterial infection and so stop taking the antibiotic. Again, the bacteria are not destroyed by the antibiotic but a selection pressure favouring bacteria that acquire resistance is created.	We rely on people acting in the best interests of the community at large. In general, we do what we feel like doing.
Create new antibiotics	This is the obvious solution – make new antibiotics faster than bacteria can acquire immunity against them.	The development of a new antibiotic takes years and the success rate is very low. This makes the development of new antibiotics not only a slow process but a very expensive one. If pharmaceutical companies fund this research, they must recoup their costs by making the drug expensive, thus restricting its availability. If governments fund the research, they must use taxpayers' money, possibly leading to unpopular tax rises.

The samples from which Figure 4.11 was produced were taken from patients in hospitals. This provides a clue about the origin of most infections by bacteria with multiple antibiotic resistance. In the UK, infections resulting from bacteria with multiple antibiotic resistance are mainly acquired in hospitals or in community-based care homes. The reasons for this are not hard to identify.

- Patients in hospitals or nursing homes are either elderly or ill, or both. They have weakened immune systems and so are more susceptible to disease.
- Many people are admitted to hospital because they are suffering from a bacterial infection. Consequently, hospitals are likely to have richer populations of pathogenic bacteria than the rest of the community.

In hospitals, bacteria can be transmitted from person to person by airborne or droplet transmission, by direct skin contact, or by contact with clothing, instruments and equipment. The medical councils of most European countries have issued advice about the.steps that hospital staff should take to reduce the risk of transmission of bacteria. Some are listed in Table 4.4.

Table 4.4 Measures taken to reduce the spread of hospital-acquired infections and some problems that might arise in meeting these guidelines

Measure	Explanation	Problems that might be encountered
Isolation of infected patients	Reduces risk of transmission between infected patient and other vulnerable patients.	Given bed shortages in many hospitals, isolation rooms are not always available.
Hand washing by staff and visitors	Effective hand washing removes bacteria from the skin and so reduces transmission by direct skin contact. This is a cheap and readily available method of reducing transmission.	It is a voluntary act that might not be followed by the thousands of people visiting patients in hospital each day.
Staff to wear disposable gloves and aprons whilst handling patients	Provided they are disposed of after handling each patient, e.g. washing or turning patient over in bed, the aprons and gloves provide a protective barrier for patients and staff.	The time spent removing and disposing of aprons and gloves when passing from bed to bed might restrict the time that hospital staff are able to spend with each patient. There might not be enough time to put on the disposable wear if an emergency occurs with one patient. Though individually cheap, the large number of aprons and gloves would lead to a considerable additional annual cost.
Screening patients on arrival for 'superbug' infections	Screening allows the isolation of infected patients, so reducing the risk of transmission to other vulnerable patients.	Screening is costly and time-consuming and isolation rooms might not be available.

Test yourself

11 Give **three** ways in which resistance to antibiotics can arise.

12 Bacteria that are thousands of years old, found in cores of ice from the Antarctic, have been shown to be resistant to antibiotics. How can this be explained?

13 Suggest how the inclusion of sub-clinical doses of antibiotic in animal feed might increase yield.

14 Antibiotics do not cause mutations that result in antibiotic resistance. Why, then, might failing to complete a course of antibiotics result in bacteria developing resistance to the antibiotic?

15 Methicillin-resistant *Staphylococcus aureus* became a problem in hospitals during the 1990s. Explain why the problem was largely restricted to hospitals.

Other pathogenic agents

Bacteria are not the only pathogenic organisms that can cause human infections. Here, we will consider one example of each of: a fungal infection; a viral infection; and an infection caused by a protoctist.

Stem rust fungus (Puccinia graminis)

Figure 4.12 Crop plant infected by stem rust fungus

Figure 4.12 shows wheat plants infected by stem rust fungus. You can see how this parasite gets its name – it infects the plant's stem (including leaves) producing a covering that is the colour of rust. The variety of stem rust fungus shown (*Ug99*) is a new variety that first appeared in Uganda in 1999. The *Ug99* strain has since spread and is currently devastating crops of wheat in countries in East Africa, the Middle East and Asia and computer models predict it will reach India, one of the world's largest producers of wheat. Since it has been estimated that 85 per cent of the world's population depend on wheat as one of their only sources of energy and 60 per cent as their main source of dietary protein, *Ug99* has the potential to cause worldwide food scarcity.

Like most fungi, the stem rust fungus produces dormant spores. These spores are carried in the air to new plants. On landing on a new plant, the spores germinate and produce threadlike structures, called **hyphae**. A mass of fungal hyphae, such as those you can see in Figure 4.12, is called a **mycelium**. Like all fungi, the stem rust fungus secretes digestive enzymes from its hyphae onto the material on which it is growing - in this case the stem of a cereal plant. The enzymes digest chemicals in the stem and the fungus absorbs the products of this digestion.

In addition to digesting its tissues, infection by stem rust fungus damages the host plant in a number of ways:

- It weakens the stem, often causing the plant to fall over. This makes mechanical harvesting impossible.
- It uses nutrients that would otherwise be stored in the plant's seeds. This reduces the harvest.
- It breaks the outer epidermis of the plant's stem. This increases the rate of water loss from the plant as well as making the stem more susceptible to infection by other plant pathogens.

Controlling the spread of stem rust fungi

Fungicides can be used to kill the stem rust fungus. These are usually expensive, reducing their availability to poor farmers in Africa and Asia. As associations between plant roots and fungi are vital to the roots' efficient absorption of inorganic ions from soil, spraying fungicides can damage delicate ecosystems.

Since, after harvesting the seeds, cereal plants are destroyed, you might think that stem rust fungus would be a problem for only one growing season. One reason that this is not the case lies in the complex life cycle of the fungus. Although the primary host of the stem rust fungus is a cereal crop, it depends on another type of plant to complete its life cycle – the barberry. Knowing this, North America began a barberry eradication programme in 1918, which continues to this day. Although getting rid of barberry plants has reduced infection rates, it has not eliminated stem rust because new spores are carried to North America by winds from the southern states of America and Mexico, in the so-called 'Puccina pathway'. In fact, spore dispersal occurs over vary large distances. In 2002, one group of scientists reported that stem rust fungus spores had dispersed up to 8000 km from the south of Africa to Australia.

The most promising way of eradicating stem rust fungus is by gene manipulation. Scientists have identified a number of genes that confer resistance to stem rust fungus and, using genetic engineering techniques that we will discuss in Chapter 7, have produced varieties of cereals resistant to some strains of stem rust fungus. Some of the most effective have only been used in Australia, so their effectiveness in other parts of the world is, as yet, unknown.

Influenza

Influenza, or flu, is caused by a virus. Figure 4.13 shows the structure of the influenza virus. It contains eight short strands of RNA (referred to as a segmented genome), surrounded by a protein capsid. Outside the capsid is an envelope with an outer lipid layer and an inner protein layer. Glycoprotein 'spikes' project through this envelope and cover the surface of the virus. The H glycoproteins (haemagglutinin) help the virus particle enter a cell of the host. The N glycoproteins (neuramidase) allow newly formed virus particles to escape from a cell of the host.

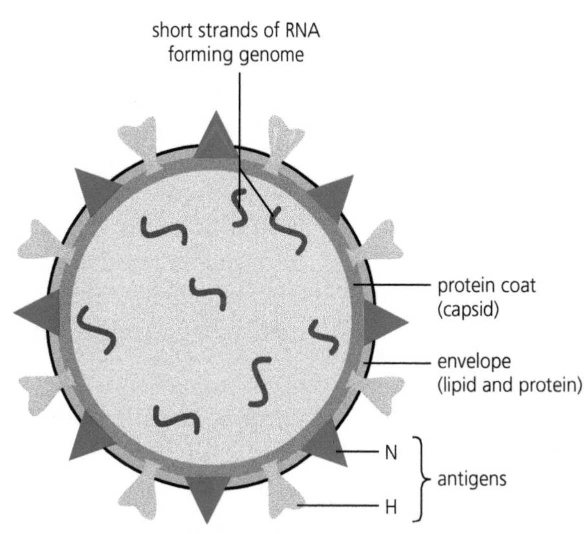

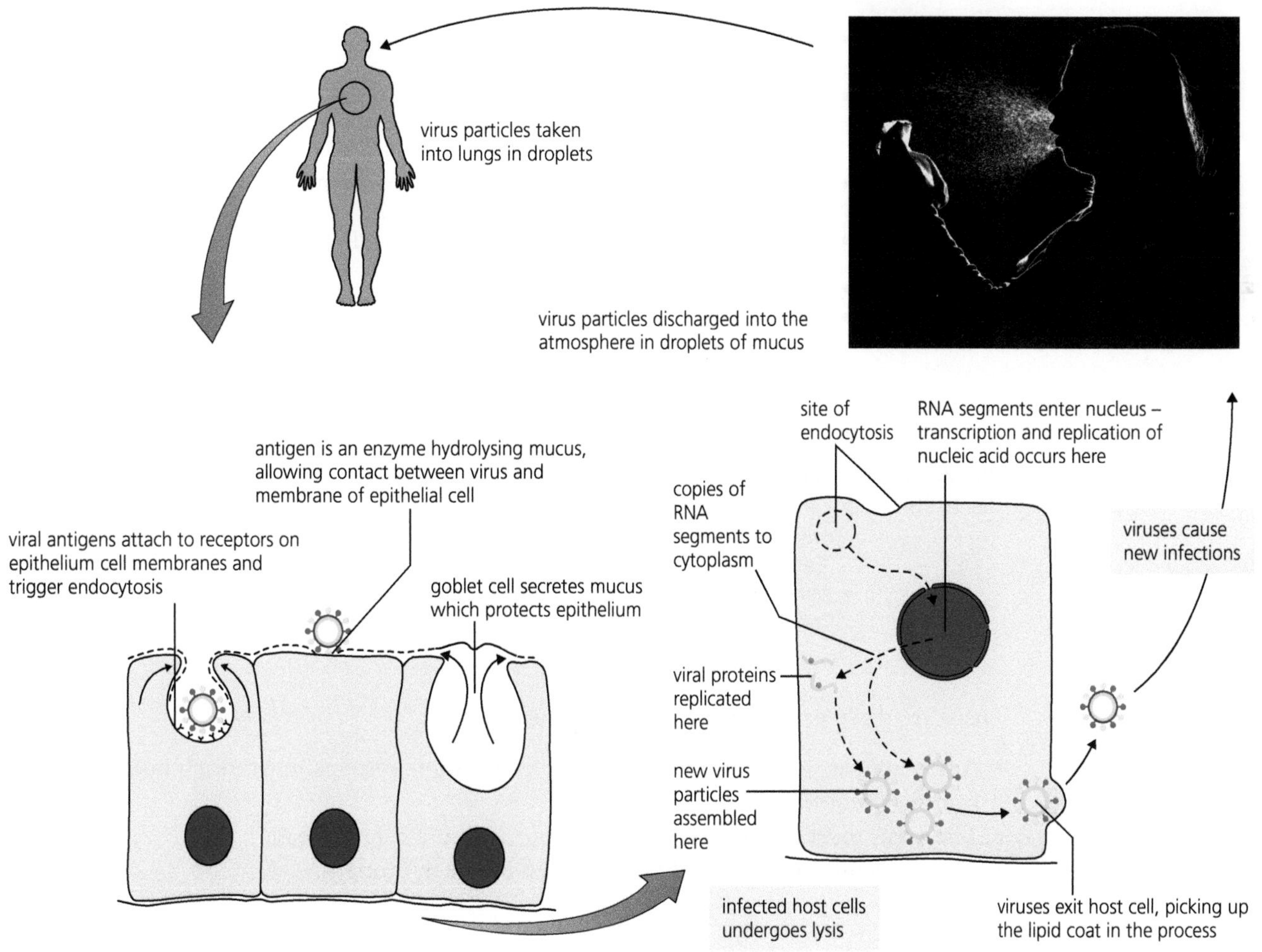

Figure 4.14 summarises the way in which the influenza virus causes infection. The events follow the lytic cycle, which you came across in *Edexcel A Level Biology 1*, Chapter 4 page 90. The influenza virus enters the human host's lungs via droplets inhaled through the nose or mouth. The virus then enters the epithelial cells lining the bronchus and bronchioles by endocytosis.

short strands of RNA forming genome
protein coat (capsid)
envelope (lipid and protein)
N
H
antigens

Figure 4.13 An influenza virus particle

virus particles taken into lungs in droplets
virus particles discharged into the atmosphere in droplets of mucus
antigen is an enzyme hydrolysing mucus, allowing contact between virus and membrane of epithelial cell
viral antigens attach to receptors on epithelium cell membranes and trigger endocytosis
goblet cell secretes mucus which protects epithelium
site of endocytosis
RNA segments enter nucleus – transcription and replication of nucleic acid occurs here
copies of RNA segments to cytoplasm
viral proteins replicated here
new virus particles assembled here
viruses cause new infections
infected host cells undergoes lysis
viruses exit host cell, picking up the lipid coat in the process

Figure 4.14 The life cycle of the influenza virus

Key terms

Antigenic variability Variation in the chemical nature of the same type of antigen resulting from frequent gene mutations and, in eukaryotes, different splicing of pre-mRNA molecules transcribed from the same gene.

Epidemic An outbreak of disease in which the number of new infections is much greater than would be expected from recent experience. The outbreak is widespread within one community at a particular time.

Pandemic A human disease that has spread worldwide.

Replication of the virus then occurs; new RNA segments are produced in the nucleus and the capsid proteins are produced in the cytoplasm of the host cell. New viral particles are assembled and the outer envelope is added as the new virus particles leave the dying host cells. As the host cells break down (lyse), toxins are released, which bring about many of the symptoms of influenza. The breakdown of epithelial cells also opens the way for secondary infections of the host's lung tissue.

Influenza viruses are classified into antigenic groups according to the H and N glycoproteins of their capsids, e.g. H5N1. When in human tissues, these glycoproteins act as antigens. The influenza virus shows great **antigenic variability**, meaning that mutations of the viral genome frequently occur that result in new forms of these antigens appearing on the outer surface of the virus. Often humans (or other hosts) have little or no resistance to these new antigens. Their immune systems will not have encountered them before, so no memory cells (see Chapter 5, page 86) with complementary receptors will be present in their lymph nodes. Because of this, influenza frequently causes a major **epidemic**.

From time to time, mutations lead to a strain of the influenza virus that is extremely virulent. A strain with just the right mix of virulence and transmissibility leads to a **pandemic** in which millions of humans die. During the last century, there were three such pandemics. These are listed and described in Table 4.5.

Table 4.5 The three flu pandemics of the 20th century

Pandemic	Profile
Spanish flu 1918–19	• Killed about 40 million people (compare this with the 10 million people killed in the First World War of 1914–18) • Nations struggled to cope; the end of the war was a time when resources were exceptionally stretched, and viruses were not understood • Vaccines were incorrectly targeted at bacteria • Civilians were put into quarantine (but troop movements continued between continents) • High levels of personal hygiene were advocated
Asian flu 1957–8	• Killed about 1 million people • Medical knowledge was more advanced than in 1918 but, also, the strain of virus was substantially less virulent • The strain of virus was identified and vaccines were produced but it was not possible to produce enough • Quarantine measures were used, again to little effect
Hong Kong flu 1968–9	• Killed about 1 million people • By this date the WHO existed and its global flu surveillance network gave early warning of an imminent pandemic as the disease spread from its origins in East Asia • Vaccines were developed quickly but not enough could be produced in time to meet the full demand

There are three pre-requisites for a flu pandemic:

- A novel virus strain, unfamiliar to human immune systems, must reach human hosts from its point of origin.
- The virus must be able to replicate in humans and cause disease.
- The virus must be efficiently transmitted between humans.

You might remember the public discussions and anxiety about the bird flu epidemic that threatened us in 2005–06. That strain was identified as H5N1. The interlocking flight paths of wild birds, migrating between seasonal feeding grounds in different parts of the globe, transmitted the infection among wild populations of birds and, occasionally, to stocks of farmed birds. Some unfortunate people, in countries as widely apart as the Far East and Eastern Europe, who had direct contact with infected birds contracted the disease. For some, this exposure proved fatal.

The H5N1 strain of 2005, although extremely virulent, failed the third of the pre-requisites listed above. It was not a strain that was quickly and easily transmitted between individual humans, and so no pandemic ensued (at that time).

The malarial parasite (*Plasmodium* spp.)

As Figure 4.15 shows, malaria is **endemic** in tropical and sub-tropical regions of the world. The WHO estimates that about 400 million people are infected with malaria, of whom 1.5 million (mostly children under 5 years old) die each year. About 80 per cent of these cases of malaria occur in sub-Saharan Africa.

Key term

Endemic A disease that occurs frequently, or is constantly present at a low level of infection, in particular areas or among people of a particular population.

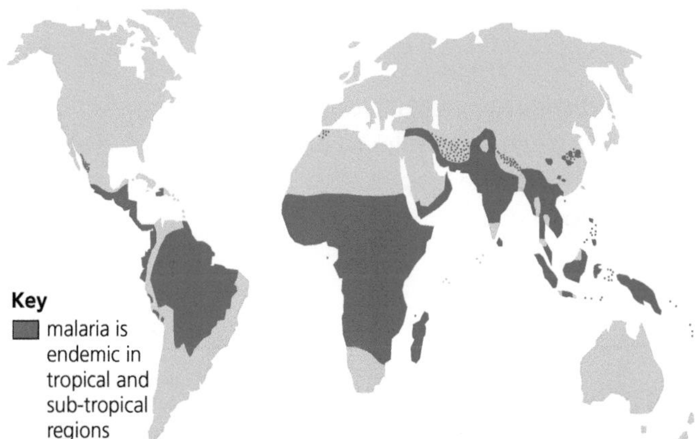

Figure 4.15 World distribution of malaria

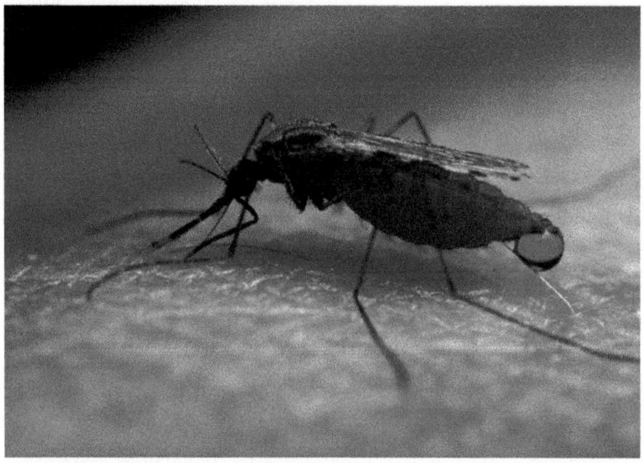

Figure 4.16 A female *Anopheles* mosquito taking a blood meal

Malaria is caused by four species of *Plasmodium*, a single-celled protoctist. About 90 per cent of malaria cases are caused by *Plasmodium falciparum*. Transmission of *Plasmodium* from an infected person to another person is by the *Anopheles* mosquito. Although the male feeds on plant juices, the female mosquito takes a blood meal from an unsuspecting human. She does this by inserting her piercing mouthparts. Using her mouthparts rather like a mini-hypodermic needle, the female inserts them through the skin into a blood vessel and injects saliva (Figure 4.16). The saliva contains anticoagulants that enable the mosquito to draw blood through its mouthparts without the blood clotting. The saliva causes inflammation and itching of the mosquito bite. If the female mosquito is infected, the saliva also contains the infective stage, called sporozoites, of *Plasmodium*.

Within a few minutes, the injected sporozoites enter cells in the liver of their new human host. As Figure 4.17 on the next page shows, each sporozoite rapidly divides to form thousands of daughter cells, called merozoites. Eventually, the infected liver cells burst, releasing their merozoites into the blood.

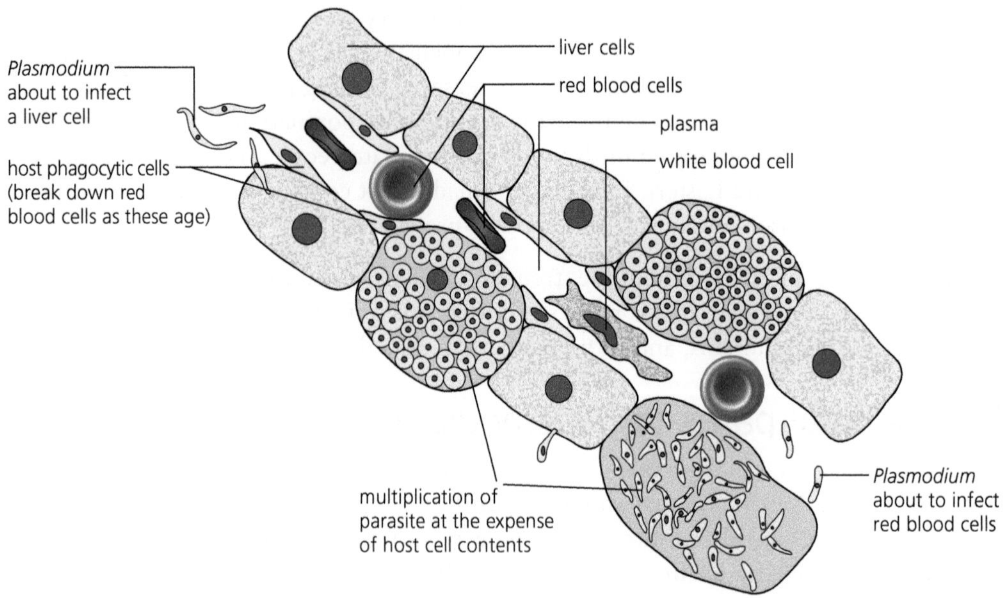

Figure 4.17 Following injection by the bite of an infected female *Anopheles* mosquito, *Plasmodium* cells enter liver cells where they divide

The released merozoites enter red blood cells, where they digest the haemoglobin as a food source. Each merozoite undergoes several cell cycles to produce between 8 and 32 new merozoites. The red blood cells burst, releasing these new merozoites into the blood, where each infects another red blood cell.

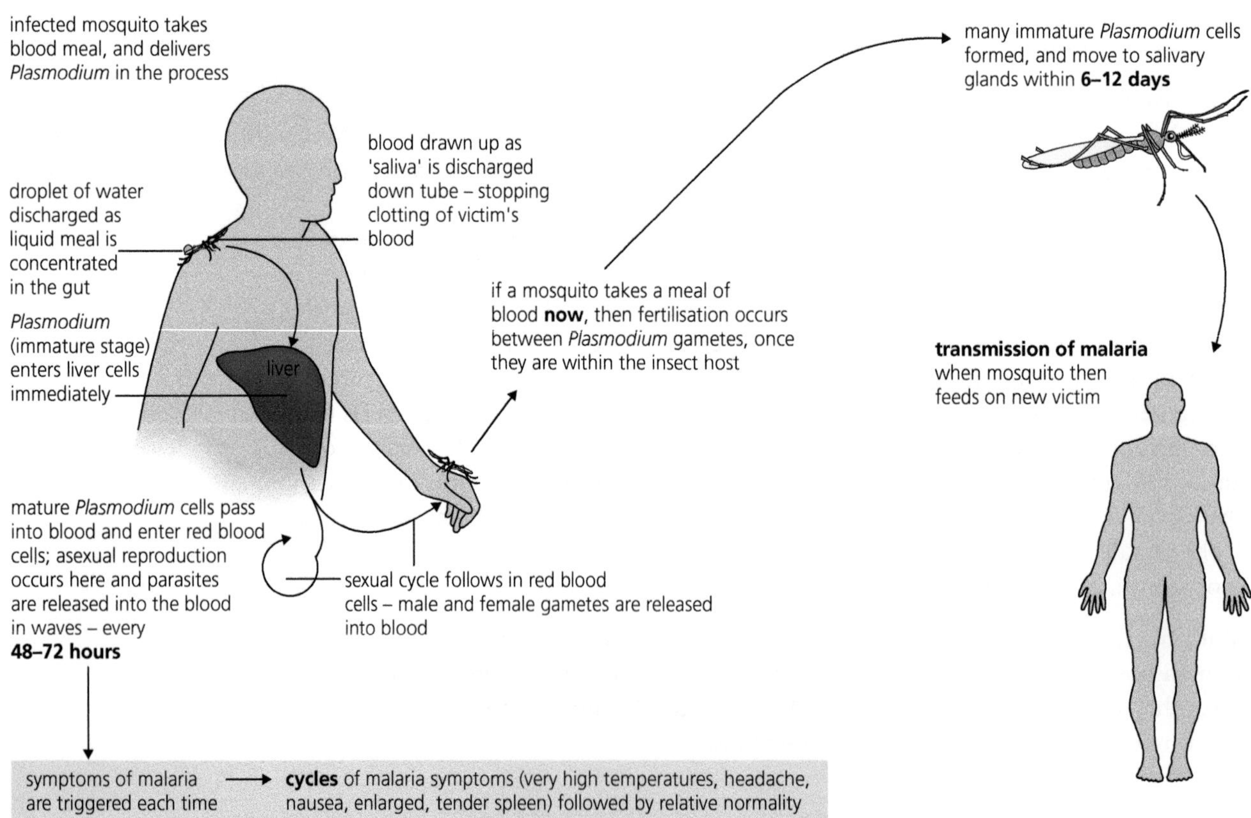

Figure 4.18 The transmission of malaria from person to person by a female *Anopheles* mosquito

A cycle of infection of red blood cells and merozoite release then follows (Figure 4.18). Every 3 to 4 days, increasing numbers of red blood cells burst, releasing merozoites that infect new red blood cells. Each red blood cell that bursts releases toxins produced during the breakdown of its contents by merozoites. These toxins cause the symptoms of malaria, including a body temperature of 40–45 °C, intense fever symptoms and a swollen spleen.

Prevention of infection includes a number of methods. Each has its drawbacks, however, as shown in Table 4.5.

Table 4.5 Some methods of preventing malaria

Method to reduce incidence of malaria	Issues
Drain wetland areas where mosquitoes breed	Not possible to drain large lakes, which are also breeding grounds for mosquitoes. Some people earn their living, or derive their main food source, from wetlands. Successful drainage might involve inter-governmental co-operation, which is not always possible.
Spray areas where malaria is endemic with insecticide to kill mosquitoes	The areas are vast, much larger than areas treated in Europe and would often involve co-operation between different governments or areas occupied by warring groups. Insecticide would damage other wildlife, including beneficial insects.
Cover beds with fine-mesh nets, coated with insecticide	This is cheap and effective, since most mosquitoes bite at night when their victims are asleep. Mosquitoes that bite during the day are, however, emerging in many areas where malaria is endemic.
Release sterilised male mosquitoes that would mate with the females but not result in viable eggs being laid	Since releasing more insects in order to reduce the number of insects is counter intuitive, local groups of people are reluctant to accept its use. Large-scale, and effective, education programmes are needed to convince people to accept such programmes.
Vaccination	A vaccine would have to be suitable for use with babies and young children, since malaria is most severe during the first years of life. Pharmaceutical companies need to charge for their products. Countries in which malaria is endemic are often poor and could not afford country-wide vaccination programmes. Plasmodium shows great antigenic variability (involving interactions between 59 different genes in the first strain of *P. falciparum* to have its genome characterised). This makes production of an effective vaccine difficult.

Test yourself

16 Suggest what data were needed to make the computer model that predicted the *Ug99* strain of stem rust fungus would spread to India.

17 Explain why the flu virus is said to have a fragmented genome.

18 Distinguish between endemic, epidemic and pandemic diseases.

19 The majority of new cases of malaria occur among children under the age of 5 years. Suggest why.

20 Malaria used to be common in southern England. The spread of malaria in England was greatly reduced when glass was introduced in windows. Suggest why glass windows had this unforeseen effect.

Exam practice questions

1 An organism that benefits by living in, or on, another organism to which it causes no harm is known as a:

A commensal **C** pathogen

B parasite **D** pest *(1)*

2 A substance derived from microorganisms that in low concentration kills or stops the growth of other microorganisms is called an:

A antibiotic **C** antigen

B antibody **D** antiseptic *(1)*

3 Influenza is caused by a:

A bacterium **C** protozoan

B fungus **D** virus *(1)*

4 *Staphylococcus aureus* is a species of bacterium commonly found on the skin. This bacterium can cause harm if it penetrates the skin, via a cut or graze or during surgery, and enters the bloodstream.

The charts show *Staphylococcus aureus* infection rates by females and males of different ages.

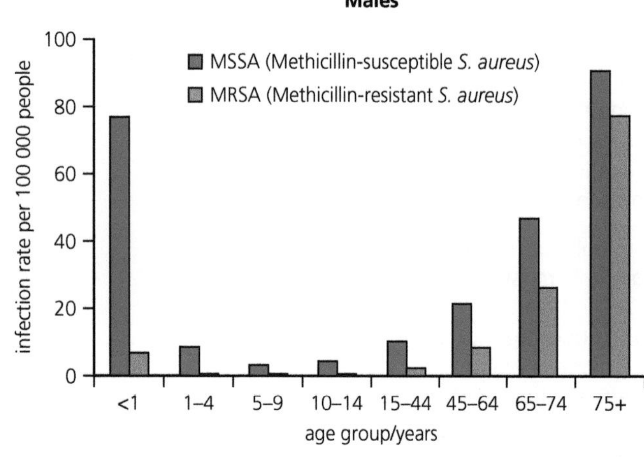

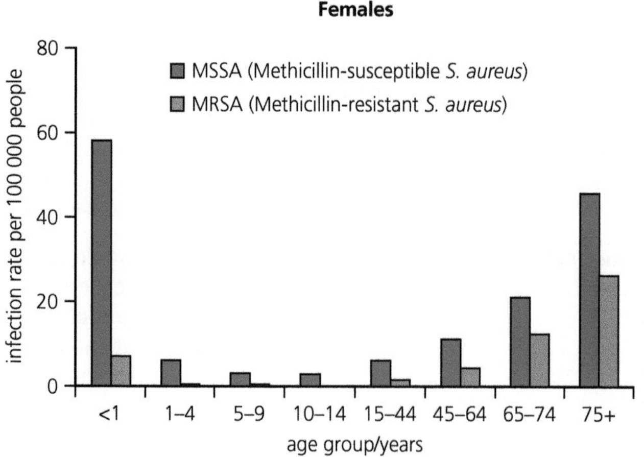

a) What can you conclude from the charts? *(3)*

b) Suggest reasons for the differences in infection rates. *(2)*

5 A hospital pathologist took samples of *Mycobacterium tuberculosis* from two patients and grew them in separate broth cultures.

He poured samples of one culture onto a number of agar plates. He then placed discs of filter paper that were impregnated with different antibiotics onto each plate.

He repeated the same procedure with samples of the second culture of bacteria and then incubated the agar plates for two days.

The diagram shows the appearance of two of his agar plates before and after incubation.

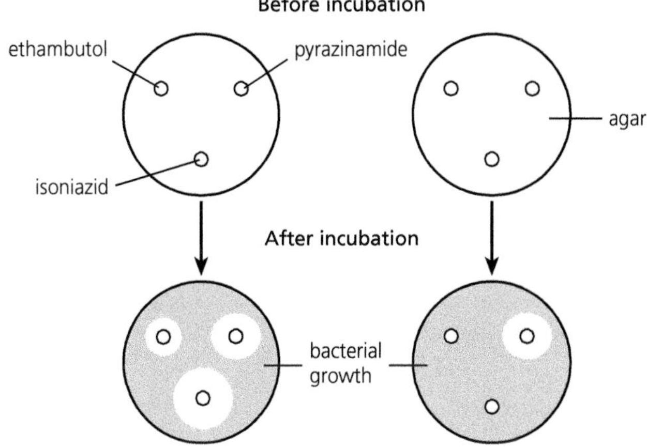

a) Explain the results of the microbiologist's investigation. *(4)*

b) When treating patients with pulmonary tuberculosis, UK doctors commonly prescribe two antibiotics to be taken every day for 6 months and a further two antibiotics to be taken every day for the first 2 months.

Explain the reasons for this treatment. *(2)*

6 Bacteraemia is a term used to describe the presence of bacteria in the blood.

The table shows data about the cases of *Staphylococcus aureus* bacteraemia reported by hospital medical laboratories in a voluntary reporting scheme in England, Wales and Northern Ireland.

Year	Number of cases of *Staphylococcus aureus* bacteraemia	
	Methicillin-sensitive (MSSA)	Methicillin-resistant (MRSA)
1993	4490	213
1994	4895	460
1995	5190	858
1996	5605	1620
1997	5609	2437
1998	5545	2858
1999	5584	3331
2000	5862	4283
2001	7168	5209
2002	7437	5529
2003	8527	6060
2004	8687	5737
2005	8622	5692
2006	8825	5393
2007	9292	4233

a) Use the data in the table to plot a graph to show the percentage of *Staphylococcus aureus* bacteraemia caused by methicillin-resistant *Staphylococcus aureus*. *(4)*

b) There has been great public concern about hospital-acquired MRSA. Can you conclude from the information in this question and from your graph that this concern is justified? Explain your answer. *(4)*

7 Mosquitoes carry a number of pathogens that infect humans. Scientists around the world have carried out numerous investigations of mosquito populations.

Tip

In an A level examination paper, you will commonly find questions that are synoptic. In other words, they require you to bring together your knowledge and understanding from different areas of the specification. Question 7 includes some synopsis.

a) In one investigation, a group of scientists investigated the behaviour of one species of malaria-carrying mosquito, *Anopheles funestus*, in two villages in Benin. They found that 3 years after the introduction of long-lasting insecticide-treated mosquito nets that completely covered sleeping areas, these mosquitoes had changed their biting behaviour.

- The proportion of bites that occurred outdoors increased from 45% to 68% ($p < 0.0001$)
- The median time for catching insects switched from 02:00 hours to 05:00 hours ($p < 0.0001$)

Explain how natural selection produced this change in behaviour of the populations of *A. funestus*. *(4)*

b) In a second investigation, a group of scientists in the Cayman Islands investigated a method of controlling a species of mosquito, *Aedes aegyti*, that transmits the virus that causes dengue fever.

They produced male mosquitoes that had been genetically modified to carry a gene that caused the death of the mosquitoes' developing larvae. They compared the size of the mosquito populations in two areas. In one, they released the genetically modified males. The other was used as a control.

They estimated the population sizes by counting the number of mosquito eggs laid in traps.

Their results are shown in the graph. The rainy season in this area lasts from April to June.

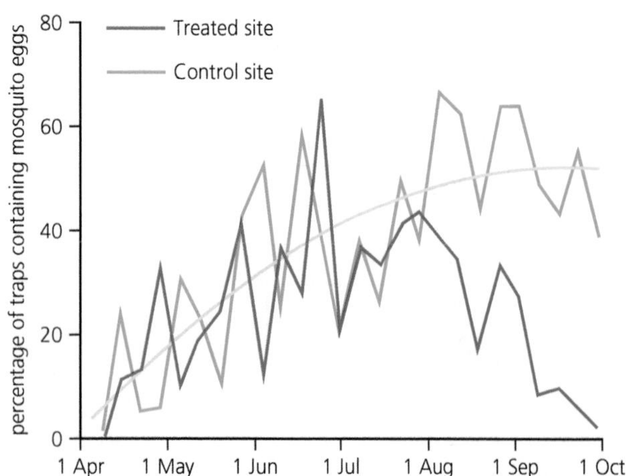

i) Do the results suggest that releasing genetically modified males could be an effective method of controlling *Aedes aegypti* populations? Explain your answer. *(3)*

ii) Consider the social and ethical issues that might arise in using the release of genetically modified males to control *Aedes aegypti* populations in areas where dengue fever is common. *(4)*

Stretch and challenge

8 The sale of counterfeit antimalarial drugs is a spreading criminal activity in some parts of the world. Consider the problems this activity could bring. You might find the following factfile a useful resource to begin your research: *Malaria – a global challenge* www.microbiologyonline.org.uk.

9 How is the UK government helping to restrict the development of bacterial resistance to antibiotics? You might find it helpful to start your research with *UK Five Year Antimicrobial Resistance Strategy 2013-2018*, available from www.gov.uk.

Response to infection

Prior knowledge

In this chapter you will need to recall that:

- communicable diseases can be passed from person to person either directly, via a vector or by airborne and waterborne pathogens. Non-communicable diseases cannot be passed from person to person
- the non-specific defence systems of the body against pathogens include: impenetrability of the skin; entrapment of pathogens by mucus in the lungs; lysosomes in tears, saliva and urine; acidic conditions in the stomach and in the vagina
- the immune system protects the human body against disease by inflammation and by the destruction of pathogens by white blood cells, including phagocytes and antibody-releasing cells
- vaccines are used to reduce, or prevent, the spread of certain diseases.

Test yourself on prior knowledge

1 Give the four main types of non-communicable disease.
2 Explain how lysozymes protect the body from pathogens.
3 Contrast the ways in which acidic conditions are produced in the stomach and in the vagina.
4 What is the role of a phagocyte?
5 Explain the difference between an antibody and an antibiotic.

You saw in the last chapter that an infectious disease is caused when another organism or virus invades the body and lives there parasitically. The invader is known as a **pathogen** and the infected organism – a human in this case – is the **host**. In this chapter you will see how your bodies respond to an infection.

First lines of defence against infectious disease

Before they can infect us, pathogens must be able to penetrate the tissues or organs of our bodies. Only then will the responses we are about to study come into play. We saw in Chapter 4, three ways that pathogens can enter our tissues – via our skin, via our lungs and via our intestine. Each of these body-environment interfaces, however, has evolved mechanisms that protect us from infection. These first lines of defence are summarised in Table 5.1.

Table 5.1 The body's first lines of defence against infection

Body-environment interface	Protective mechanisms
Surface of body	Our skin is normally an effective physical barrier. It has many layers of cells, making it difficult to penetrate. Cells in the outer layer are dry and filled with a tough, indigestible protein, called keratin, providing a hostile environment for bacterial growth. The waxy sebum that makes our skin supple contains fatty acids that are toxic to many microorganisms.
Lungs	The linings of our trachea, bronchi and bronchioles secrete copious volumes of mucus in which microorganisms become trapped. The constant beating of cilia – hair-like extensions of surface membranes of cells lining these organs – moves the mucus to the pharynx, and we swallow it. You can see a mucus-secreting cell and ciliated cells from the lining of the trachea in Figure 5.1.
Intestine	Our saliva contains lysozymes – enzymes that hydrolyse components of bacterial cells. Other enzymes operating in the intestine, including lipases and proteases, together with the very low pH of the stomach, contribute to the destruction of pathogens. Our intestines are also colonised by hundreds of different species of commensal bacteria, such as *Escherichia coli,* which we encountered in Chapter 4. Competition with the bacteria in this well-established community further reduces the ability of pathogenic bacteria to become established in our intestines.

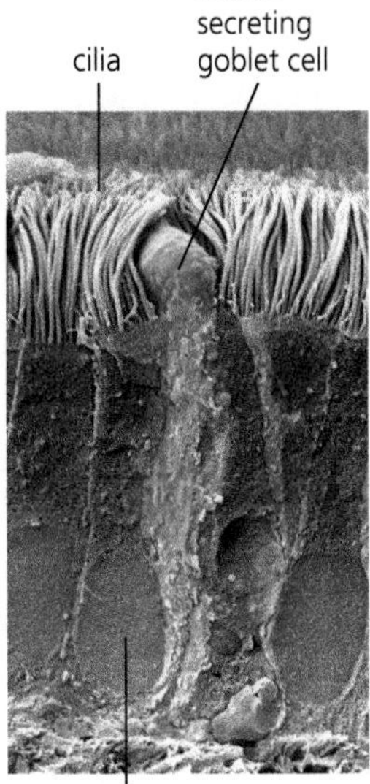

Figure 5.1 Scanning electron micrograph of cells lining the trachea. You can see a mucus-secreting cell in the middle of this SEM and the cilia that move mucus out of the lungs are clearly visible

Despite these and other defences, many pathogens do enter your body. Once inside, these pathogens are much more difficult to eradicate. Not only must your body distinguish them from your own cells, it must then destroy them without causing damage to your own tissues.

Non-specific inflammatory response

If a pathogen gets into your body, an **inflammatory response** is our second line of defence. This type of response is non-specific, meaning that it is the same for all pathogens. If you look at Figure 5.2 on the next page, you can follow the several processes involved in this inflammatory response.

Inflammation is the rapid, localised response of our tissues to damage. It is triggered when damaged cells release 'alarm' chemicals, including histamine and prostaglandins. These chemicals have the following immediate effects in the wounded area:

- The smooth muscle of arterioles relaxes, increasing blood flow to that area.
- Cells in the walls of capillaries draw away from one another (diapedesis), so that the capillaries become 'leaky', forming more tissue fluid than usual. This extra tissue fluid causes local swelling of the infected area.
- Sensory neurones become more sensitive.

The initial outcome is that the volume of blood in the damaged area is increased and we suffer local **oedema**.

> **Key term**
>
> **Diapedesis** A localised response to damage in which cells lining capillaries move apart creating gaps, through which plasma can leave the capillaries. This leads to the local production of increased volumes of tissue fluid (oedema). It also allows phagocytes to leave the capillaries.

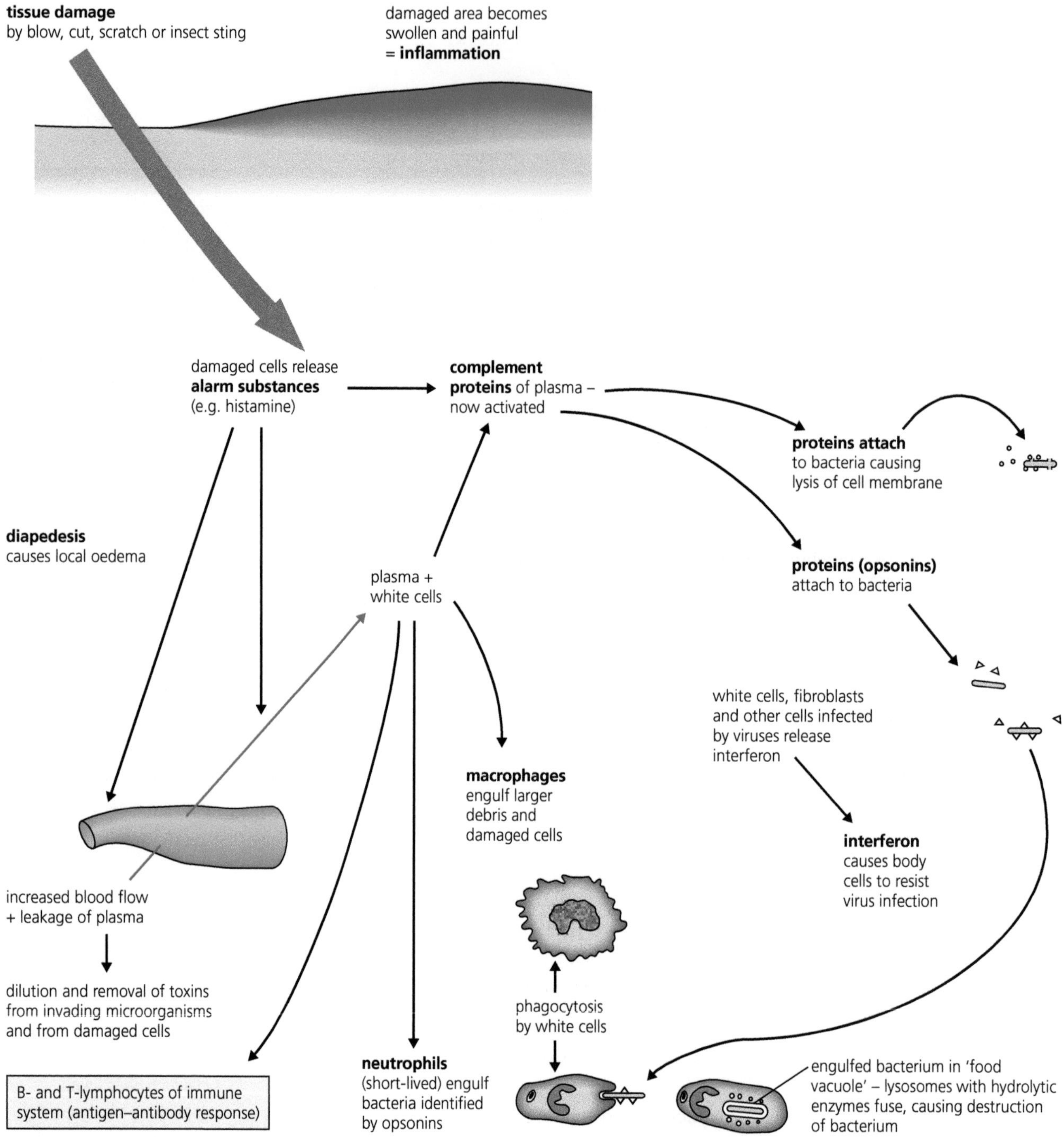

tissue damage
by blow, cut, scratch or insect sting
damaged area becomes
swollen and painful
= inflammation
damaged cells release
alarm substances
(e.g. histamine)
complement
proteins of plasma –
now activated
proteins attach
to bacteria causing
lysis of cell membrane
diapedesis
causes local oedema
plasma +
white cells
proteins (opsonins)
attach to bacteria
white cells, fibroblasts
and other cells infected
by viruses release
interferon
macrophages
engulf larger
debris and
damaged cells
interferon
causes body
cells to resist
virus infection
increased blood flow
+ leakage of plasma
dilution and removal of toxins
from invading microorganisms
and from damaged cells
phagocytosis
by white cells
B- and T-lymphocytes of immune
system (antigen–antibody response)
neutrophils
(short-lived) engulf
bacteria identified
by opsonins
engulfed bacterium in 'food
vacuole' – lysosomes with hydrolytic
enzymes fuse, causing destruction
of bacterium

Figure 5.2 The processes involved in the non-specific inflammatory response to infection

neutrophil
eosinophil
lymphocyte
monocyte
erythrocytes
basophil

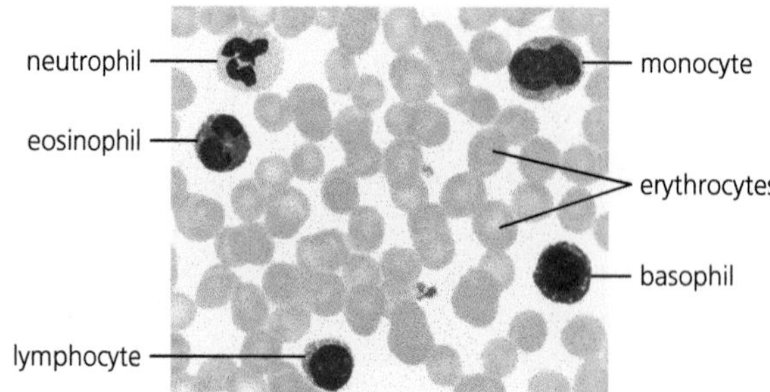

Figure 5.3 A blood smear showing red blood cells (erythrocytes) and all five types of white blood cells (leucocytes)

The increased blood flow also results in more leucocytes being brought to the infected area. Figure 5.3 shows a blood smear containing red blood cells (erythrocytes) and different types of white blood cell (leucocytes). Macrophages and neutrophils are phagocytic leucocytes. Figure 5.4 shows how one of these phagocytes engulfs and destroys a foreign cell. In doing so, it must be able to distinguish foreign cells from the body's own cells. Phagocytes are aided in this by a group of proteins, called **complement proteins**. These have several effects including:

- attracting more phagocytes to the site of infection
- binding to, and forming pores in, the surface membranes of foreign cells, leading to the lysis of these cells
- binding to surface membranes of foreign cells, thus aiding the attachment of the surface membrane of a phagocyte to a foreign cell. These proteins are called **opsinins** and the process they invoke is opsonisation.

Key term

Leucocytes The generic term for all types of white blood cell.

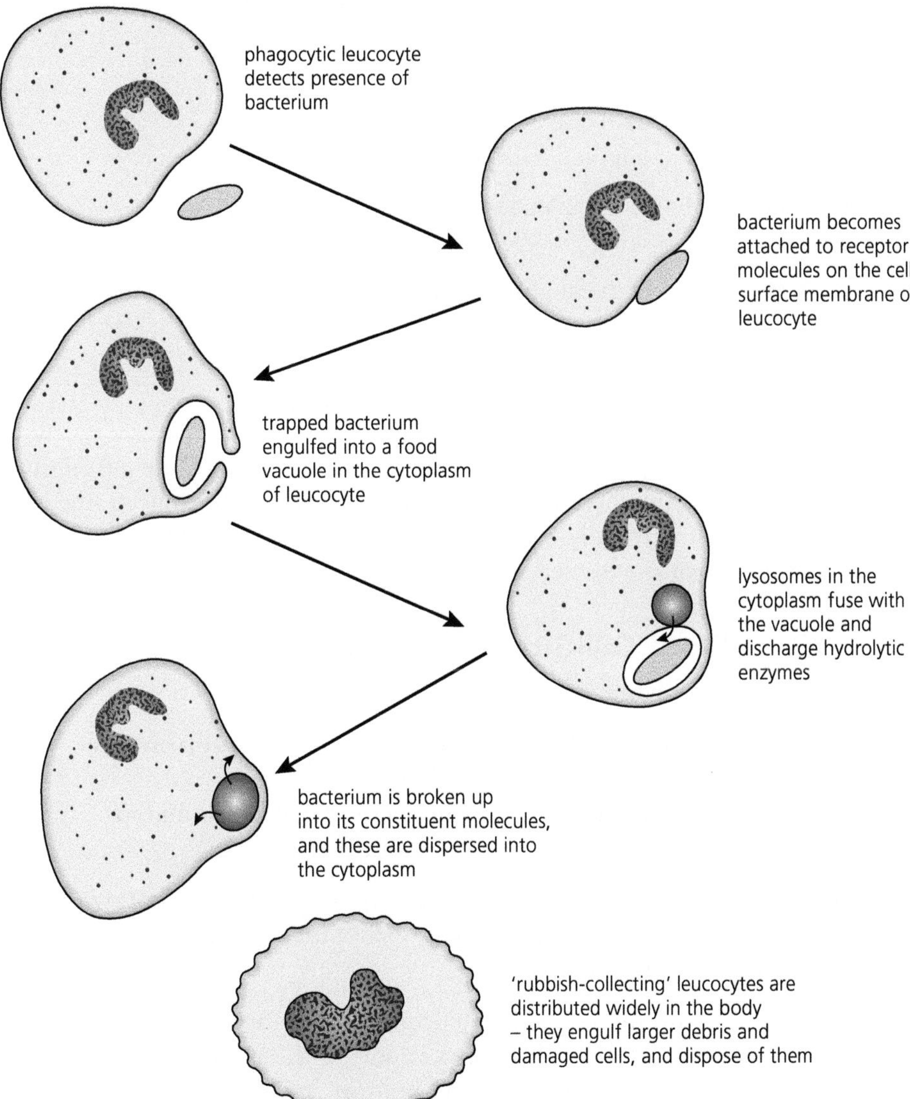

Figure 5.4 Phagocytosis of a bacterium

Key term

Cytokines A group of small proteins released by one type of cell that affect the behaviour of other cells.

A group of cytokines, called **interferons**, are released by cells infected by viruses. These interferons bind to neighbouring, healthy cells and trigger synthesis of antiviral proteins. As a result, viral replication is slowed or halted.

As previously stated, all the above responses to infection are referred to as non-specific responses because they help to destroy any invading pathogen. In contrast, the immune responses discussed next are triggered by, and directed towards, specific pathogens.

Test yourself

1 The frequent smoking of tobacco destroys the cilia lining the trachea and bronchi. What is the disadvantage of this?
2 Suggest why some people suffer diarrhoea after they have taken a course of antibiotics.
3 Explain each of the following effects that you might feel following a cut in your skin.
   a) The affected area of your skin becomes hot.
   b) The affected area of your skin swells.
   c) The affected area of your skin feels painful.
4 Explain why the action of macrophages is said to be 'non-specific'.
5 State the role of interferons.
6 Neutrophils are short-lived. Suggest **one** advantage of this.

The specific immune response

You have seen above that two types of leucocyte, the macrophages and the neutrophils, react in a non-specific way to any foreign cell that enters your body. In contrast, the next group of leucocytes you will study, the **lymphocytes**, are specific. Lymphocytes make up 20 per cent of the leucocytes in your bodies.

Lymphocytes – the B cells and T cells

Key term

Multipotent stem cells Cells that retain the ability to divide by mitosis and give rise to more than one type of daughter cell, in this case blood cells.

You have two distinct types of lymphocyte, based on the ways they function.

- B lymphocytes (**B cells**) secrete antibodies (producing the **humoral immune response**).
- T lymphocytes (**T cells**) attack infected cells (producing the **cell-mediated immune response**) and, as you shall see shortly, assist B cells.

B cells and T cells are both produced by multipotent stem cells in the marrow of certain bones. As Figure 5.5 shows, the lymphocytes then migrate to different locations where they undergo separate maturation processes.

T cells leave the bone marrow and complete their maturation in the **t**hymus gland. **B** cells do not leave, but complete their maturation within the **b**one marrow where they were formed. In both cases, during their maturation, any lymphocytes that would react against the body's own cells are selectively destroyed. Mature B cells and T cells eventually rejoin the blood system. Many are stored in lymph nodes throughout the body.

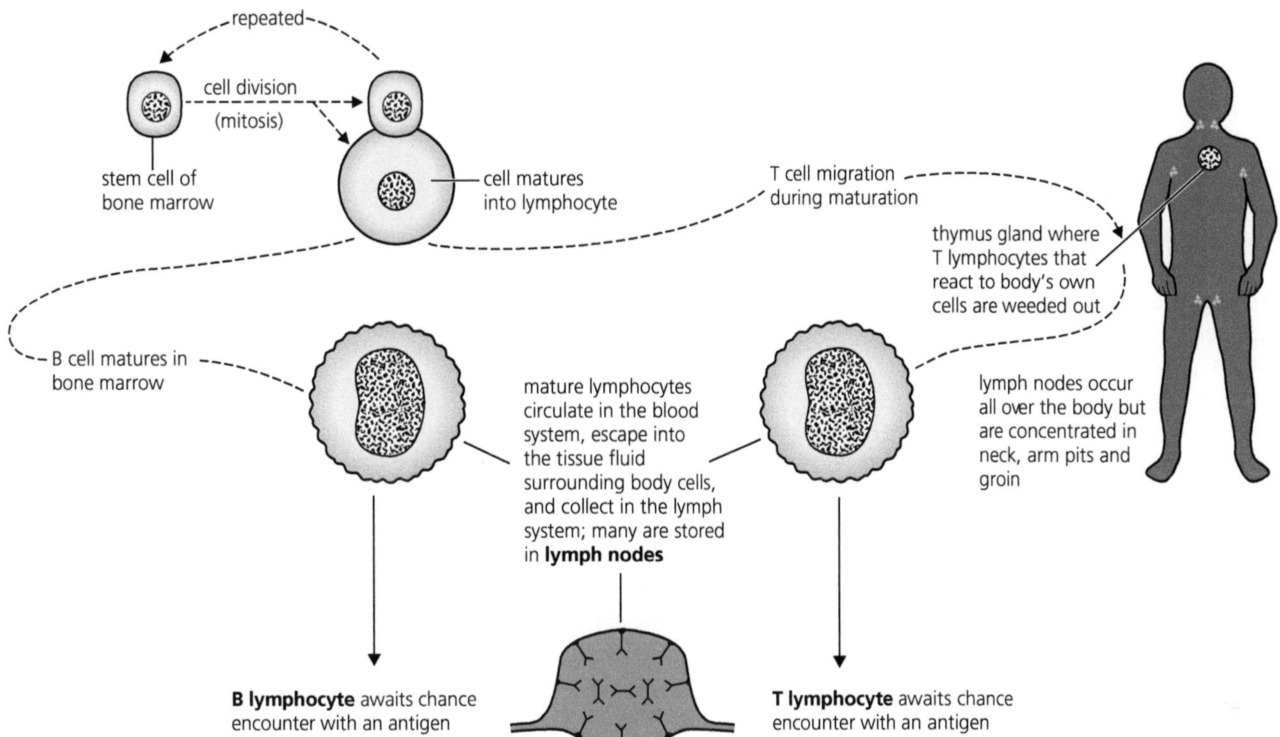

Figure 5.5 The formation of B cells and T cells

How B cells and T cells recognise 'non-self' antigens

In *Edexcel A level Biology 1* Chapter 9, page 184, you learnt about the fluid-mosaic model of cell surface membranes. Lodged within this structure are molecules of glycoproteins. The glycoproteins are highly variable and help to identify cells – so are called cell 'markers'. Cells in each tissue of your body have markers that are different from those in other tissues; cells from other organisms, including other people, have markers that are different from yours.

The glycoproteins that identify cells are known as the **major histocompatability complex (MHC) proteins** (in humans, these proteins are often called human leucocyte antigens – HLA). These MHC proteins are encoded by genes on the short arm of your chromosome 6. So, each individual's MHC proteins are genetically determined, and are features you inherit. For reasons that are too complex for an A level study, the MHC genes generate almost unbelievable variation in the proteins they encode. To give you a rough idea of this variation, there are at least eight different MHC genes and each gene has multiple alleles, with the number of alleles of a single gene ranging from 13 to 661. Unless you have an identical twin, your MHC proteins are unique.

Key term

Major histocompatability complex (MHC) proteins Glycoproteins found on the cell surface membrane of a cell that are unique to that individual. In humans, these are often called human leucocyte associated antigens, or HLA.

Tip

Proteins on the surface membrane of any cell act as antigens - either self-antigens or non-self antigens. Consequently, you will often see MHC proteins described as MHC antigens.

Key term

Antigen Any molecule that triggers an immune response. Small molecules, like amino acids, sugars and triglycerides, do not trigger an immune response. Antigens are large molecules, like proteins, glycoproteins and lipoproteins. The surface membrane of cells is coated with antigens.

B cells and T cells both have molecules on their cell surface membranes that are antigen receptors. These receptors are complementary to **antigens** – molecules that trigger an immune response – that might occur on the surface membranes of other cells. Each lymphocyte has only one type of antigen receptor. Consequently, each lymphocyte can 'recognise' only one type of antigen. A conservative estimate is that each of us can produce several million different antigen-receptor molecules. You might wonder how this is possible. Again, the mechanism is too complex for an A level study, but several genes are involved in encoding each antigen-receptor molecule and these genes mutate and genetically cross-over and recombine at a very fast rate.

Tip

It is easy to confuse the complementary fit of antibody to antigen with the induced fit model of enzyme action. In an exam, take care not to refer to the 'active site of an antibody'. It is incorrect terminology and will not gain you marks.

As you will see shortly, the antigen-receptor molecules of a B cell are released as antibodies. Figure 5.6 shows the structure of the two most common antibodies. Both are a type of protein, called an **immunoglobulin** (abbreviated to Ig). You can see that the smaller of the two, called IgG, has four polypeptides linked by disulfide bridges (–S–S–) to form a Y-shape, and that two of these polypeptides are large (the heavy chains) while two are small (the light chains). You can also see that it is only two small parts of the antibody molecule that bind to a complementary antigen. The larger molecule in Figure 5.6 is called IgM. You can see that it is formed by five IgG-like molecules held together.

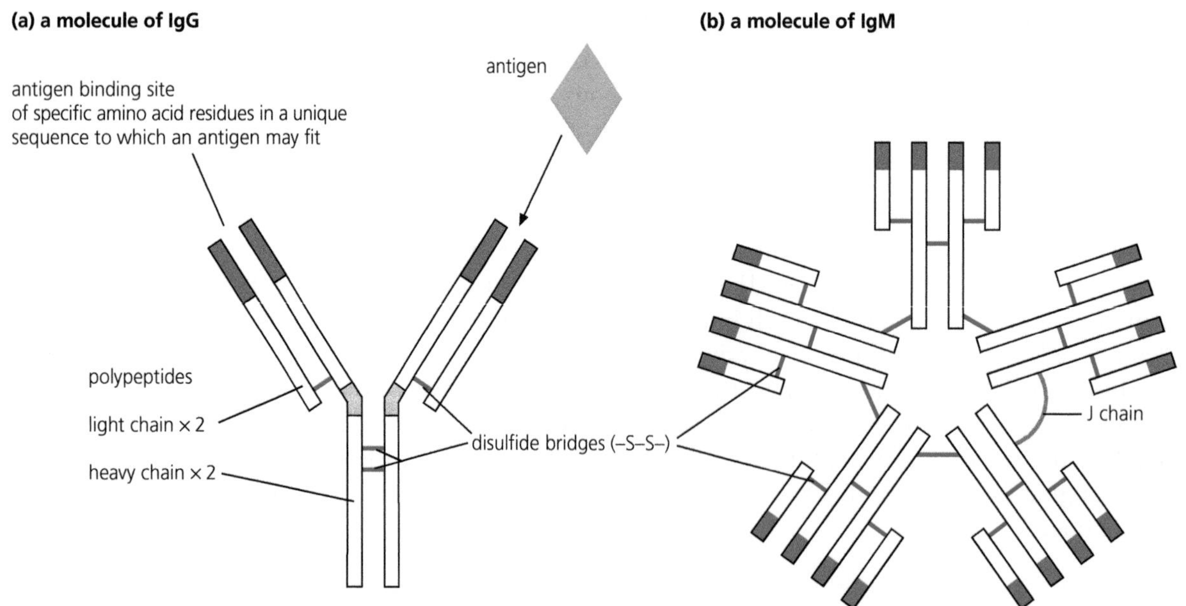

Figure 5.6 The structure of the two most common antibodies: (a) is a molecule of IgG, (b) is a molecule of IgM

The antigen-receptor molecules on the surface of T cells are chemically similar to the Y-shaped IgG immunoglobulin shown in Figure 5.6 but have only two polypeptide chains instead of four. Like those of B cells, these receptors are almost infinitely variable, but each T cell has only one type on its surface. Unlike B cells, T cells do not release their antigen-receptor molecules.

As you saw in Figure 5.5, B cells and T cells mature in different parts of the body. It is during this maturation that any randomly produced cells with antigen-receptor molecules complementary to your own MHC proteins (so-called **self-antigens**) are destroyed. Otherwise, your B cells and T cells might attack your own body cells. That said, Type I diabetes and rheumatoid arthritis are just two examples of human disorders caused by an immune reaction to the body's own cells.

Key term

Self-antigen A glycoprotein on the cell surface membrane of one of the body's own cells. The ability of this glycoprotein to cause an immune response is reduced by the body destroying any randomly produced lymphocytes with a complementary antigen-receptor.

Test yourself

7 The thymus gland is relatively much larger in young people than in adults. Suggest the advantage of a large thymus gland during childhood.

8 Explain how the names of B cells and T cells relate to their development.

9 B cells and T cells are produced by multipotent stem cells. What does 'multipotent' mean?

10 One effect of an IgM antibody is to clump pathogens together. This makes phagocytosis by macrophages easier. Suggest how the structure of IgM molecules enables them to clump pathogens together.

The humoral immune response

The humoral immune response involves the release of antibodies by B cells. It depends, however, on an interaction between B cells, T cells and macrophages. Figure 5.9 demonstrates this interaction. It might help you to follow the numbered stages in this diagram as you read the associated steps in the account of the humoral immune response on the next page.

1 A pathogen enters the body; its surface has non-self antigens on it. Completely at random, this pathogen collides with a B cell that has an antigen receptor on its surface membrane that is complementary to a non-self antigen on the pathogen. The antigen binds to the antigen-receptor molecule on the B cell. At the same time, other cells of the pathogen are attacked by non-specific macrophages that engulf cells of the pathogen.

2 The B cell engulfs the antigen and digests it. The B cell then displays fragments of the antigen on its cell surface membrane, bound to its own MHC proteins.

3 Meanwhile, the macrophage (step 1) also digests the antigen-carrying pathogen it has engulfed. It too displays fragments of the antigen bound to the MHC proteins on its cell surface membrane. This is called antigen presentation by a macrophage.

4 The **antigen-presenting macrophage** comes into contact with a T cell that has an antigen-receptor protein complementary to one of the pathogen's antigens now displayed on the macrophage. The two briefly bind. This activates the T cell, which is now called an **activated T helper cell**.

> **Key term**
>
> **Antigen-presenting macrophage** A type of white blood cell, a macrophage, that has engulfed and digested a pathogen and is displaying antigens from the pathogen on its own cell surface membrane.

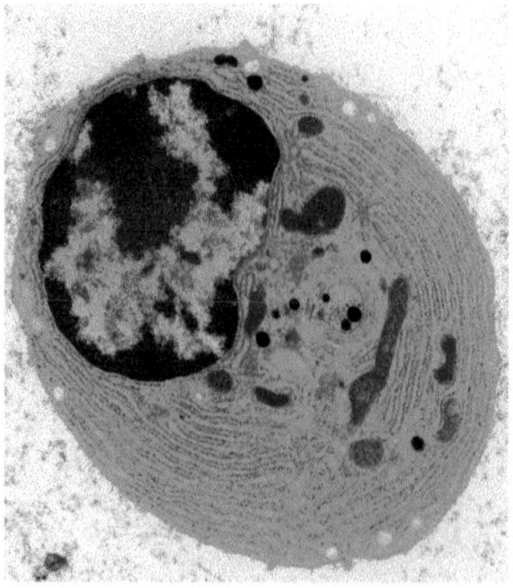

5 An activated T helper cells now binds to a B cell displaying the same antigen on its cell surface membrane (step 2 above). This, in turn, activates the B cell.

6 Stimulated by the secretion of cytokines from the activated T helper cell, the activated B cell immediately divides very repeatedly by mitosis, forming a clone of cells called **plasma cells**. The transmission electron micrograph in Figure 5.7 shows the plasma cell is packed with rough endoplasmic reticulum. It is here that the antibody is mass-produced, and is then exported from the plasma cell by exocytosis. The antibodies are normally produced in such numbers that the antigen is overcome.

The production of an activated B cell, its rapid cell division to produce a clone of plasma cells, and the resulting production of antibodies that react with the antigen, is called **clonal selection**. Sometimes several different antibodies react with one antigen – this is polyclonal selection.

7 After these antibodies have destroyed the foreign matter and the threat of disease that it introduced, the antibodies disappear from the blood and tissue fluid, along with the bulk of the specific B cells and T cells responsible for their formation. However, a few of these specifically activated B cells and T cells are retained in the body as **memory cells**. In contrast to plasma cells and activated T cells, these memory cells are long-lived. In the event of a re-infection of the body by pathogens carrying the same antigen, these memory cells make possible the early and effective response shown in Figure 5.8. This is the basis of natural immunity (see below).

Figure 5.7 Transmission electron micrograph of a plasma cell. Note that the cytoplasm is almost filled by rough endoplasmic reticulum – the site of antibody production

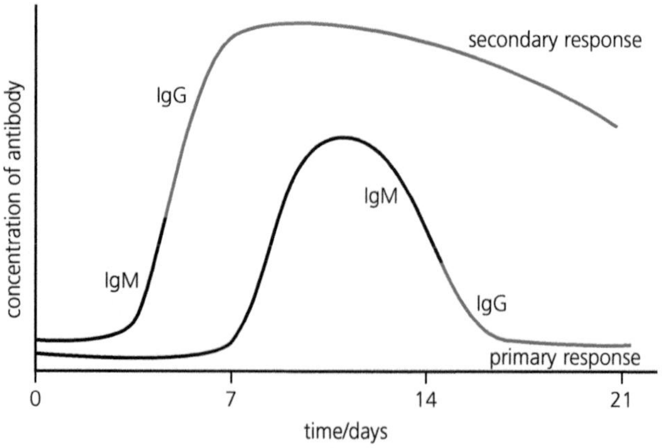

Figure 5.8 Profile of antibody production during infection (primary response) and re-infection (secondary response)

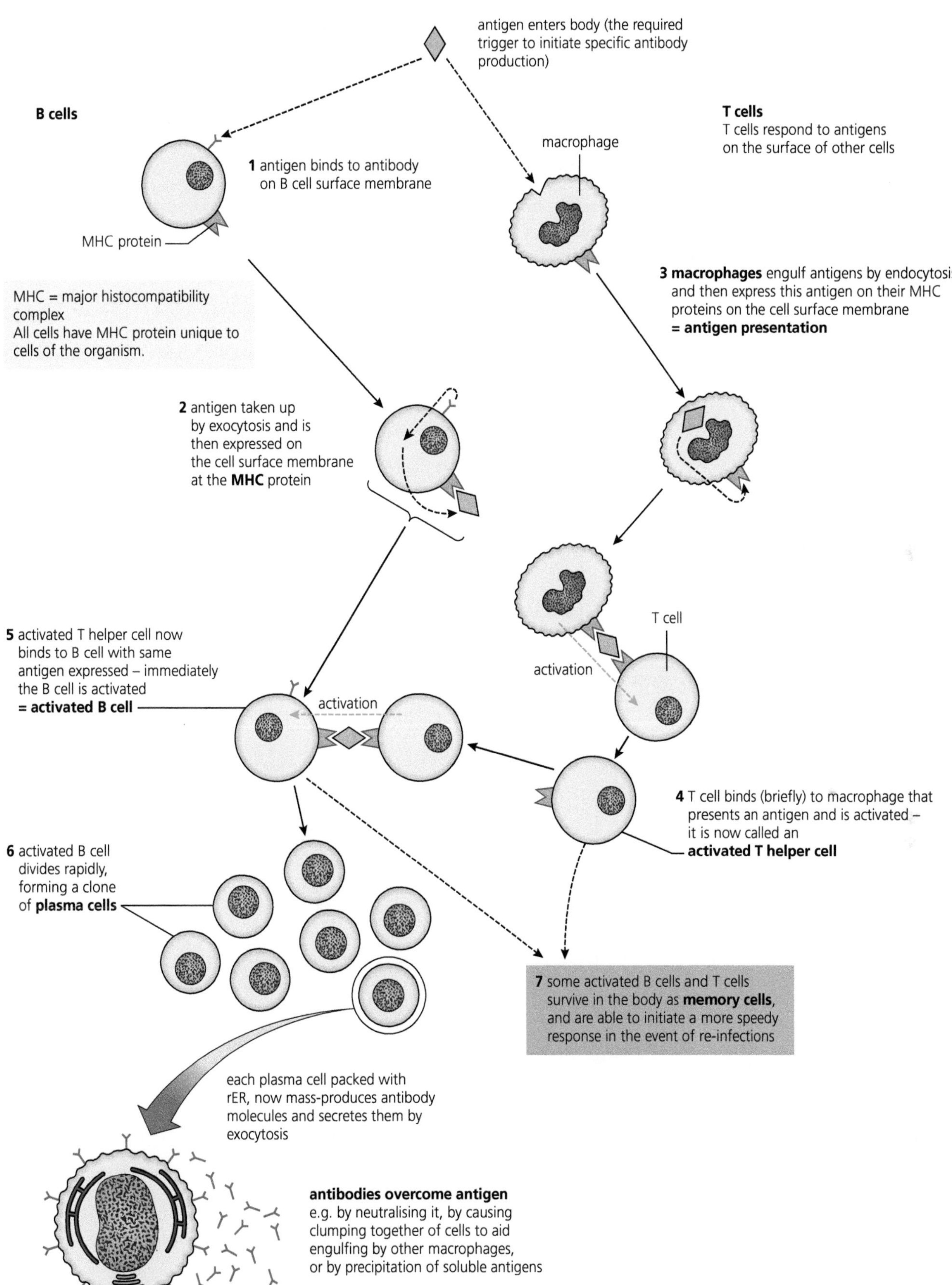

Figure 5.9 The steps involved in antibody release

The cell-mediated immune response

The cell-mediated immune response is brought about by the activity of T cells. As you saw in Figure 5.9, T cells will bind to an antigen-presenting cell that has antigens from a pathogen on its surface. In doing so, an activated T helper cell is formed that releases cytokines. The release of cytokines stimulates the activated T cell to divide repeatedly to form a clone. Within this clone are three types of T cell:

- T killer cells – destroy body cells infected by viruses.
- T helper cells – release cytokines that stimulate production of B cells.
- Memory cells – remain in the body and bring about the secondary response.

Figure 5.10 provides an overview of the complex roles of B cells and T cells in the immune system.

Test yourself

11 State the importance of antigen-presenting cells in the specific immune response.

12 Give **two** ways in which the antigen-receptor molecules of a T cell differ from those of a B cell.

13 Name **two** processes that enable us to produce millions of different cell-surface antigen-receptor proteins.

14 How does the role of a T helper cell differ from that of a T killer cell?

15 Use Figure 5.9 to suggest why the secondary response is much faster than the primary response to infection.

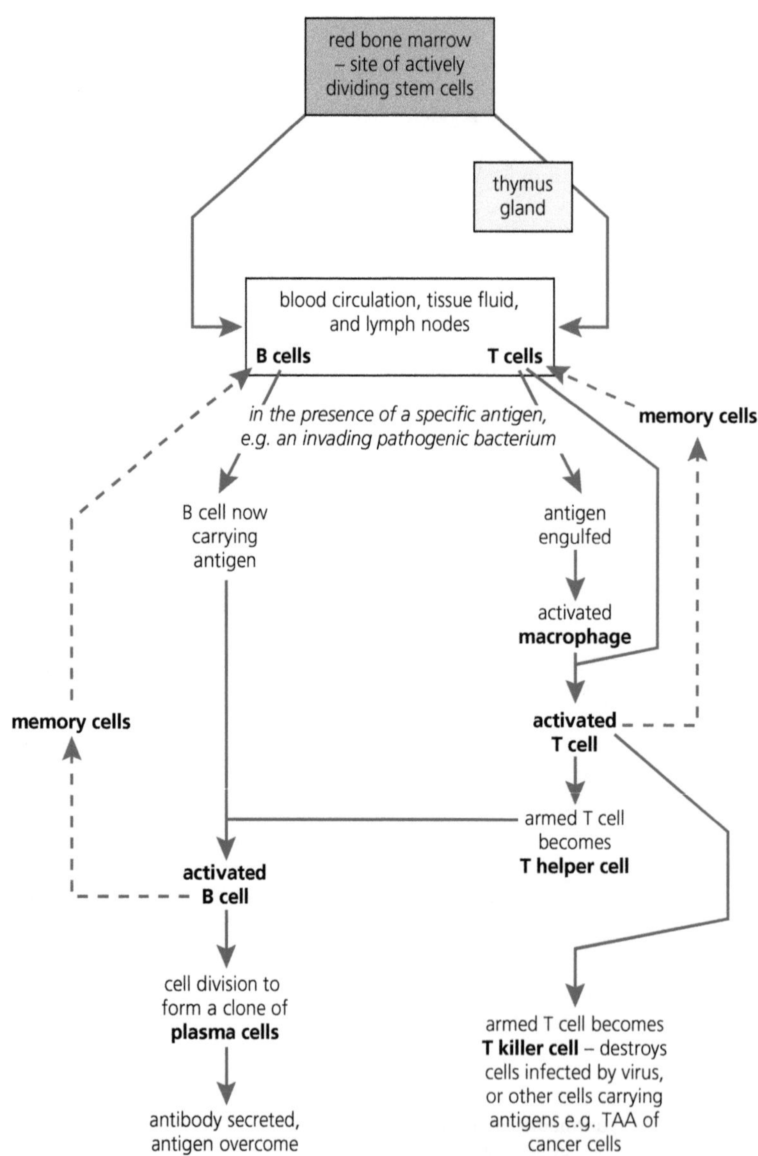

Figure 5.10 A summary of the roles of B cells and T cells in the specific immune response

Activity

Testing the clonal selection hypothesis

Scientists propose explanations for the observations they make. These explanations are referred to as **hypotheses**. In order to test the validity of a hypothesis, scientists use it to make predictions and then devise experiments to test these predictions. If their results are always consistent with their predictions, they become confident in the validity of their hypothesis.

The **clonal selection hypothesis** proposes an explanation for the way that you produce antibodies against non-self antigens. You can summarise this hypothesis in the following way:

- Your immune system randomly produces millions of different types of B and T cells. Each type has a unique protein receptor on its surface membrane. When, and only when, one of these cells binds with a complementary antigen, it is stimulated to produce large numbers of cells that are identical to itself and, consequently, to each other. A group of identical cells is called a **clone**. Thus, in response to the presence of a particular antigen, a B or T cell is selected and it forms a clone. Within the clone, each cell has the identical protein receptor on its surface.

Scientists tested this hypothesis by injecting two rats, **R** and **S**, with antigens from two different strains of pneumococcal bacteria. They injected each rat twice, with the second injection made 28 days after the first.

1 Why do you think the scientists injected each rat twice, with a gap of 28 days between injections?
2 Which type of immune cell do you think the scientists were attempting to stimulate?

The scientists injected rat **R** with antigens from strain **X** of the pneumococcal bacterium (type **X** antigens) and injected rat **S** with antigens from strain **Y** of the pneumococcal bacterium (type **Y** antigens).

3 What do you think the scientists were predicting would happen as a result of these injections?

The scientists now needed a way to find out which cells had been stimulated to produce a clone. The method they chose was rather neat. They coated inert beads with the type **X** antigens and put the beads into two glass columns (labelled 1 and 3 in Figure 5.11 on the next page). They then did the same with type **Y** antigens and put these beads into another two glass columns (labelled 2 and 4 in Figure 5.11). They reasoned that if they washed samples of lymphocytes through these columns, those lymphocytes that had the appropriate complementary protein receptors would bind to the antigens on the inert beads. As a result, they would stay in the glass column and not emerge at the bottom.

4 Why was it important that the beads they used were inert?

One week after the second injection, the scientists removed a sample of blood from each of the two rats and separated the lymphocytes from the rest of the blood. They put half of each sample into a column of inert beads coated with type **X** antigen and half into a column of inert beads coated with type **Y** antigen. They then washed the samples through the columns and collected any lymphocytes that passed through the column.

5 What type of fluid would they have used to wash the lymphocytes through the columns?

Finally, the scientists tested the lymphocytes that had passed through each column to see whether they could make antibodies against either of the two antigens used in the experiment. Figure 5.11 summarises the method they used and their results.

Now look at their results in Figure 5.11 and see if you can interpret them. Start with the results from column 1. First you need to make sure you have understood what they show by describing them.

6 How would you describe the results from column 1?
7 Can you explain both parts of that description?
8 Explain why cells emerging from column 2 give large quantities of anti-X antibody but no anti-Y antibody.

You should be able to use similar arguments to explain the results from columns 3 and 4. You can then **evaluate** the experiment. This means you ask whether the experiment was a valid test of the clonal selection hypothesis. If the results had not been consistent with this prediction, they would have cast serious doubt on the hypothesis. By using two rats injected with different antigens of the same bacterium, and by using columns with only type X antigen or only type Y antigen, the scientists had built a **control** into their experiment. Without further details, you must assume that the scientists made sure that the conditions under which the rats were reared were kept constant. You could criticise the scientists for using only two rats, since this was a small sample size. In fact, they used large groups of rats. This account has been simplified to help you understand what was done. Therefore, you can conclude that the experiment was a valid test of the clonal selection hypothesis.

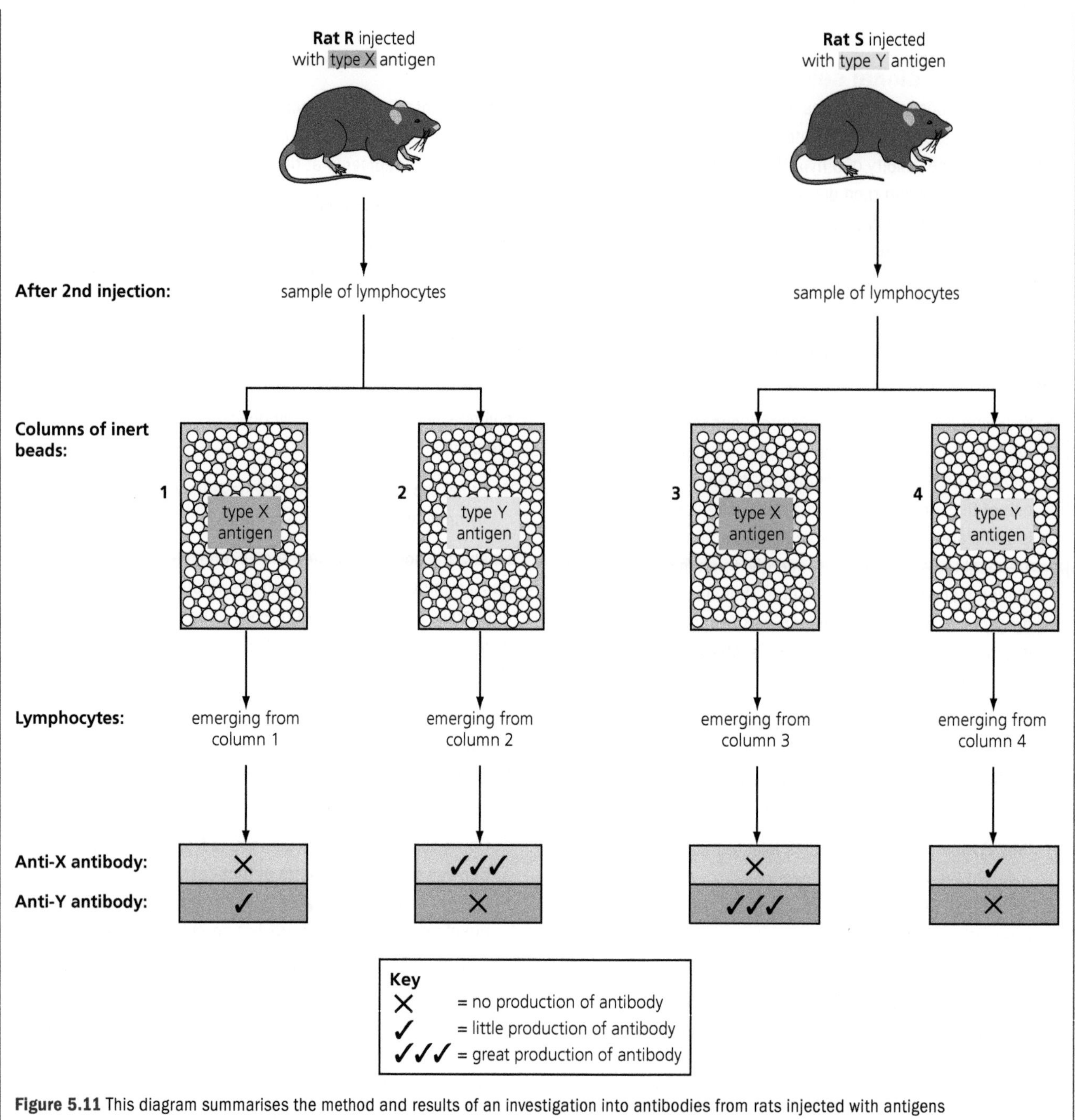

Figure 5.11 This diagram summarises the method and results of an investigation into antibodies from rats injected with antigens

Types of immunity and vaccination

As you have seen, your immune system protects you from the worst effects of many of the pathogens that might infect you. This immunity might be acquired actively or passively and by natural or artificial means. Table 5.2 explains the difference.

Table 5.2 Types of immunity displayed by humans

	Artificial immunity	Natural immunity	Longevity of immunity
Active immunity (antibodies made by subject)	Antibodies made following administration of a vaccine B cells, T cells and memory cells are made	Production of antibodies following infection by, and recovery from, a disease B cells, T cells and memory cells are made	Long-lasting since memory cells are maintained throughout life
Passive immunity (antibodies not made by subject but made by another organism)	Antibodies administered by injection (immunisation)	Mother's antibodies cross placenta to fetus or ingested by baby in mother's breast milk (especially in the colostrum, the first-formed milk)	Fades with time, since the recipient has not made memory cells and the antibodies received are themselves treated as non-self antigens and destroyed by recipient's active immunity

Vaccination

Vaccination is the deliberate administration of antigenic material to stimulate the recipient to develop active immunity against a pathogen. The active agent of the vaccine can be:

- the intact pathogen that has been inactivated (to stop it being able to cause infection) or attenuated (to reduce its ability to cause infection)
- purified components of the pathogen that have been found to have antigenic properties but do not cause disease
- toxoids – modified toxins – to develop active immunity against toxin-based diseases
- genetically engineered DNA – which can be designed to stimulate particular target cells of the immune system.

Vaccines are administered either by injection or by mouth. They cause the recipient's immune system to make antibodies against the pathogen without suffering an infection, and then to retain the appropriate memory cells. Active artificial immunity is established in this way. In terms of antibody production, the response caused by any later exposure to the antigen is exactly the same as if the immunity had been developed after the body overcame an earlier infection.

Vaccination has been so successful that some formerly common and dangerous diseases have become very uncommon occurrences in many human communities. For example, vaccination has led to the worldwide eradication of smallpox and to a great reduction in the incidence of measles, polio and tetanus. You can find the recommended schedule of vaccinations for children brought up in the UK on www.doh.gov.uk.

Herd immunity

Interestingly, it is not essential for everyone in a population to be immunised against a contagious disease in order to control that disease. The **herd immunity** theory proposes that the risk of someone who is susceptible to a contagious disease becoming infected gets less, the greater the proportion of people in that community who are

> **Key term**
>
> **Herd immunity** Provided a large enough proportion of a population are immune to a particular contagious disease, the likelihood of the causative pathogen being transmitted to a member of the population who is not immune is negligible.

immune. To take an extreme example, if all but one person in the UK is immune to a contagious disease, the chance of the one susceptible person being exposed to the pathogen within the UK is almost nil.

You might be wondering what proportion of a population could avoid vaccination but still be protected by the majority who have been vaccinated. This depends on the nature of the disease-causing organism and its method of spread. Table 5.3 provides information about a number of contagious diseases with which you might be familiar.

Table 5.3 The threshold percentage of the population needed to achieve herd immunity

Disease	Transmission route	Percentage threshold to achieve herd immunity
Diphtheria	Via saliva	85
Measles	Airborne	83 to 95
Mumps	Droplet	75 to 86
Rubella	Droplet	83 to 85
Smallpox	Social contact	83 to 85

Looking at Table 5.3, you might think it will be safe for you to avoid immunisation or vaccination. Remember, though, you are dealing with a mathematical model concerned with reducing the likelihood of infection. The likelihood of being knocked down by an automobile is low, but you still look both ways before crossing a road. It is best to restrict failure to immunise or vaccinate to those who might be harmed by it, such as people with weakened, or compromised, immune systems.

Given the success of vaccination programmes in eradicating contagious diseases, the public has sometimes become casual about the threat such diseases still pose. During your GCSE science course, you probably studied the effects of a newspaper article suggesting a link between a vaccine to protect against measles, mumps and rubella (the combined MMR vaccine) and autism. As a result of this article, so many parents declined the invitation to have their children vaccinated with the MMR vaccine that the proportion of vaccinated children fell below the herd immunity threshold in some communities. One such community was the area around Swansea, a city in South Wales. Here, it is believed the percentage of vaccinated children fell to almost 80 per cent. The result was the 2013 measles epidemic amongst children in the Swansea area. You might also recall that doctor who proposed this link to autism was subsequently discredited and left the UK.

Test yourself

16 Anti-venom, injected into someone who has been bitten by a snake, does not confer long-lasting protection against further bites by the same species of snake. Suggest why.

17 Suggest **one** factor that determines whether a vaccine is administered by injection or by mouth.

18 It is said that the only smallpox viruses left in the world are in sealed cultures in a few laboratories. Suggest why these laboratories keep samples of this virus.

19 Does vaccination lead to long-term protection against measles? Explain your answer.

20 Define the term 'herd immunity'.

Exam practice questions

1 Antibodies are released by:

A lymphocytes **C** neutrophils

B macrophages **D** plasma cells *(1)*

2 As an emergency treatment, a person who is showing symptoms of rabies can be injected with a rabies immunoglobulin. This is an example of:

A active, artificial immunity **C** passive, artificial immunity

B active, natural immunity **D** passive, natural immunity *(1)*

3 MHC proteins are:

A antibodies **C** antigen receptors

B antigens **D** toxoids *(1)*

4 IgG and IgM are two types of antibody produced by humans.

a) IgG is found in the bloodstream and in body tissues. IgM is found only in the bloodstream. Suggest a reason for this difference in distribution. *(3)*

b) Compare the primary and secondary responses to infection. *(4)*

5 The human immunodeficiency virus (HIV) is a retrovirus that causes autoimmune deficiency syndrome (AIDS). It infects T helper cells that have $CD4^+$ proteins on their cell surface membranes.

a) What is meant by the term 'retrovirus'? *(1)*

b) Suggest why HIV infects only T helper cells that have a $CD4^+$ protein on their cell surface membranes. *(2)*

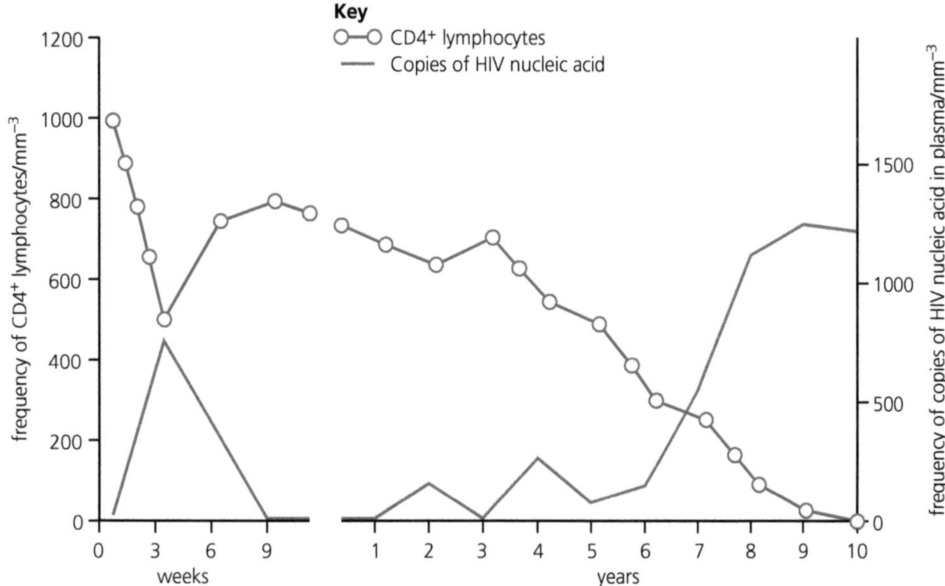

c) The graph shows the frequency of T helper cells with $CD4^+$ proteins and of copies of HIV nucleic acid in the plasma of someone infected with HIV but who receives **no** medical treatment.

i) Suggest an explanation for the data between 9 weeks and 3 years in the graph. Use evidence from the graph to justify your explanation. *(3)*

ii) The person represented in the graph probably died in year 10. Explain why death would occur. *(4)*

6 Dogwhelks are snails that are common on rocky shores around the UK. Their main diet comprises barnacles and mussels.

Dogwhelks show variation in their shell colour. A geneticist investigated a theory that dogwhelks developed coloured shells only if they ate mussels.

A dogwhelk feeds by rasping small particles from its prey, using a file-like tongue. This made it difficult for the geneticist to identify the contents of the guts of dogwhelks by examining them with a light microscope.

Instead he used a technique called the enzyme-linked immunoabsorbent assay (ELISA). The flow chart summarises this technique.

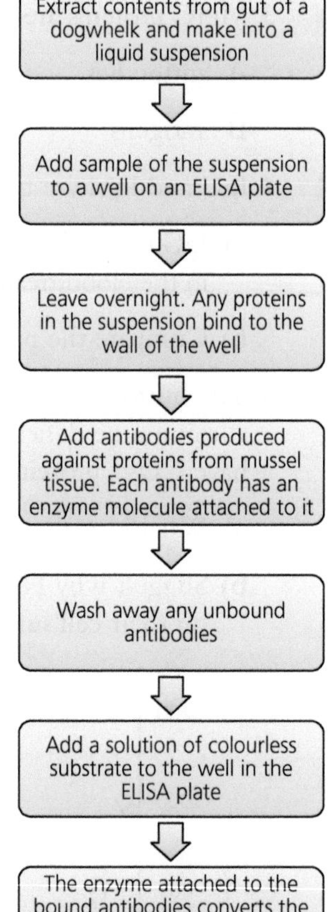

a) The antibody against mussel proteins was produced from samples of blood taken from rabbits that had been injected with a suspension of mussel tissue. The geneticist was not allowed to carry out this part of the experimental procedure because he did not hold a UK Home Office licence. Suggest why the UK Home Office has a licensing system. *(1)*

b) The rabbits were injected twice with a 21-day gap between injections. Suggest an explanation for this procedure. *(2)*

c) Explain why the antibodies against proteins from mussels would bind only to these proteins. *(1)*

d) Explain why the unbound antibody was washed away (step 5). *(2)*

e) Describe how the geneticist could tell if a dogwhelk had **not** eaten mussels. *(1)*

f) Suggest an appropriate null hypothesis for the geneticist's investigation. *(1)*

g) How could the geneticist use the ELISA technique to test his null hypothesis? Justify the choice of any statistical test you include in your method. *(4)*

Stretch and challenge

7 Against which diseases are vaccines currently in use, worldwide? Is there a pattern in the distribution of these diseases?

8 Monoclonal antibodies are widely used in biological and medical research.

a) How are monoclonal antibodies made?

b) Outline how they are used.

6 Cell control

Prior knowledge

In this chapter you will need to recall that:

- after their formation in a multicellular organism, most new cells differentiate, becoming specialised to carry out a particular function. In doing so, the majority lose the ability to divide by mitosis
- the genetic code comprises 64 different sequences of three nucleotide bases (base triplets), each of which encodes a specific amino acid
- the genetic material of living organisms consists of one or more molecules of DNA
- the sequence of bases in an organism's DNA determines the sequence of amino acids in the polypeptides and proteins that the organism is capable of producing
- the sequence of bases that encodes a polypeptide is called a gene
- polypeptides are made by ribosomes that assemble amino acids in a sequence that is prescribed by the base sequence of a gene
- protein synthesis involves transcription and translation
- whereas all the DNA of prokaryotes encodes polypeptides, much of the DNA of eukaryotes does not encode polypeptides.

Test yourself on prior knowledge

1. State the name given to a group of specialised cells that are derived from a common ancestor and perform the same function.
2. The genetic code is described as universal and degenerate. Explain what this means.
3. Are all genes the same length? Explain your answer.
4. Describe the difference between a polypeptide and a protein.
5. Outline the processes of transcription and translation during polypeptide production.
6. Give **two** locations in a chromosome of DNA that do not code for the amino acid sequence of polypeptides.

Key term

Genome The genetic material of an organism. For a prokaryote, this includes all of its genes; for a eukaryote it includes all of its genes plus the base sequences of DNA that do not encode polypeptides.

Introduction

You saw in Chapter 6 of *Edexcel A level Biology 1* that the zygote of a multicellular animal or plant is a single diploid cell. It contains the organism's genome – the complete set of chromosomes that are characteristic of that species – and gives rise to all the somatic cells of that animal or plant. Since the zygote divides by mitosis, you know that all the somatic cells derived from it are clones of the zygote, that is they contain copies of the same genome.

Figure 6.1 Two stages in the life cycle of a dragonfly

Figure 6.2 Part of a deadly nightshade plant (*Atropa belladonna*)

Look at Figure 6.1. Unless you are already familiar with these animals, you probably think they are different species. The animal in the photo on the left lives in freshwater, has no wings and, as you can see, is a carnivore (it is eating a fish). The animal in the photo on the right does not live in water, has wings and does not feed on other animals. In fact, they are the same species. The animal on the left is a young stage dragonfly (a larva) and the animal on the right is the adult dragonfly. They look and behave completely differently yet, within one life cycle, have exactly the same genome. A rather less dramatic example of how a single genome can produce different results is shown by the deadly nightshade in Figure 6.2. You can see some of the organs this plant possesses – the stem, leaves, flower and fruit. Although these organs look very different, the cells within them have the same genome.

These examples present the same puzzle. If their cells contain copies of the same genome, how can animals and plants produce tissues and organs that are structurally and functionally so different? It is this puzzle that this chapter will seek to explain.

Factors affecting gene expression

There are many ways in which the transcription or translation of genes can be regulated in eukaryotes. Here we will examine five:

- DNA regulatory sequences and transcription factors
- post-transcriptional modification of mRNA
- destruction of mRNA
- DNA methylation
- histone acetylation.

(The last three: epigenetic modifications)

Key term

Epigenetic modification Stable, long-term changes in the ability of a cell to transcribe its genes. Though these changes are inherited, they are not caused by changes to the base sequence of the DNA.

Three of these methods are said to be epigenetic modifications. This means that they are inherited by daughter cells but, in contrast with chromosome and gene mutations, are not the result of changes to the DNA. You will see later how these epigenetic modifications are important in the process of cell differentiation.

Regulatory sequences and transcription factors

Biologists used to think that all the DNA in the nucleus of a eukaryotic cell formed genes that encoded polypeptides. One of the many surprises brought about by genome sequencing (considered in Chapter 7) was the realisation that this is not so. Biologists now believe that only about 2 per cent of the DNA in the nuclei of human cells forms genes. Because the function of the other 98 per cent – the non-coding DNA – was not known, it used to be called 'junk DNA'. In fact, nothing could be further from the truth and we now know that this non-coding DNA plays an important role in regulating which of our genes are 'switched on' or 'switched off'.

Every gene is associated with short base sequences of non-coding DNA that regulate whether the gene is transcribed ('switched on') or not. There are two types of these **regulatory sequences** that stimulate the transcription of genes:

- **Promoters** – short base sequences that lie closed to their target genes. They initiate transcription by enabling RNA polymerase to bind to the gene they regulate.
- **Enhancers** – short base sequences that lie some distance from their target genes. They stimulate promoters causing an increase in the rate of transcription of the genes they regulate.

Figure 6.3 represents one of these regulatory sequences and its **target gene**, that is, the gene it regulates. The sequence represented in Figure 6.3 is a **promoter**. Notice that it is close to, and 'upstream' of, its target gene.

Key terms

Regulatory sequence
A sequence of DNA nucleotides that controls whether its target gene is transcribed or not.

Transcription factor
A protein, or assembly of several proteins, that regulate the production of mRNA. A specific transcription factor binds to a promoter region upstream of its target gene and either promotes or inhibits the binding of RNA polymerase to the target gene.

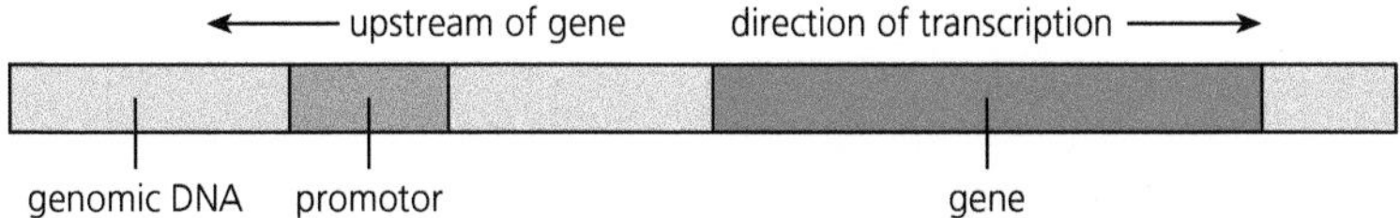

Figure 6.3 In eukaryotes, every gene is controlled by promoters, one or more non-coding sequences of bases. The promoter shown here is close to, and 'upstream' of, its target gene

As you know from *Edexcel A level Biology 1*, Chapter 3, in order for a gene to be transcribed, the enzyme RNA polymerase must be able to attach to it. If this enzyme cannot, the gene will not be transcribed, in other words no mRNA will be produced from it and so none of the polypeptide the gene encodes will be produced. In eukaryotes, RNA polymerase cannot initiate transcription itself. Attachment of RNA polymerase to a gene is regulated by the gene's promoter. Figure 6.4 shows how. One or more, specific proteins, called **transcription factors**, bind to the promoter. Once they have done so, an RNA polymerase binds to the transcription factor complex and becomes activated to begin the synthesis of mRNA from a unique point on the target gene.

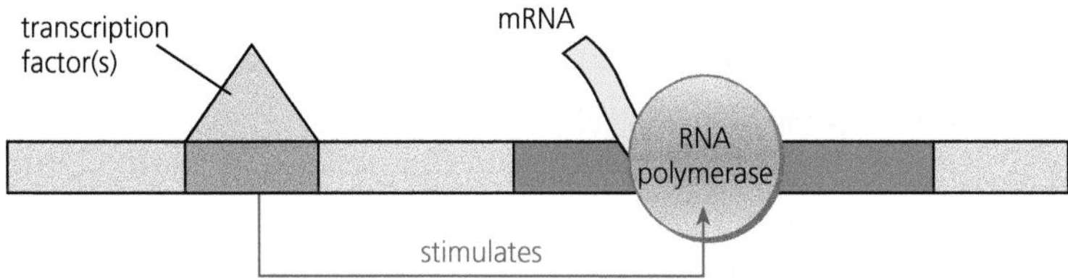

Figure 6.4 How transcription is initiated in a eukaryotic cell. One or more transcription factors bind with a promoter. By binding with the transcription factor complex, a molecule of RNA polymerase is activated and can begin to transcribe the promoter's target gene, producing mRNA

Test yourself

1 Explain why the term 'genome' is applied differently to a bacterial cell and a plant cell.
2 What is meant by the term 'epigenetic'?
3 Explain why the term 'junk DNA' is now thought to be inappropriate.
4 Give the relationship between the direction of transcription and 'upstream'.
5 Describe the role of a promoter in DNA transcription.

Activity

Glucocorticoids as transcription factors

Glucocorticoids are a class of steroid hormones. Steroids are all derived from cholesterol. Figure 6.5 shows how molecules of a glucocorticoid enter human cells.

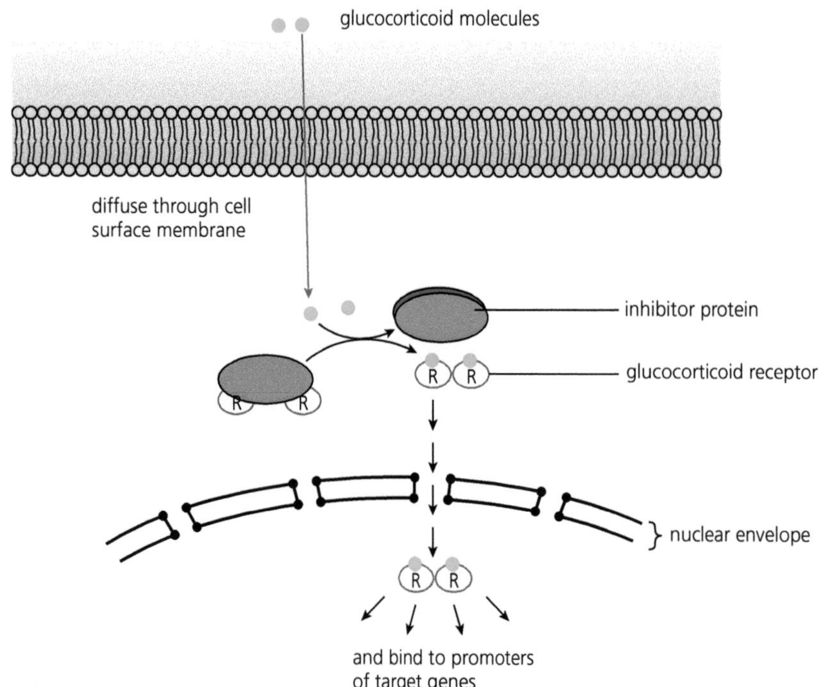

Figure 6.5 The action of glucocorticoids in human cells

1 Explain why molecules of glucocorticoids can diffuse through the surface membranes of human cells.
2 Write a word equation to represent the reaction that forms a glucocorticoid-receptor complex in the cytoplasm.
3 Suggest one advantage of the glucocorticoid receptors normally being bound in the cytoplasm to protein inhibitor molecules.
4 How do molecules of the glucocorticoid-receptor complex get into the nucleus?
5 Once inside the nucleus, the glucocorticoid-receptor complexes activate the expression of their target genes. Outline how they will do this.
6 One effect of glucocorticoids is to up-regulate the expression of genes encoding anti-inflammatory proteins. Use your knowledge from previous chapters to suggest the function of anti-inflammatory proteins.

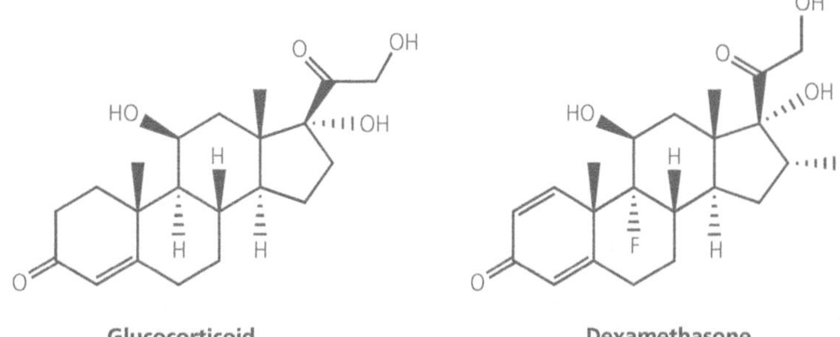

Figure 6.6 The molecular structure of glucocorticoid and the drug, dexamethasone

7 Figure 6.6 shows the generalised structure of a glucocorticoid and of a drug called dexamethasone. This drug binds to glucocorticoid receptors more powerfully than does glucocorticoid itself. Suggest why dexamethasone might be prescribed and use Figure 6.5 to suggest why dexamethasone might bind more powerfully to glucocorticoid receptors.

Post-transcriptional modification of mRNA

An essential feature of science is that when new evidence shows that a long-held hypothesis is not true, we change our ideas. The one-gene-one-polypeptide hypothesis, first proposed by Francis Crick, one of the Nobel prize-winning discoverers of the structure of DNA, is one such hypothesis. We now know that a single gene can give rise to more than one polypeptide.

In organisms with simple genomes, such as bacteria, mRNA molecules can be produced by continuous transcription through several adjacent genes. These organisms then process these 'multigenic transcripts' to generate two or more different polypeptides. This is not common in eukaryotes; the vast majority of eukaryotic genes are transcribed individually. To understand how eukaryotes can produce more than one polypeptide from a single gene, you need to revise your knowledge of the structure of a gene from *Edexcel A Level Biology 1*.

> **Tip**
>
> It might help you to remember that **e**xons are **e**xpressed regions of DNA whilst **i**ntrons are **i**ntragenic regions that are not expressed.

Figure 6.7 reminds you that the genes of eukaryotic cells contain base sequences that do encode mRNA, called exons, and sections that do not, called introns. During transcription, the exons and introns are all copied into the base sequence of an RNA molecule, called pre-mRNA. This molecule is then spliced; the introns are removed and the exons are rejoined, to form mature mRNA. This mRNA migrates to the cytoplasm where its base sequence is translated by ribosomes.

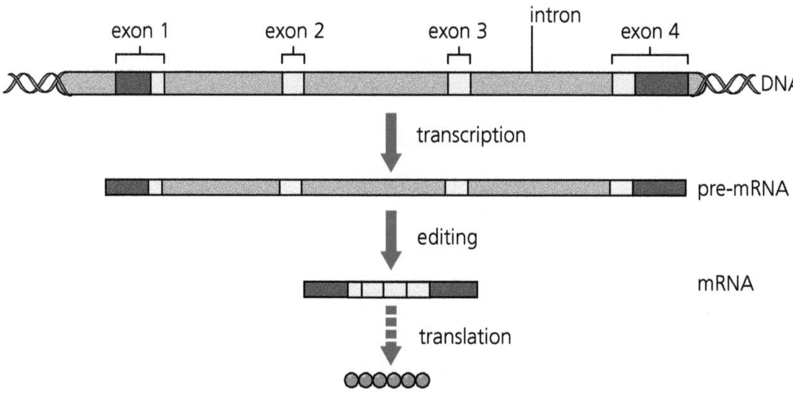

Figure 6.7 Editing of pre-mRNA in eukaryotic cells

Figure 6.8 shows how alternative splicing of pre-mRNA can produce mature RNA with different combinations of exons. Alternative splicing is known to be common in humans. The functions of the different mature mRNA molecules, called **isoforms**, are not always clear. Sometimes, they result in polypeptides that are tissue-specific, as is the case with the gene represented in Figure 6.8. Here, pre-mRNA transcribed from the calcitonin gene is spliced:

- in thyroid tissue to produce the polypeptide **calcitonin**. This polypeptide comprises 32 amino acid residues and acts as a hormone, stimulating a reduction in the concentration of calcium ions in the blood.
- in nervous tissue to produce the polypeptide calcitonin-gene-related peptide (**CGRP**). This polypeptide comprises 37 amino acid residues and is a powerful vasodilator that is involved in the transmission of pain. High levels of CGRP have been associated with migraine.

> **Key term**
>
> **Isoforms** Different arrangements of exons in mature messenger RNA transcribed from a single eukaryotic gene. The different isoforms result in different polypeptides being produced from one gene.

So, you can see that alternative splicing of pre-mRNA produced from a single gene can result in the production of completely different polypeptides. The one-gene-one-polypeptide hypothesis, at one time a core dogma of cell biology, does not hold true.

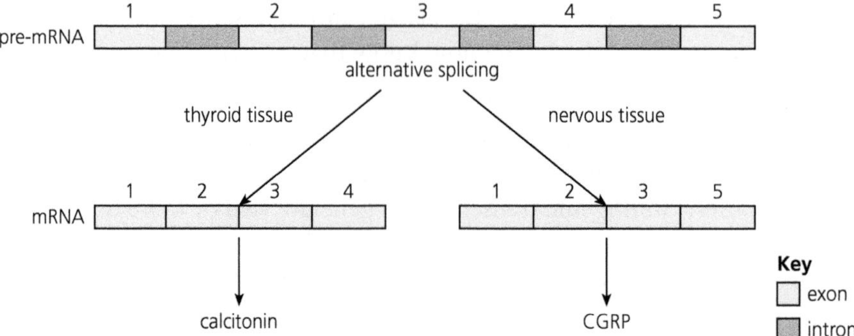

Figure 6.8 Alternative splicing of pre-mRNA results in tissue-specific products of the calcitonin gene

Test yourself

6 Distinguish between an exon and an intron.

7 Several different enzymes are involved in the editing of pre-mRNA. Suggest the different functions of **two** of them.

8 In Chapter 5, you saw that a small number of major histocompatability complex genes encoded a vast number of antibodies. Suggest how they do this.

9 How does a bacterium produce a multigenic transcript and how does it then use it?

10 Suggest the difference in the number of nucleotides in a molecule of mRNA encoding calcitonin and one encoding CGRP. Explain your answer.

Destruction of mRNA

In *Edexcel A level Biology 1*, Chapter 3, you learnt about the roles of three types of RNA during protein synthesis: messenger RNA (mRNA), ribosomal RNA (rRNA) and transfer RNA (tRNA). Recently, RNA molecules have been discovered that were thought to be too small to have any important function. Although these small RNA molecules are non-coding, they have been found to play an important part in gene regulation. They ensure that mRNA can be destroyed, so that it never gets translated.

The destruction of mRNA involves these short regulatory RNA molecules binding to a protein to form a complex called an **RNA-induced silencing complex** (**RISC**). As you might expect, these small RNA molecules are themselves encoded by genes. There are two types of this regulatory RNA:

- **microRNA** (**miRNA**) – short, single-stranded RNA
- **small inhibitory RNA** (**siRNA**) – short, double-stranded RNA.

Let's see how these two types of RNA lead to the destruction of mRNA.

miRNA

In 2013, scientists estimated there to be over 2000 different miRNA molecules in human cells and that they are involved in regulating over 60 percent of the mRNA that encodes polypeptides. Each is produced as a long, precursor molecule in the shape of a hairpin bend. You can see the shape of the precursor molecule, called pri-miRNA in Figure 6.9.

You can also see in Figure 6.9 how, under the action of a RNA hydrolase, one of these molecules is hydrolysed to form much shorter, single-stranded miRNA molecules.

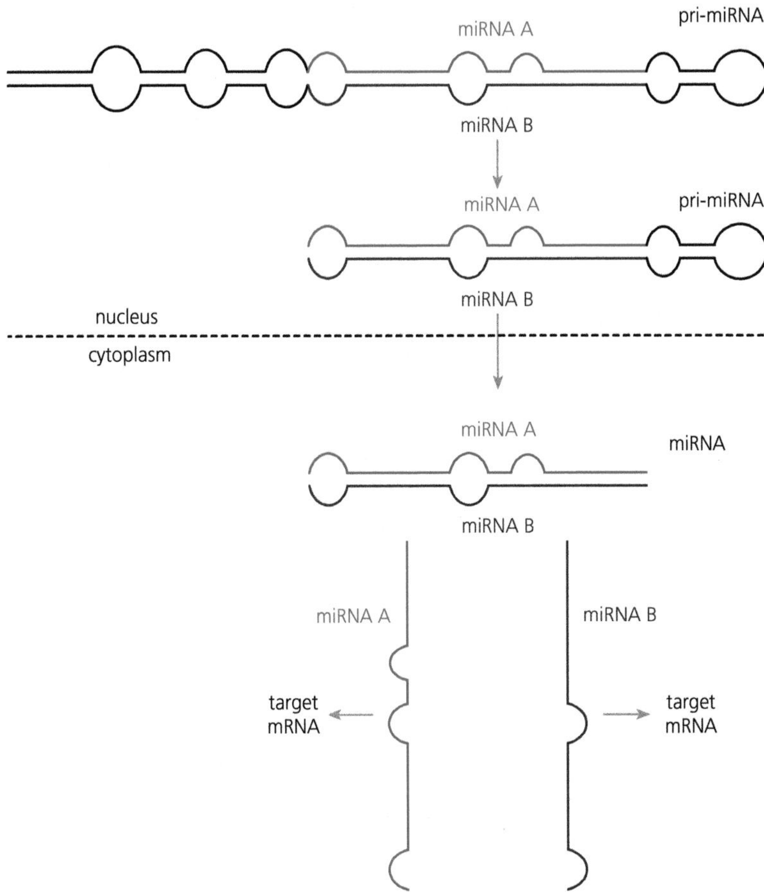

Figure 6.9 The formation of mature microRNA (miRNA) molecules from precursor miRNA (pri-miRNA)

As described above, each miRNA molecule binds to a protein to form a RISC. The miRNA within the RISC will bind to mRNA in the cytoplasm. The two RNA molecules bind by the formation of hydrogen bonds between base pairs – cytosine (C) with guanine (G) and adenine (A) with uracil (U). The sequence of bases in the two strands is not, however, fully complementary. Figure 6.10 shows how 'bulges' appear in the two strands where the bases are not complementary. These bulges prevent the mRNA being transcribed.

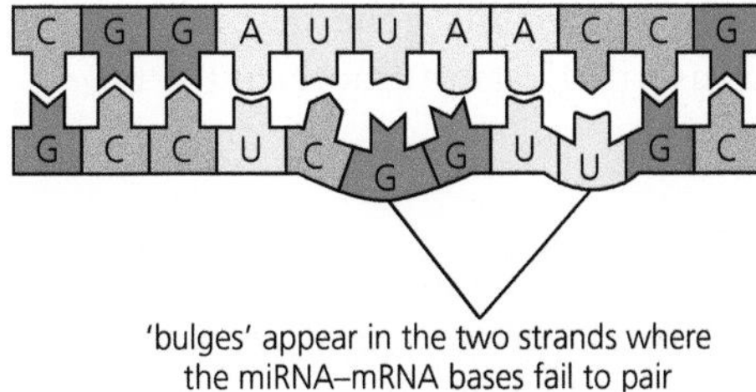

Figure 6.10 'Bulges' appear in the two strands where the miRNA-mRNA bases fail to pair. These 'bulges' prevent the transcription of the mRNA

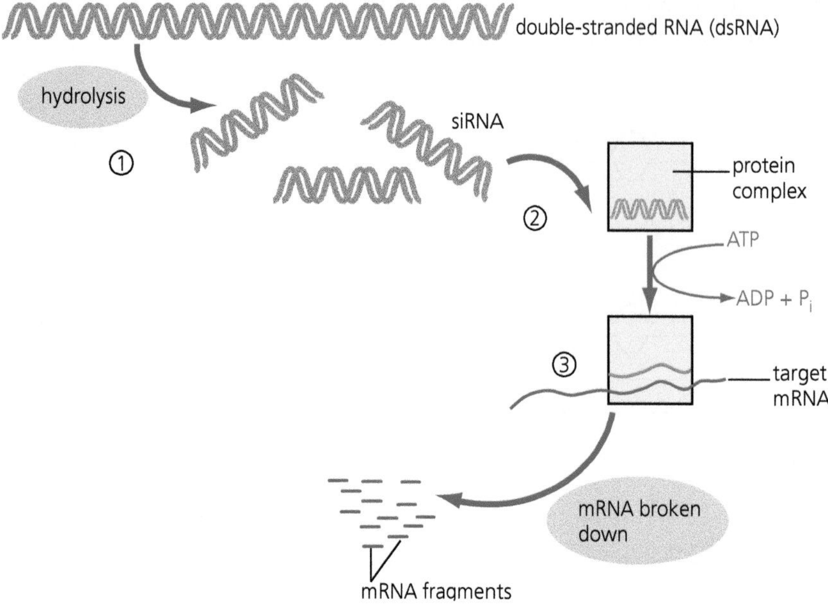

Figure 6.11 Small inhibitory RNA (siRNA) combines with a protein complex, leading to the destruction of messenger RNA (mRNA)

siRNA

Unlike miRNA, siRNA are formed as molecules of double-stranded RNA (dsRNA). A RNA hydrolase, called a **dicer**, hydrolyses these dsRNA molecules into lengths of about 20 base pairs. These are the siRNA molecules shown as step 1 in Figure 6.11.

Like the miRNA, these double-stranded siRNA molecules bind with a protein complex to form a RISC (step 2 in Figure 6.11). Proteins within the RISC unwind the RNA and remain bound to one of the strands. It is this single-stranded, antisense RNA, bound within the RISC, that binds to the target mRNA (step 3 in Figure 6.11). Again, this binding is by hydrogen bonds between complementary base pairs. Finally, the bound mRNA is hydrolysed by yet another RNA hydrolase.

DNA methylation

DNA molecules are very long polymers of nucleotides. Each nucleotide contains the pentose deoxyribose, a phosphate group and a purine or pyrimidine base. Two of these bases – adenine and cytosine – can be methylated, that is a methyl group (CH_3) can be added to one of their carbon atoms. In vertebrate animals, methylation of DNA bases is restricted to the pyrimidine base, cytosine. If you look at Figure 6.12, you can see how a cytosine residue is methylated: the hydrogen atom on carbon-5 is replaced by a methyl group. Only about 3 per cent of the cytosine residues in human DNA are methylated. Most are found where the cytosine is linked by a phosphodiester bond to a guanine residue, represented **CpG**.

Figure 6.12 Methylation - the addition of a methyl group (CH_3) - of a cytosine molecule

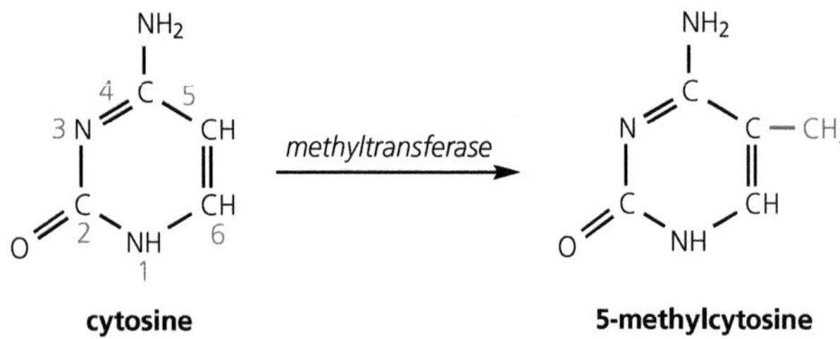

Repeated CpG sequences are common in the DNA near promoters. The presence of methylated DNA, especially near promoters, prevents the activation of RNA polymerase, described above. This means that the target genes of these promoters are effectively silenced. Conversely, before any gene that is silenced can be transcribed, its promoter must be demethylated.

Histone acetylation

Eukaryotic DNA is different from prokaryotic DNA in a number of ways. One is that eukaryotic DNA is associated with a protein called histone. You can see this represented in Figure 6.13, in which the 'thread' of DNA is wound around 'beads' of histone.

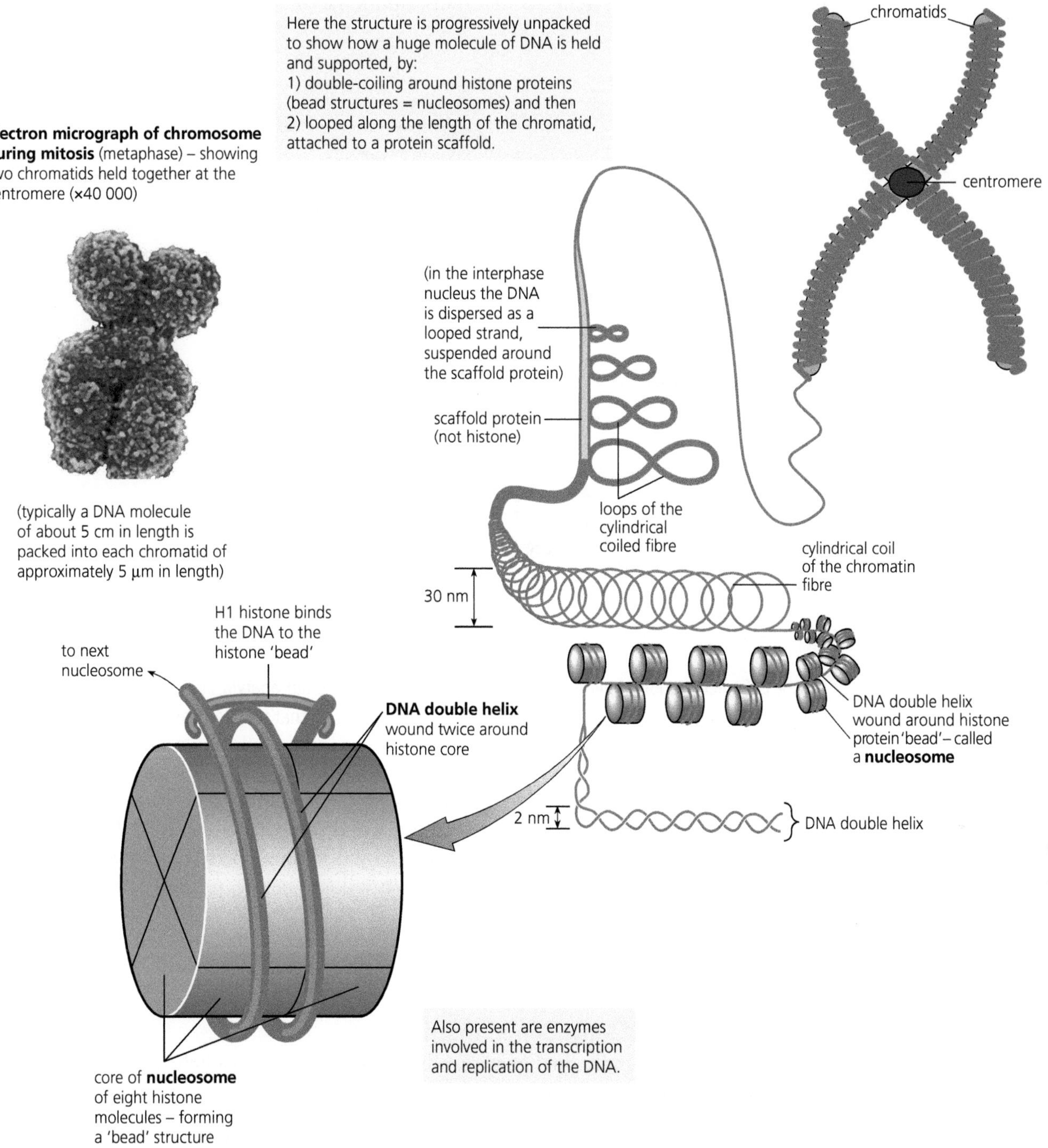

Figure 6.13 The packaging of DNA in the chromosomes of eukaryotic cells

Figure 6.14 on the next page, represents a tiny part of a chromosome. The 'tails' of the histone molecules contain the amino acid leucine. This amino acid residue can be acetylated, that is an acetyl group ($COCH_3$) can be transferred on to it from acetyl coenzyme A. When this happens, the binding of the histones changes and they become more loosely packed. You can see the change from (a) to (b) in Figure 6.14. In part (a) the histones are so tightly packed that transcription factors and RNA polymerase cannot gain access to this region. After acetylation in part (b), however, the loosening of the histones frees the gene, so that transcription factors and RNA polymerase can now gain access to it.

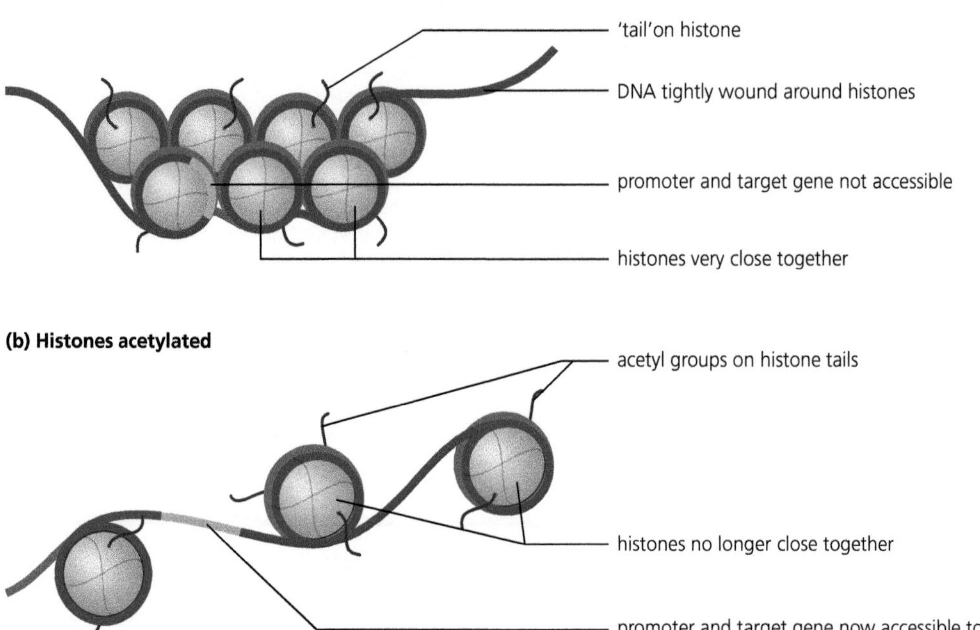

Figure 6.14 Acetylation of histones. In (a), the histone 'tails' are not acetylated and the histones are bound tightly together. In (b), the histone 'tails' are acetylated so that the histones are held together more loosely, exposing the gene in this region of DNA

The different patterns of genes 'switched on' and genes 'switched off' that you have seen above are stabilised by DNA methylation and histone acetylation, so that these patterns are passed on to daughter cells formed by division. The stabilisation of these patterns is called **gene imprinting** and, as you will see below, is important both in inheritance and in cell differentiation.

Extension

X chromosome inactivation in mammals

You have seen that individual genes can be 'switched on' or 'switched off'. In mammals, an entire chromosome can be 'switched off'.

In mammals, sex is determined by sex chromosomes. A female has two X chromosomes whereas a male has one X chromosome and a much smaller Y chromosome. Having the 'wrong' number of chromosomes almost always results in abnormal development. So how do mammals avoid abnormal development when the cells of females and males have different numbers of X and Y chromosomes?

Since the Y chromosome contains very few genes, and those that are present are mostly genes regulating male sexual function, it appears that a female can develop normally without a Y chromosome. The X chromosome, however, contains a large number of genes and they are genes that control many processes that are vital to both sexes. You can, therefore, consider that a female mammal has too many X chromosomes for normal development.

The solution is that one of the X chromosomes is inactivated, so that each somatic cell in the body of a female mammal has one activated X chromosome and one inactivated X chromosome. The inactivation occurs early in embryonic development, probably in cells in the inner cell mass of a blastocyst (see *Edexcel A Level Biology 1*, Chapter 6, page 126). It also occurs at random in each cell of the blastocyst. In one embryonic cell it might be the paternal X chromosome that is inactivated, and in an adjacent embryonic cell it might be the maternal X chromosome that is inactivated. Whichever, once the inactivation has been established, it is perpetuated. When a cell in the blastocyst divides by mitosis, its daughter cells inactivate the same X chromosome and so on, as they and their daughter cells divide. Consequently, a female mammal is a mosaic of clones derived from these embryonic cells. Within each clone, the same X chromosome (maternal or paternal) is inactivated but between clones the 'choice' is random. The mechanisms involved in X inactivation – how the cell 'counts' the number of X chromosomes and then how it inactivates one of them – remain hotly debated and research in this area is on-going.

This pattern of mosaic clones can be observed in some conditions in which inheritance involves a sex-linked gene, in other words one carried on the X chromosome. One clone might express one allele of a sex-linked gene and another clone might express another allele of the same sex-linked gene. This can result in a female showing patches of tissue that are either affected or unaffected by an inherited disease, such as X-linked hypohidrotic ectodermal dysplasia (XLHED), with symptoms including missing sweat glands.

Test yourself

11 Give the roles of mRNA, rRNA and tRNA.

12 What type of molecule is an RNA-induced silencing complex?

13 Describe how the structure of a molecule of siRNA differs from that of mRNA, miRNA, tRNA.

14 Explain the notation 'CpG'.

15 Ultimately, DNA methylation and histone acetylation have the same effect. Describe that effect.

Stem cells in humans

Textbooks of human histology identify over 200 different cell types in the human body. Like all multicellular organisms, each human starts life as a single-celled zygote. This cell has the ability to produce every one of the different cells identified in textbooks of human histology. In the early stages of development, the human zygote divides to form a ball of cells, the blastocyst (see *Edexcel A Level Biology 1*, Chapter 6, page 126). The blastocyst contains an inner mass of cells that will form the embryo and an outer layer of cells that will form part of the placenta. Because

> **Key term**
>
> **Totipotent** A term used to describe an unspecialised cell that can divide repeatedly to renew itself and to give rise to any cell in the body of an organism including, in mammals, cells of the placenta and umbilical cord.

the human zygote can form any human cell, including those of the placenta and umbilical cord, it is said to be **totipotent**.

During development, the zygote repeatedly divides by mitosis. The new cells quickly begin to differentiate to form different tissues. As they do so, their ability to divide and to produce different types of progeny cells becomes restricted. The extent of this restriction is described in Table 6.1.

Table 6.1 The different levels of cell 'potency', i.e. ability to divide and produce different types of progeny cell. Mature cells, such as nerve cells, have no 'potency'

Description of cell	Ability to divide and produce different progeny cells	Human examples
Totipotent	Can divide to replicate itself and produce any cell (including placenta and umbilical cord in mammals)	Zygote
Pluripotent	Can divide to replicate itself to produce any cell (but not placenta or umbilical cord in mammals)	Cell from inner mass of blastocyst
Multipotent	Can divide to replicate itself and to produce the different cells in one type of tissue	Haematopoietic cells in bone marrow that give rise to red blood cells, white blood cells and platelets
Unipotent	Can divide to replicate itself and to produce only one type of cell	Cells in the germinal epithelium of the skin

The cells in mature, fully differentiated tissues generally lack the ability to self-renew. This presents a problem in diseases that involve the death of cells.

Following a myocardial infarction, some of the muscle cells of a sufferer's heart die. These dead muscle cells cannot renew themselves. One way in which this can be treated is by a heart transplant – using surgery, the healthy heart of a dead donor replaces the faulty heart of the recipient. Of course, the heart muscle cells of the donor heart will have non-self antigens, so the recipient must be given immunosuppressant drugs to prevent its rejection.

> **Key term**
>
> **Stem cells** Cells that are capable of self renewal and of producing new cells that differentiate into other specialised types of cell.

What if, instead of replacing whole organs, surgeons could transplant cells that have the ability to divide and replace the dead cells? You might be familiar with one way in which this is commonly done. People with cancer of their bone marrow are routinely treated with a bone marrow transplant. Initially, chemotherapy is used to destroy the sufferer's own, faulty bone marrow. They are then given an intravenous transplant of healthy bone marrow from a donor. Again, there is a danger that the new tissue will be rejected, so the donor's bone marrow must be carefully matched to that of the recipient. The new transplanted bone marrow contains haematopoietic cells – multipotent cells that can divide to produce all the blood cell types that the recipient needs (see Table 6.1). Used in this way, the multipotent cells are acting as **stem cells**.

Types of stem cell and their uses

You have seen above how stem cells in the bone marrow of a donor can be successfully used to replace those in the bone marrow of a recipient. In theory, this use of transplanted stem cells could be extended to enable sufferers of many diseases to produce new, healthy cells. In reality, the use of stem cells involves moral judgements that have led to governmental restrictions.

Embryonic stem cells

Looking at Table 6.1, you might think that the best type of human stem cell to use in transplants would be pluripotent cells. These cells can produce any type of cell in the body of a child or adult. They are also present in a discrete area – the inner cell mass of a blastocyst – so would be easy to harvest. Since they are found only in the human embryo, these cells are known as **embryonic stem cells**. Herein lies an obstacle to their widespread use; taking embryonic stem cells destroys the embryo. Many people believe that it is unethical to destroy an embryo, since it has the potential to become a human being. Many religious, moral and humanistic organisations also share the same viewpoint.

Governments have to take a balanced view when faced with two conflicting issues, in this case the ability to extend life or relieve suffering by using embryonic stem cells and the understandable objections raised by a large proportion of the population to destroying embryos. In the UK, the use of embryonic stem cells is enshrined in law. Scientists are allowed to use embryonic stem cells only if:

- they are from unused embryos produced by *in vitro* fertilisation (IVF) during the treatment of infertile couples
- the unused embryos are donated specifically for this purpose by both members of the couple receiving fertility treatment
- the use of the stem cells is registered with a professional organisation, called the UK Stem Bank.

An organisation called the Human Fertilisation and Embryology Authority regulates the donation of embryos and, on behalf of the UK government, ensures that the UK law is upheld.

Adult stem cells

Small groups of stem cells are found in the tissues of adults, where their general function is to repair damaged cells or replace lost cells. These cells are called **adult stem cells**. Unlike embryonic stem cells, these cells are either unipotent or, at best, multipotent. As we saw above, the bone marrow is a rich source of adult stem cells but far fewer are found in tissues such as the brain. This makes it much more difficult to harvest adult stem cells from most tissues. One advantage that adult stem cells have over embryonic stem cells, however, is that it is possible to use a patient's own stem cells, removing the risk of rejection.

Induced pluripotent stem cells (iPS cells)

Earlier in this chapter, you saw that epigenetic modifications accompany the progressive restriction of specialised cells to divide. Totipotent cells in the embryo give rise to pluripotent cells in the blastocyst that give rise to multipotent cells in some tissues and fully differentiated somatic cells in most cases. About 10 years ago, scientists working in Japan found they could reverse this process.

Fibroblasts are relatively unspecialised cells found in connective tissue. They are capable of differentiating into adipose tissue (fat), bone, cartilage and smooth muscle (such as that found in arteries and the intestine). The Japanese research scientists reasoned that if they could find the genes that were vital to embryonic stem-cell function, they could induce an embryonic state in adult cells. They identified a number of such genes, all of which encoded transcription factors. One-by-one, they transferred these genes into mouse fibroblasts using retroviruses as vectors. They then devised a technique that enabled them to find which cells had become pluripotent. Eventually they found four genes, called *Oct4*, *Sox2*, *cMyc*, and *KLf4*, that induced the change from an adult fibroblast to a pluripotent stem cell.

Since the 'reverse differentiation' of these fibroblast cells was induced, they called these cells **induced pluripotent stem cells (iPS cells)**. As these cells have the properties of self-renewal and the ability to differentiate into all adult cell types, but do not involve destruction of an embryo, they are considered to be as useful as, but more ethically appropriate than, embryonic stem cells.

Using pluripotent stem cells to treat medical conditions and test drugs

Pluripotent stem cells have the potential to be used in two ways in medical research.

- They can be used to produce new types of cell that directly replace those affected by disease. Diseases with the potential to be treated in this way include Crohn's disease, Type I diabetes, osteoarthritis and Parkinson's disease.
- They can be used to generate new types of cell that can be used in the laboratory to test the effectiveness and safety of new drugs, reducing the need for animal (and human) experimentation.

We must treat the potential uses of pluripotent stem cells with caution, though.

- Scientists do not yet fully understand how to control the differentiation of pluripotent stem cells into the 200-plus cell types catalogued in human histology textbooks.
- Two of the genes that transformed adult fibroblasts to iPS cells (namely *cMyc* and *KLf4*) are oncogenes – genes with the potential to cause cancer.
- The conversion of adult cells to iPS cells is currently slow and the success rate is very low.
- Growing cells for therapies will require specialist systems and research centres, so access to therapies may be limited to areas with suitable facilities.
- The safety aspects of introducing pluripotent stem cells into recipients must be trialled. Clinical trials take a long time and are expensive.

In the meantime, doctors must manage the expectations of their patients suffering diseases that have the potential to be cured using pluripotent stem cells, who read and hear of these 'miracle cures' in the media.

Tip

Some of these Test yourself questions rely on you being able to make links between the content of this chapter and other parts of the specification, i.e. they are synoptic.

Test yourself

16 Lymphoid cells are found in bone marrow. They are able to divide, producing lymphocytes. Are lymphoid cells totipotent, pluripotent, multipotent or unipotent? Explain your answer.

17 During a myocardial infarction, some of the muscle cells of a sufferer's heart, die. Explain what causes the death of these cells.

18 What is a human stem cell?

19 Explain why immunosuppressant drugs must be given to people who have received a bone marrow transplant.

20 What is an oncogene?

Exam practice questions

1 The fertilised egg cell of a multicellular organism is:

A multipotent | **C** totipotent

B pluripotent | **D** unipotent *(1)*

2 Which one of the following is a form of epigenetic modification?

A alternative pre-mRNA splicing

B continuous transcription

C DNA methylation

D DNA replication *(1)*

3 An exon is:

A a sequence of bases found at the end of a gene

B a sequence of bases found at the end of a molecule of mRNA

C a sequence of bases that is extracted from a pre-mRNA molecule

D a sequence of bases that is copied into a mature mRNA molecule *(1)*

4 Copy and complete the table by placing a tick in each box where the statement is true. *(4)*

Statement	Type of cell		
	Adult stem cell	Embryonic stem cell	Induced pluripotent stem cell
Can be grown in suitable laboratory conditions			
Could be used with patient's own cells			
Use in UK is restricted by law			
Involves removal of epigenetic modification			

5 In mammals, the somatic cells of females have two X chromosomes and those of a male have one X chromosome.

A gene encoding ginger-coloured fur is located on the X chromosome of domestic cats. If an X chromosome carries the gene for ginger fur, the cat will produce ginger fur.

A male cat with ginger fur mated with a female with black fur. The female had four kittens, two male and two female. Neither of the males had ginger fur. Both the females were tortoiseshell, with patches of ginger fur and patches of black fur.

a) What is meant by a 'somatic cell'? *(1)*

b) What name is given to the position of a gene on a chromosome? *(1)*

c) Use the information in the passage to explain why neither of the male kittens had ginger fur. *(2)*

d) Use your understanding of cell regulation to explain why the female kittens had patches of ginger fur. *(4)*

6 The intestinal bacterium, *Escherichia coli*, is able to absorb the disaccharide lactose and hydrolyse it to its constituent monosaccharides. To do this, it uses two enzymes:

- lactose permease, encoded by a gene *lacY*, speeds up the absorption of lactose by the cells of *E. coli*
- β-galactosidase, encoded by a gene *lacZ*, speeds up the hydrolysis of lactose.

(A) No lactose present

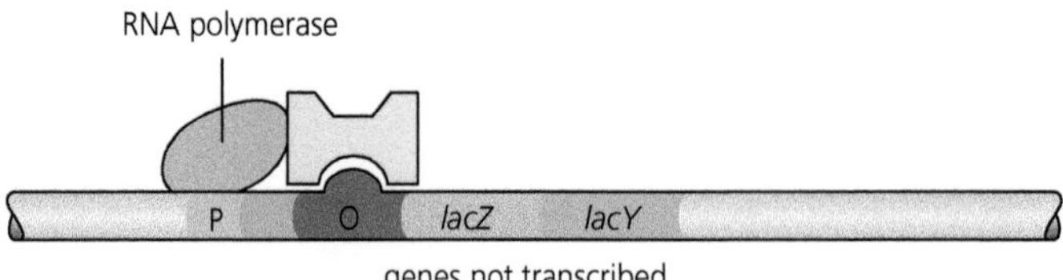

(B) Lactose present

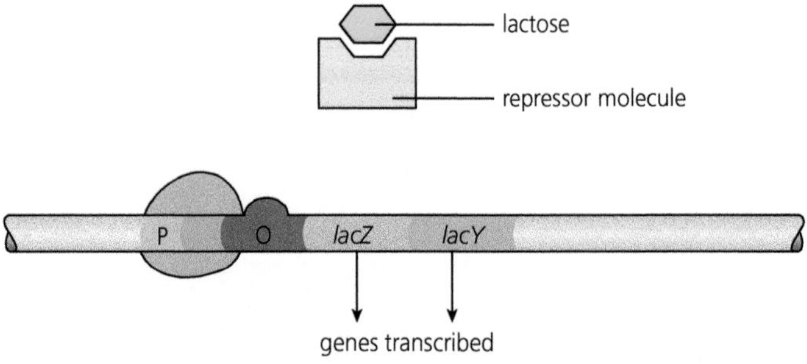

The diagram represents a section of DNA from *E.coli* with these genes and operator (O) and promoter (P) regions. In (A), neither *lacY* nor *lacZ* is transcribed. In (B) both *lacY* and *lacZ* are transcribed.

a) Use the diagram to suggest the function of the operator and promoter regions of DNA. *(3)*

b) Use the diagram to explain why lactose permease and β-galactosidase are produced only when lactose is present. *(3)*

c) Suggest **one** advantage that cells of *E. coli* gain by producing these enzymes only when lactose is present. *(1)*

d) Compare and contrast the method of gene regulation represented in the diagram with gene regulation in eukaryotes. *(3)*

7 Insulin is a peptide, 51 amino acids long, that acts as a hormone. It is produced in small groups of cells in the human pancreas, called the islets of Langerhans. It stimulates the uptake of glucose from the blood by cells mainly in the liver.

a) How many RNA nucleotides would you expect to be present in the mRNA translated to produce insulin? Explain your answer. *(1)*

Type 1 diabetes results from the autoimmune destruction of cells in the islets of Langerhans that normally produce insulin. It has been proposed that human embryonic stem cells could be used to replace the missing insulin-producing cells in the pancreas of a person suffering Type I diabetes.

b) What are the main features of embryonic stem cells? *(2)*

c) Suggest why the method proposed is likely to be more successful than transplanting insulin-producing cells from the pancreas of an adult donor. *(3)*

d) Discuss any ethical issues that are raised by this proposal. *(4)*

Stretch and challenge

8 Huntington's disease is an inherited human disorder caused by the presence of huntingtin protein in brain cells. Its presence causes the progressive death of brain cells and, ultimately, the death of the sufferer.

Outline how the use of siRNA might one day make it possible to provide a treatment for Huntington's disease.

9 Produce a 10-minute presentation in which you provide a balanced view of the use of stem cells in medicine.

7 DNA profiling, gene sequencing and gene technology

Prior knowledge

In this chapter you will need to recall that:

- a nucleotide is formed from three sub-units – a pentose, a phosphate group and an organic base
- a strand of DNA is a polymer of nucleotides, held together by phosphodiester bonds
- a DNA molecule comprises two antiparallel strands of DNA held together by hydrogen bonds between complementary bases, adenine with thymine and cytosine with guanine. Although individually weak, the sheer number of hydrogen bonds in a DNA molecule holds the two strands together very strongly
- during DNA replication, a helicase enzyme separates the two strands in a DNA molecule exposing their bases. Free nucleotides pair with these exposed bases, adenine with thymine and cytosine with guanine, and DNA polymerase catalyses the formation of new phosphodiester bonds in the 5-prime to 3-prime direction of the developing strand
- a DNA molecule contains some base sequences that encode the sequence of amino acids in polypeptides and some that do not. Some of the latter sequences regulate gene expression
- during transcription, part of the DNA base sequence is transcribed into a sequence of bases in a molecule of RNA. In eukaryotes, this sequence is then edited to form mRNA. The mRNA is used by ribosomes during the production of polypeptides
- the genetic code is universal and is based on a combination of three bases encoding one amino acid
- bacteria are prokaryotic. They have a single, circular DNA molecule that is not associated with proteins (in other words is naked) and might have many smaller circular molecules of DNA, called plasmids
- viruses are non-living particles that can infect living cells. During infection, viral nucleic acid is inserted into a host cell. The host cell then replicates the viral nucleic acid, produces the encoded polypeptides and assembles new virus particles
- the DNA sequence that encodes a functional polypeptide is known as a gene and its location is known as its locus.

Test yourself on prior knowledge

1. Name the pentose sugar in a DNA nucleotide.
2. Give the RNA sequence that is complementary to the DNA base sequence AGTCAT and name the bases in your sequence.
3. Identify **two** events that might occur during the editing of newly transcribed RNA in a eukaryotic cell.
4. Explain why mRNA is not edited in prokaryotic cells.
5. During DNA replication, new bases can only be added to a growing DNA strand in a 5-prime to 3-prime direction. What property of DNA polymerase does this exhibit?

Introduction

Some of the content of your A level biology course has changed very little over many years. In complete contrast, this part of the course is one of the fastest changing topics of biology. The techniques that are used have developed, and are continuing to develop, rapidly and most have become automated in the process. The results gained using these new techniques have brought about rapid changes in our understanding of the field of molecular genetics and with it new understanding of how we can better identify and treat diseases.

There are, however, a few techniques that remain at the heart of gene technology. Two of these are used to make multiple copies of DNA and are summarised in Figure 7.1. One, called the polymerase chain reaction (PCR), was originally carried out in glassware (in other words *in vitro*), although glassware has now largely been replaced by plastic; the other involves the use of living cells (*in vivo*) to replicate fragments of DNA that have been inserted into their genomes.

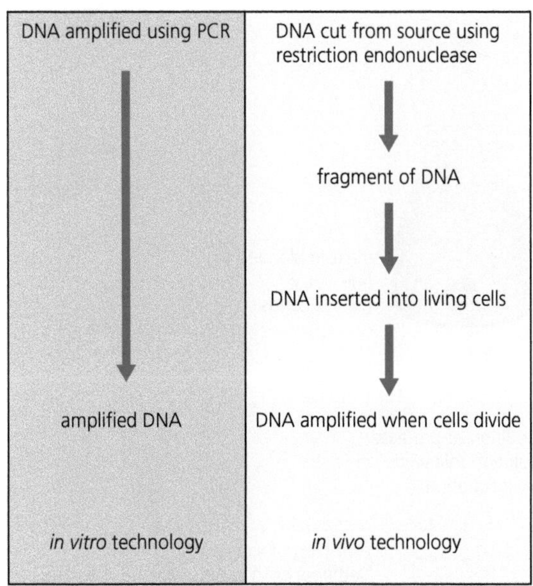

Figure 7.1 Producing multiple copies of DNA fragments can be achieved using *in vitro* and *in vivo* methods

Both techniques are used with purified samples of DNA. Figure 7.2 on the next page, shows how samples of DNA can be extracted from tissue by mechanically breaking up the cells, filtering off the debris, and breaking down cell membranes by treatment with detergents. The protein framework of the chromosomes, often called a scaffold, is then removed by incubation with a protease. The DNA, now existing as long threads, is isolated from this mixture of chemicals by precipitation with ethanol, and is thus 'cleaned'. The DNA strands are then re-suspended in an aqueous, pH buffered medium, ready for use.

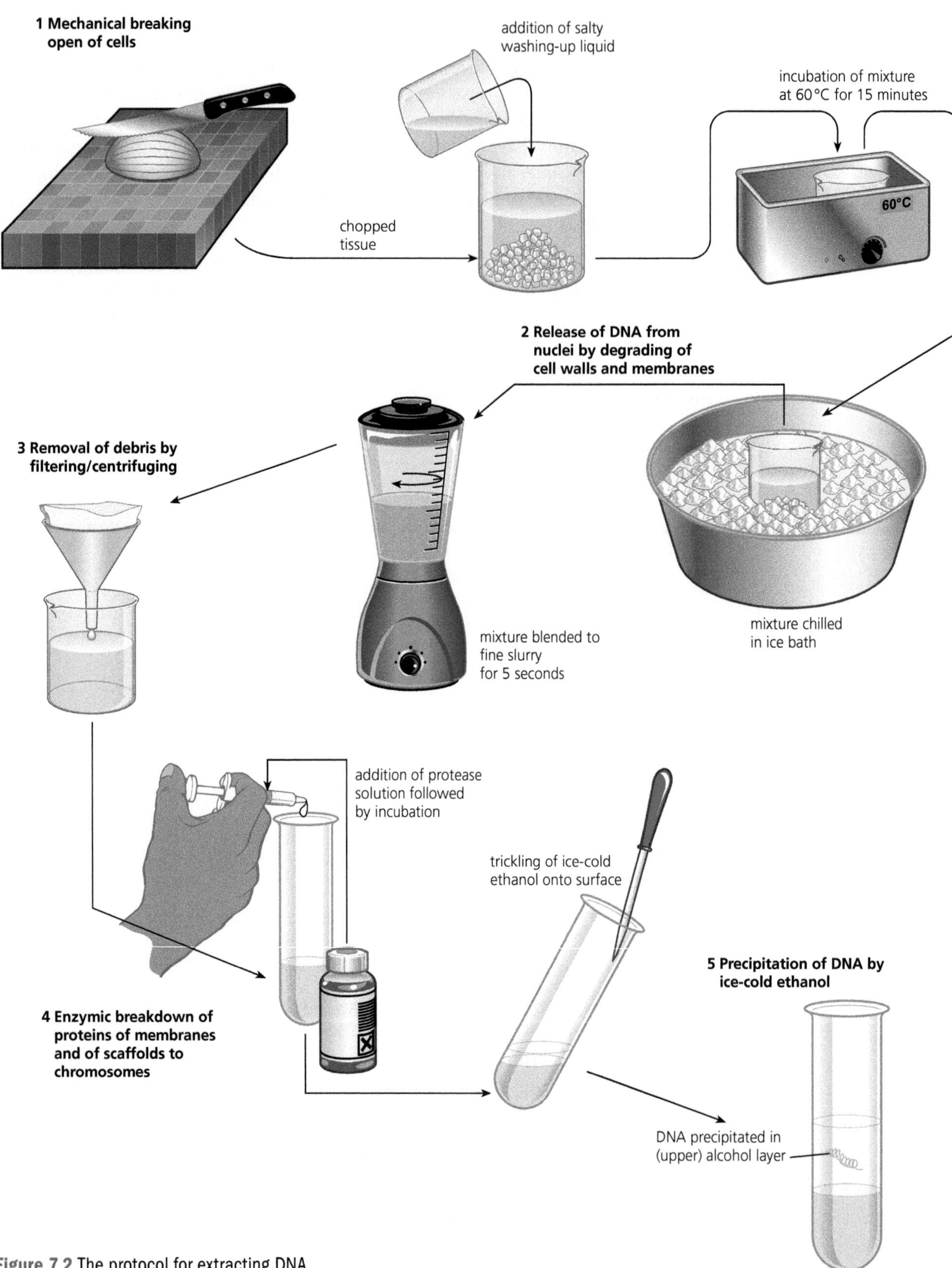

Figure 7.2 The protocol for extracting DNA

In-vitro amplification of DNA – the polymerase chain reaction (PCR)

Since the 1980s, the PCR has revolutionised molecular genetics because it allows the rapid analysis and cell-free cloning of DNA. Consequently, it has become a technique that is fundamental to other aspects of gene technology.

The PCR is used to amplify target DNA sequences that are present within a DNA source. By amplify, we mean it produces multiple copies. The technique involves mixing DNA containing the target sequence with a mixture of reagents in a plastic PCR tube and placing the tube in a machine called a **thermal cycler**.

The reagents include:

- primers – short lengths of single-stranded DNA (called oligonucleotides) that are complementary to the base sequence of part of the 3' (3-prime) ends of the strands of target DNA to be copied
- free DNA nucleotides, each with an adenine, cytosine, guanine or thyme base
- thermostable DNA polymerase.

Key terms

Amplify target DNA sequences Produce a very large number of copies of a fragment of DNA. This can be done by transferring the target DNA into bacterial cells and allowing the resulting bacterial culture to produce the new DNA copies - the *in-vivo* method. More often, it is done using the polymerase chain reaction (PCR)- the *in-vitro* method.

Primer A short strand of nucleotides (an oligonucleotide) with a base sequence that is complementary to the short base sequence at the beginning (3' end) of a strand of DNA to be copied. Primers are used as a starting point for the action of DNA polymerase, both in the semi-conservative replication of DNA we considered in Book 1 and in the PCR.

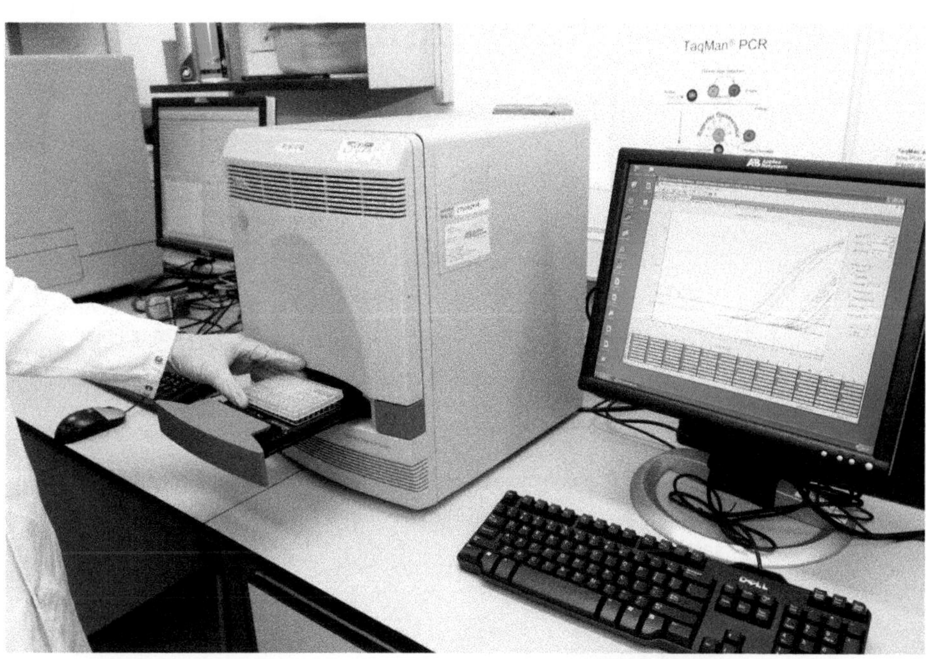

Figure 7.3 Loading a PCR machine. This demonstrates how much of the laboratory work involved in molecular genetics has become automated. The scientist has loaded the contents of each tube. Her next direct involvement will be to interpret the computer printout

You can see a PCR machine being loaded in Figure 7.3. It is a thermal cycler that varies the temperature at which the tubes are incubated in a pre-programmed way. Figure 7.4 on the next page, summarises these temperatures and the reactions that occur during one cycle of the PCR. You can follow these steps in Figure 7.4 as you work through the example that follows.

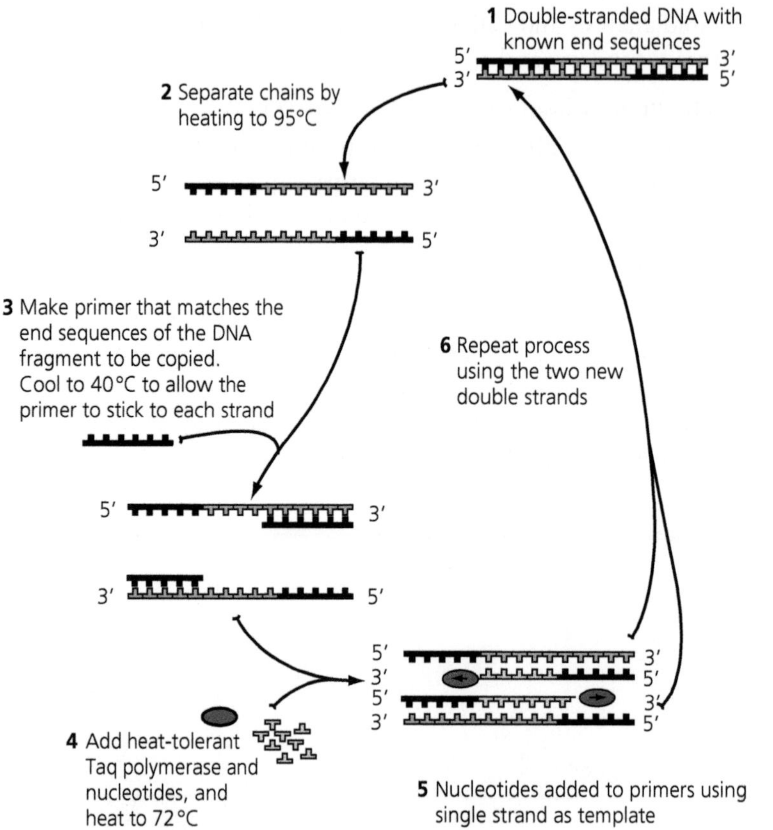

Figure 7.4 The polymerase chain reaction (PCR). The purine and pyrimidine bases are represented by A (adenine), C (cytosine), G (guanine) and T (thymine). The letter N represents a DNA nucleotide in a satellite repeat sequence

Example

Using the PCR to amplify specific fragments of DNA

The PCR is used to amplify fragments of DNA.

1 Suggest why it might be unhelpful to attempt to amplify the whole of a cell's genome at one time.

At the start of the PCR in Figure 7.4, we have a fragment of double-stranded DNA. In step 1, the two strands of this DNA are separated by heating to 93 °C.

The thermal cycler then cools the separated strands to 55 °C. This enables hydrogen bonds to reform. In theory, the separated strands could join again to reform the original double-stranded DNA.

2 Explain why heating to this temperature causes the two strands to separate.

3 Use information in Figure 7.4 to explain what prevents the two separated DNA strands joining back together.

4 Use your knowledge of DNA structure and replication to explain why the primer is made to be complementary to a base sequence at the 3' end of each template DNA strand.

5 Suggest a disadvantage of using very short primers.

6 Use the information in Figure 7.4 to estimate the distance between any two of these primers on the template DNA.

7 If molecular biologists wish to transfer a human gene to a bacterial cell, they need to amplify the gene plus its promoter. Explain why.

8 Which enzyme, used in natural DNA replication, is missing from the PCR process?

Answers

1 Molecular biologists usually want to amplify only part of a DNA molecule, for example, a gene that they wish to transfer to another organism or store in a gene library. A whole genome contains too much genetic material to be amplified during a single PCR amplification.

2 The two strands are held together by hydrogen bonds between nucleotides with complementary bases – adenine with thymine and cytosine with guanine. Although the very large number of hydrogen bonds in a molecule of DNA strongly holds the two strands together, each individual hydrogen bond is relatively weak. A temperature of 93 °C causes sufficient random thermal motion to break the hydrogen bonds, so that the two strands separate.

3 Figure 7.4 shows that an **excess** of primer is added to the original mixture. If you think back to what you learnt about competitive inhibitors in *Edexcel A Level Biology 1*, you will recall that the likelihood of a competitive inhibitor binding to the active site of an enzyme becomes less the more substrate you add. The same principle holds true here as well. If the concentration of primer is very high, the single-stranded DNA is more likely to collide with, and bind to, a primer than it is to its former complementary DNA strand.

You can see in Figure 7.4 that the primer is complementary to part of the base sequence at the 3' end of each template DNA strand.

4 The primer is a sequence of nucleotides that attaches to the end of each strand of DNA that is to be used as a template for the production of a new strand. The enzyme DNA polymerase must attach to such a primer before it can start to build a complementary DNA strand. Like all enzymes, DNA polymerase is specific in the way it binds to its substrate. Here, it can only add a new nucleotide to the 3'–OH end of the growing strand, so must grow the new strand in a 5' to 3' direction. Since the new strand will be antiparallel to the existing strand, the 5' end of the new strand will be complementary with the 3' end of the template strand.

You can also see in Figure 7.4 that the primer is several nucleotides long. Since scientists need to know the base sequence of the template DNA, you might expect it to be easier for them to produce primers that are very short.

5 A primer is a sequence of nucleotides that attaches to template DNA wherever it collides with a base sequence that is complementary to its own. A sequence of, say, only two bases is likely to occur several times within a short length of template DNA.

DNA polymerase starts copying when it attaches to a primer. It then moves along the template DNA but stops when it reaches another primer. With primers only short distances apart, a large number of very short lengths of DNA would be amplified.

6 The primer shown in Figure 7.4 is six nucleotides long. Since there are four different DNA nucleotide bases (A, C, G and T), the sequence shown (ATCCGG) is likely to occur, by chance, every 4^6 nucleotides, in other words every 5120 nucleotides.

The PCR techniques can be used to amplify individual genes. One reason for amplifying a gene is to transfer it into another cell, as described later in this chapter.

7 As you saw in Chapter 6, the promoter region regulates transcription of its target gene. Simply transferring a gene into another cell might not result in that gene being transcribed unless its promoter was also present. Bacteria are prokaryotic and have a method of regulating genes that is different from that of eukaryotic cells, such as human cells. Consequently, the promoter must be transferred along with its target gene.

Figure 7.4 represents a single cycle of the PCR. In one cycle, the two strands of a fragment of DNA are separated and each is used as a template for the production of a new, complementary strand. This mimics the natural semi-conservative DNA replication that you learnt about in *Edexcel A level Biology 1*.

8 In natural DNA replication, the two strands are separated by DNA helicase. In the PCR, the same effect is achieved using a high temperature that 'melts' the DNA.

The thermal cycler is normally programmed to repeat the cycle shown in Figure 7.4 many times. At each cycle the number of DNA copies doubles, so that after *n* cycles there will be 2^n copies of the DNA.

Test yourself

1. Describe how an *in-vitro* technique differs from an *in-vivo* technique.
2. Explain the role of a DNA primer.
3. List the groups of reagents used in the PCR.
4. Calculate the number of copies of a single fragment of DNA that will be produced after ten cycles of the PCR.
5. The thermostable enzyme used in the PCR is obtained from certain species of bacteria. Suggest the advantage these bacteria gain by possessing thermostable enzymes.

Key term

DNA profiling The general use of DNA tests to establish the identity of, or relationships between, individuals. The technique involves the production of copies of the DNA fragments, their separation using electrophoresis and a comparison of the location of these fragments on the developed electrophoresis gel.

The PCR is used in DNA profiling

The term **DNA profiling** refers to the general use of DNA tests to establish the identity of an individual or the relationship between individuals. This is possible because, with the exception of clones (including human identical twins), the sequence of bases in the DNA of each sexually reproducing organism is unique.

You learnt in *Edexcel A level Biology 1* that most of the DNA in eukaryotes does not code for polypeptides. Some of this non-coding DNA occurs within genes – as introns. There are non-coding sequences between genes as well. Within the latter non-coding regions, we find some short sequences of bases that are repeated many times. They are called **short tandem repeats**, but are often known as 'satellite' regions of DNA. Actually, there are two types of satellite regions:

- **minisatellites** of about 20–50 bases, possibly repeated 50 to several hundred times
- **microsatellites** of 2–5 bases repeated 5 to 15 times.

We all inherit a distinctive combination of these apparently non-functional 'repeat regions', half from our mother and half from our father. The same satellites occur at the same loci on both chromosomes of a homologous pair, but the number of times the sequence is repeated on each chromosome of a homologous pair will differ. It is for these reasons that each of us has a unique sequence of nucleotides in our DNA (except for identical twins). It is the microsatellites that we use as the genetic markers in DNA profiling.

> **Key term**
>
> **Short tandem repeats** (also known as satellite regions of DNA) In eukaryotes, these are regions of non-coding DNA that contain a particular sequence of nucleotide bases repeated many times over. The number of repeats at a particular locus is unique and so can be used as a genetic marker.

The steps involved in DNA profiling

1 Using the PCR to amplify the desired DNA

A sample of DNA is obtained from cells of the individuals being investigated. The PCR is used to amplify those sections of each DNA sample that are known to contain the desired satellite regions

You have seen above how the PCR can be used to amplify DNA. By choosing appropriate primers, only those parts of the DNA that contain the satellite regions to be used as genetic markers will be amplified. Since these satellite regions from different individuals will contain different numbers of tandem repeats, the amplified sections of DNA will be of different lengths.

> **Tip**
>
> Although here we are looking at the use of electrophoresis to separate fragments of DNA produced by the PCR, the technique is also used to separate fragments of DNA produced by restriction endonucleases, considered later in this chapter.

2 Separating DNA fragments by electrophoresis

Electrophoresis is used to separate these amplified sections and the result is converted into a pattern of bands similar to a bar code.

The process of electrophoresis separates particles according to their charge and size. The apparatus used to do this is shown in Figure 7.5 on the next page. Its major component is a tank. You can see that each end of the tank has an electrode connected to a power source. This generates an electrical field from one end of the tank to the other. Since the phosphate group on each nucleotide gives DNA an overall negative charge (PO_4^{3-}, fragments of DNA will migrate to the positive electrode (anode) of an electrical field. You can also see in Figure 7.5 that the samples containing DNA fragments are placed in wells cut into a sheet of gel that is supported on a glass plate. The gel is made from either agarose (a very pure form of agar) or polyacrylamide. These gels contain tiny pores that allow them to act like molecular sieves. When migrating from the cathode to the anode through these gels, small molecules move faster, and so travel further, than larger molecules.

The small diagram at the bottom right-hand side of Figure 7.5 represents the separated fragments of DNA. The wells are at the top of the diagram so the fragments have migrated down the page. Those fragments with fewer tandem repeats in their satellite regions have travelled further than those with more tandem repeats.

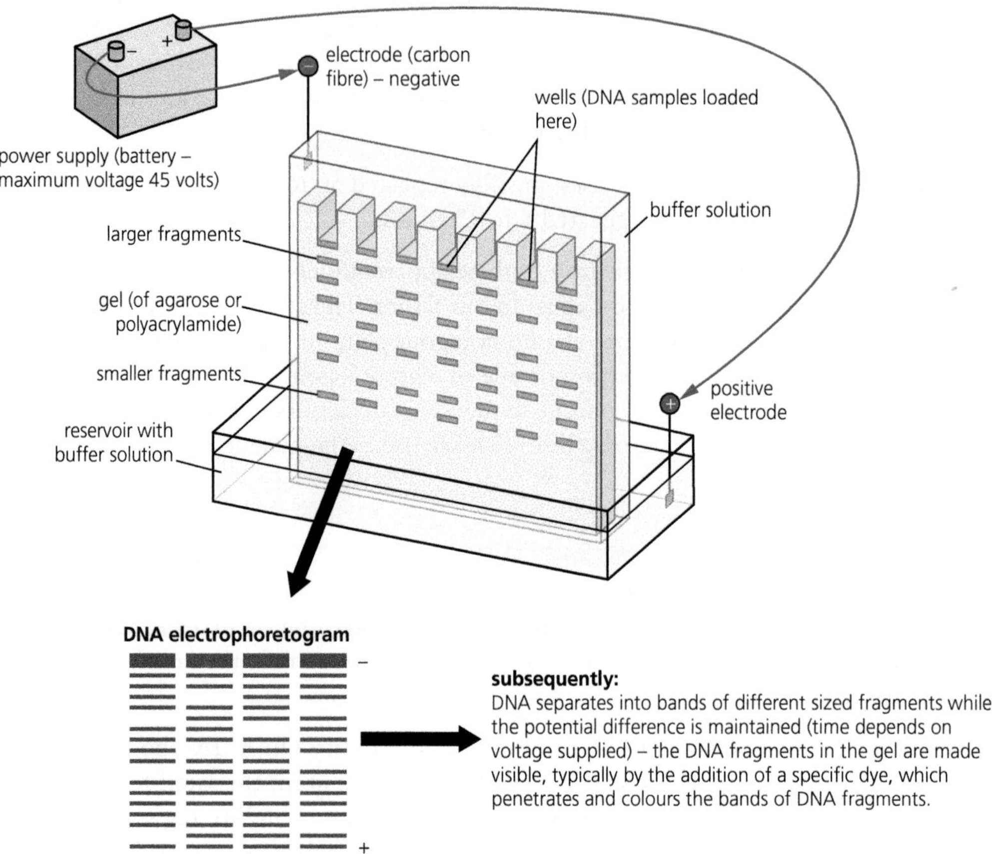

Figure 7.5 Separation of fragments of DNA using electrophoresis

> **Key term**
>
> **Radioactive DNA probe** A short length of DNA (an oligonucleotide) that has been produced with a base sequence that is complementary to that within a DNA fragment you wish to locate. The probe has also been labelled using a radioactive isotope so that it will cause 'fogging' of a photographic plate during autoradiography, allowing its location to be identified.

3 Visualising the DNA fragments using Southern blotting

After first separation by electrophoresis, however, none of the DNA bands is visible at all. They must be visualised by, for example, adding a dye that binds only to DNA and is visible in white light. The more commonly used method for visualising the results of electrophoresis uses the Southern blotting technique.

The principles of the Southern blotting technique are shown in Figure 7.6. The gel with the separated DNA fragments is placed on a blotting paper 'wick' fed by alkaline buffer solution. A nylon membrane is placed on the gel and more blotting paper is applied above, topped by a weight.

- As the buffer solution is drawn up through the gel, contact between the alkali and the DNA fragments breaks the hydrogen bonds between complementary bases, so that the DNA fragments become single-stranded.
- The compression caused by the weight, transfers a copy of the DNA fragments from the gel to the nylon membrane, which is much more robust than the gel.

As Figure 7.6 shows, the nylon membrane is then removed, and the distribution of DNA fragments detected by the application of selected **radioactive DNA probes**, followed by autoradiography.

Southern blotting (named after the scientist who devised the routine):

- extracted DNA is cut into fragments with restriction enzyme
- the fragments are separated on electrophoresis gel
- fragments are made single-stranded by treatment of the gel with alkali.

1 Then a copy of the distributed DNA fragments is produced on nylon membrane:

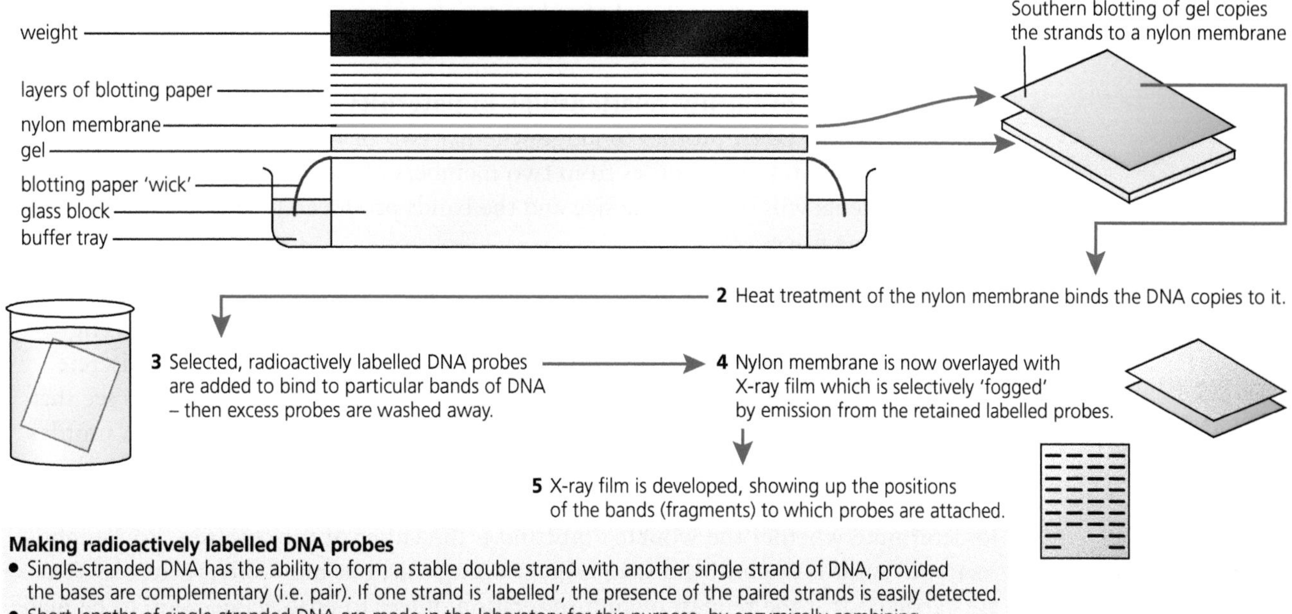

Making radioactively labelled DNA probes

- Single-stranded DNA has the ability to form a stable double strand with another single strand of DNA, provided the bases are complementary (i.e. pair). If one strand is 'labelled', the presence of the paired strands is easily detected.
- Short lengths of single-stranded DNA are made in the laboratory for this purpose, by enzymically combining and then adding selected nucleotides one at a time, in a precise sequence.
- Consequently, the base sequence of probes is predetermined and known.
- All the nucleotides used contain radioactive phosphorus (^{32}P), or carbon (^{14}C) in the ribose of the nucleic acid backbone so the subsequent positions of the probes (and the location of a complementary strand of DNA, e.g. on a nylon membrane) can be located by autoradiography.

What a probe is and how it works

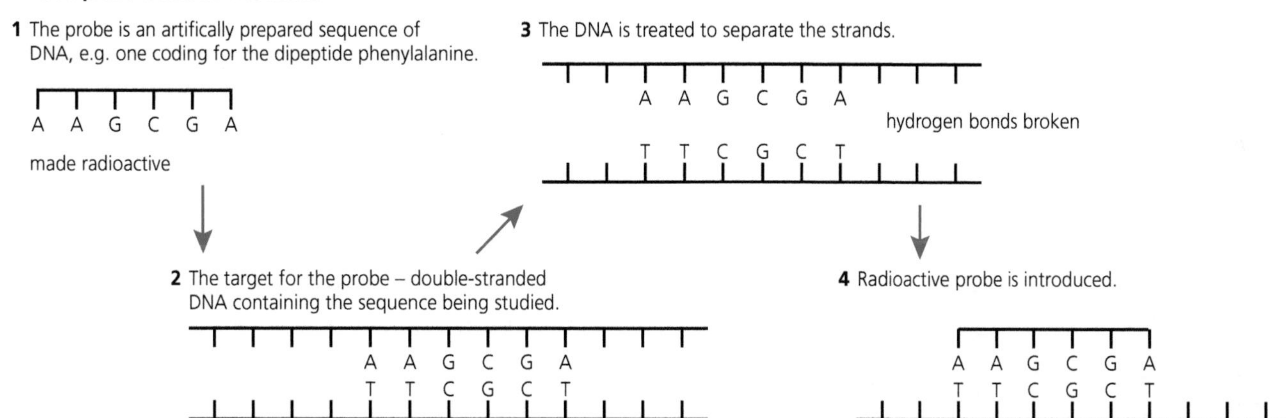

Figure 7.6 Visualisation of separated DNA fragments using Southern blotting and labelled probes

Test yourself

6 Explain why electrophoresis separates fragments of DNA.

7 Why are microsatellites useful in DNA profiling?

8 Explain how scientists can control which part of a genome is amplified by the PCR.

9 The Southern blot technique is used to transfer a DNA profile from a gel to a nylon membrane. Give the advantage of this.

10 Outline how autoradiography is used to show the location of DNA probes in a DNA profile.

The use of DNA profiling

DNA profiles, amplifying several fragments of DNA containing different repeated sequences, result in a unique banding pattern for every individual. There are many uses of these DNA profiles. Here, we will consider just two:

- The identification of family relationships in paternity testing.
- The identification of individuals in forensic science.

1 Identification of family relationships in paternity testing

Scientists can use DNA profiles to judge whether two or more individuals are genetically related. DNA profiles from two members of a clone will be identical – the DNA fragments will be the same size and the bands produced by electrophoresis will be in the same places.

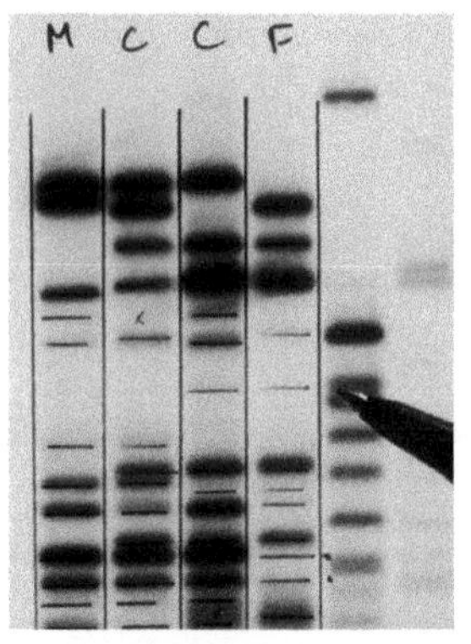

Figure 7.7 DNA profiles used to establish family relationships

Figure 7.7 shows four DNA profiles. The dark bands show the fragments of DNA that have migrated down the gel during electrophoresis and been visualised using the Southern blotting technique. You can see that the actual bands are less discrete than those represented diagrammatically in Figures 7.5 and 7.6. You can also see that they differ in thickness and in intensity. The first three bands show the DNA profiles of a woman (labelled M) and her two children (labelled C). The final profile is that of the woman's husband (labelled F). The reason the profiles have been produced is to determine whether the woman's husband is the father of the children. You might imagine that it is the husband who is questioning this, but more often paternity tests are carried out at the request of an immigration officer to enable a decision to be made about whether a child should be granted immigration status.

Because a child inherits half its DNA from its mother and half from its father, the bands in a child's DNA profile that do not match any of those in its mother's profile must come from the child's father. Look at the DNA profiles in Figure 7.7. Identify those bands in the children's profiles that are **not** present in the mother's profile. Are they present in the profile of her husband? If so, the profiles provide evidence that the husband is likely to be the father of both children. If not, they provide evidence that he is not the father of one or both.

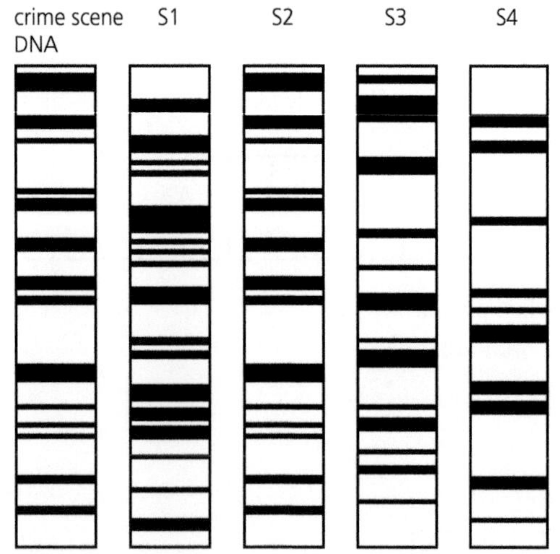

Figure 7.8 DNA profiling in forensic science – profiles from a specimen taken at the scene of a crime and from four different suspects

Similar DNA profiling also has applications in studies of wild animals, for example, to ensure that only unrelated animals are used in captive breeding programmes of animals in danger of extinction.

2 Identification of individuals in forensic science

DNA profiles produced from biological specimens taken from the scene of a serious crime, such as a rape attack or murder, can provide reliable evidence. Following amplification by the PCR, samples such as a few hair roots or a tiny amount of blood or semen will provide sufficient DNA to carry out DNA profiling. Specimens might also be collected from people suspected of being present at the crime. The greatest care has to be taken to ensure the authenticity of the samples – there must be no possibility of contamination if the outcome of subsequent testing is to be helpful in a court case.

DNA profiling of these samples can help to eliminate innocent suspects as well as to identify a person, or people, who might be responsible for the crime. You can see an example of this use of DNA profiles in Figure 7.8, in which five DNA profiles are represented. The one on the left of the diagram is from a crime scene. The others, S1 to S4, are from four people the police suspect might have been at the scene of the crime. Having produced them, the role of a forensic scientist is to advise the police, and then possibly a jury, what these profiles show about the guilt of the suspects.

DNA profiles can also help forensic scientists to identify corpses otherwise too decomposed for recognition, or isolated body parts remaining after bomb blasts or other violent incidents.

Test yourself

11 Which of the DNA fragments in Figure 7.7 is the largest? Explain your answer.

12 Can you conclude from Figure 7.7 that the man (F) is the father of both his wife's children? Explain your answer.

13 TV programmes often show crime scene investigators searching a crime scene for clues dressed in normal clothing. Explain why this is unlikely to happen in reality.

14 What can you conclude from Figure 7.8 about the involvement of the suspects in the crime? Explain your answer.

Using gene sequencing

What is gene sequencing?

Gene sequencing involves finding the sequence of bases in a DNA molecule. This can be done at the level of an entire genome or at the level of a single fragment of DNA.

The Human Genome Project was launched in 1990. In 2001, it achieved its objectives when the US National Institutes of Health published the first draft of the base sequence of the entire **human genome**. One of the outcomes of this research was the realisation that much of the genome did not code for polypeptides; something we considered in the last chapter. Another outcome is that scientists can use the known base sequence of a gene to predict the amino acid sequence of the polypeptide it encodes. A third outcome is the ability to screen samples of DNA for links to genetically determined conditions. This has enabled medics to screen DNA from:

- potential parents to find if one of them is a carrier of an inherited disorder and, if so, offer genetic counselling
- fetal cells to find if a child would suffer a debilitating inherited disorder and, if so, the potential to offer the parents an abortion
- embryos prepared for *in vitro* fertilisation (IVF), so that embryos with 'faulty' genes can be rejected and not used for implantation into the mother's uterus
- sufferers of several diseases, including cancer, to help to identify the most suitable drug to alleviate their condition; a science known as pharmacogenomics.

Of course, each of these techniques raises questions about the ethics and cost-effectiveness of these procedures. Since the Human Genome Project began, the techniques used in genetic screening have evolved. The first sequence took 11 years to complete (1990 to 2001) and cost an estimated US$ 2.3 billion. In March 2015, should you have so chosen, you could have had your own genome sequenced in a period of weeks at a cost of £125. Since 2008, a new technique has been developed that enables scientists to sequence the bases of mRNA (called RNA-seq). This has shown that over 90 per cent of human genes contain one or more alternative splicing variants (see page 100).

Key terms

Gene sequencing Finding the sequence of bases in a DNA molecule. This can be done at the level of an entire genome or at the level of a single fragment of DNA.

Human genome All the genetic material in a human cell. This includes all of the genes plus the base sequences of DNA that do not encode polypeptides.

A method of gene sequencing based on the PCR

In this method, developed by Fred Sanger, the DNA strand to be sequenced is used as a template in the PCR. You have seen that the PCR will, in a short space of time, produce a very large number of copies of the target DNA. The neat trick in this technique is to stop replication when particular nucleotides are incorporated – a process known as **termination sequencing**.

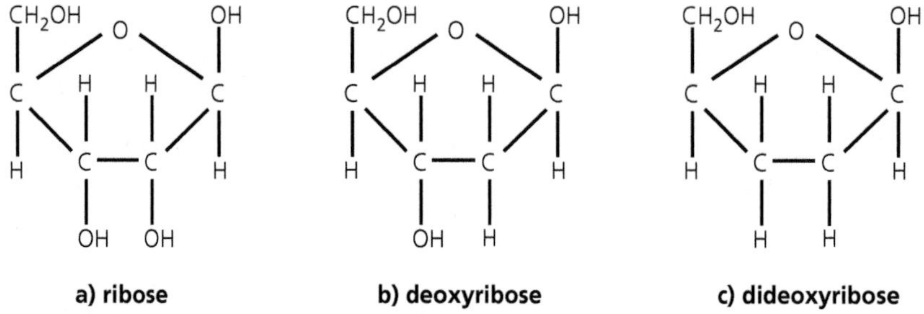

a) ribose b) deoxyribose c) dideoxyribose

Figure 7.9 A molecule of (a) ribose, (b) deoxyribose and (c) dideoxyribose

Look at Figure 7.9. It shows two pentose sugar molecules you should recognise – ribose and deoxyribose. It also shows a third molecule that you might not have seen before, dideoxyribose. Just as deoxyribose is a ribose molecule with an oxygen atom missing from carbon-2, dideoxyribose is a ribose molecule with two oxygen atoms missing, one from carbon-2 and one from carbon-3. A nucleotide that contains dideoxyribose (a dideoxynucleotide) can pair with a nucleotide that has a complementary base, but because it lacks the OH group on carbon-3, DNA polymerase cannot add another nucleotide to it. This means that the incorporation of a dideoxynucleotide into a developing DNA strand stops the further development of that strand by the PCR.

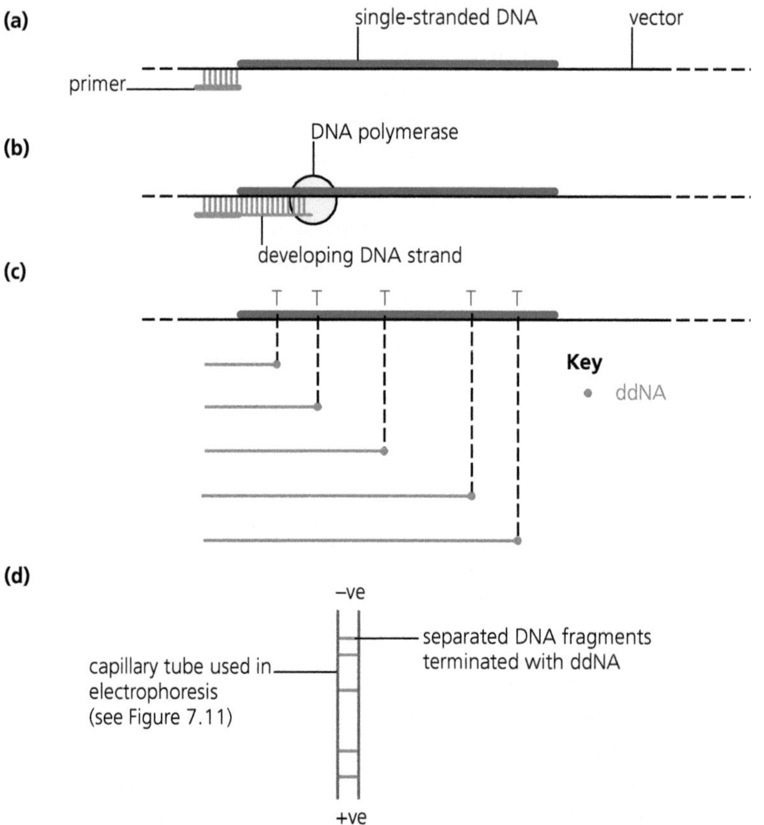

Figure 7.10 The principles involved in one type of gene sequencer. The four stages are described on the next page

Figure 7.10 represents what happens when you use dideoxynucleotides in the PCR. As described earlier, the reagents used include a heat-resistant DNA polymerase, short primers with base sequences complementary to the start of the section of DNA to be copied and an excess of nucleotides, each carrying one of the four bases. This time, however, four reaction mixtures are used, each containing a percentage of dideoxynucleotides carrying adenine, cytosine, guanine or thymine. Each of these dideoxynucleotides is labelled with a different fluorescent dye.

- In Figure 7.10(a) you can see an early stage in one PCR cycle, with a single strand of DNA with a primer attached.
- In Figure 7.10(b), DNA polymerase has attached to the primer and begun to produce a complementary strand, adding nucleotides with bases that are complementary to the bases on the template strand.
- In Figure 7.10(c) the development of five new strands has been terminated when DNA polymerase added a dideoxynucleotide-carrying adenine (ddNA). You can deduce that thymine (T), the base complementary to adenine, must occur at the positions shown on the template strand.
- The fragments of different length are then separated by electrophoresis. Figure 7.10(d) shows one way in which this can be automated. Instead of using gel in a tank, as shown in Figure 7.5 on page 120, gel within a capillary tube has been used. By using a laser, detector and computer, the bands of labelled adenine-carrying dideoxynucleotide can be detected as they migrate through the gel in the capillary and displayed on a screen as shown in Figure 7.11.

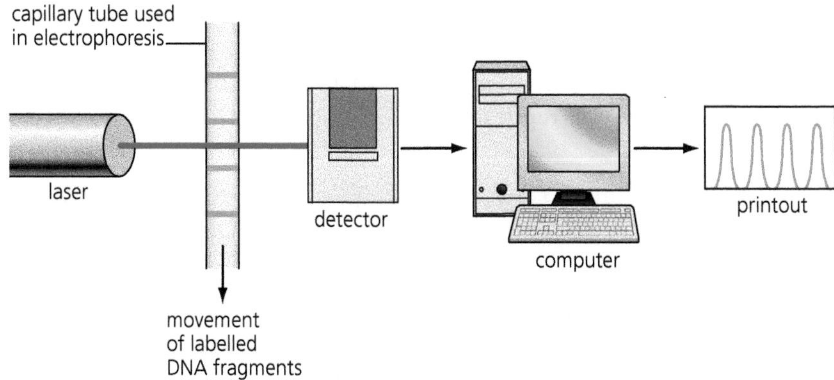

Figure 7.11 Using electrophoresis through a capillary tube allows the automation of electrophoresis. In this case, DNA fragments labelled with green fluorescent dye result in green peaks on a computer screen

By compiling the results of labelled dideoxynucleotide with each of the four bases, an automated sequencer produces results like those in Figure 7.12. Scientists can use this to determine the entire base sequence of the original DNA fragment. Typically, the method described is accurate for sequences up to a maximum of 700–800 base pairs long. Modifications of the same method, in which overlapping fragments are sequenced and the overlaps arranged by computer software, allow sequencing of larger genes and, in fact, whole genomes.

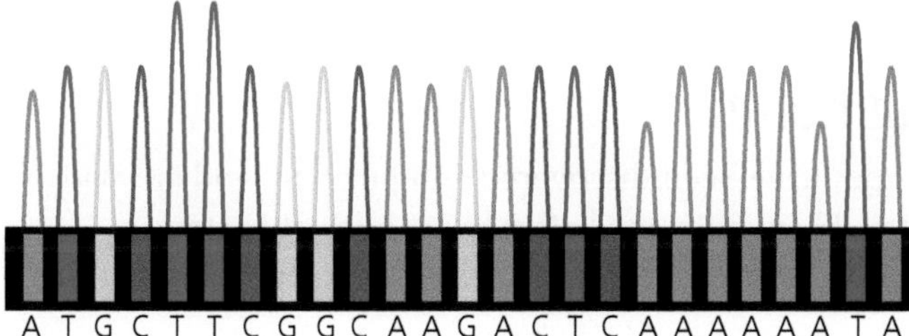

Figure 7.12 Part of the printout of a gene sequencing machine. The bases have been labelled with fluorescent dye: adenine green, cytosine blue, guanine yellow and thymine red

Test yourself

15 Define the term 'genome'.

16 DNA sequencing has found that an intron, 729 base pairs long, found in the *ZFY* gene on the X chromosome was identical in the X chromosomes from a random sample of men. The same intron showed a great deal of genetic variation within samples of X chromosomes from chimpanzees, gorillas and orang-utan. Suggest an explanation for the lack of variation in humans.

17 Does knowledge of the genome enable scientist to predict the proteome (full complement of proteins) that a bacterium and a person can make? Explain your answer.

18 In March 2015, someone in the UK could have had their own genome sequenced in a period of weeks at a cost of £125. Suggest **one** disadvantage that might arise from the knowledge of a person's genome.

19 If a dideoxynucleotide is incorporated into a developing DNA strand, DNA polymerase cannot add further nucleotides to this strand. Explain why.

20 Give the base sequence of the template strand used to obtain the results in Figure 7.12.

In-vivo amplification of DNA

Key term

Restriction endonuclease
An enzyme (often abbreviated to 'restriction enzyme') that hydrolyses the phosphodiester bonds at a particular base sequence in a DNA molecule, cutting the DNA into double-stranded fragments. The specific base sequence is called the **recognition sequence** of that endonuclease.

As outlined in the introduction on page 113, DNA can be amplified by inserting fragments into the genomes of living cells. Those cells that successfully incorporate the new DNA into their genomes will replicate it and pass it on to their daughter cells when they divide.

The way in which DNA to be replicated is extracted from tissue samples was described in Figure 7.2 on page 114. As Figure 7.13 on the next page shows, the extracted DNA can then be cut into fragments by the addition of a **restriction endonuclease** (often called a restriction enzyme). An endonuclease is an enzyme that hydrolyses the bonds in the sugar-phosphate backbone of both strands in a DNA molecule, producing double-stranded fragments of DNA. Restriction endonucleases occur naturally in bacteria, where they protect against viruses that infect the bacterium (bacteriophages) by cutting the viral DNA into small pieces, thereby preventing the lytic cycle (*Edexcel A level Biology 1*, Chapter 4, page 90). Restriction enzymes were so named because they restrict the multiplication of phage viruses.

DNA may be extracted from tissue samples by mechanically breaking up the cells, filtering off the debris, and breaking down cell membranes by treatment with detergents. The protein framework of the chromosomes is then removed by incubation with a protein-digesting enzyme (protease). The DNA, now existing as long threads, is isolated from this mixture of chemicals by precipitation with ethanol, and is thus 'cleaned'. The DNA strands are then re-suspended in aqueous, pH buffered medium. They are now ready for 'splicing' into fragments.

Cutting DNA into fragments

Many different restriction enzymes have been discovered and purified, and are widely used to produce DNA fragments. They are named after the species of bacterium from which they are extracted, for example the enzyme *Eco*R1 is extracted from the bacterium *Escherichia coli*. Where more than one restriction endonuclease has been extracted from the same species of bacterium, roman numerals are added, e.g., *Pvu*II was the second restriction endonuclease to be extracted from the bacterium *Proteus vulgaris*. Like all enzymes, restriction endonucleases are specific in the substrates to which they bind. In this case, each binds only to one particular base sequence, called its **recognition sequence**. You can see the recognition sequences of six different restriction endonucleases in Figure 7.13. One distinctive feature is that they are palindromic, that is they read the same in either direction.

> **Key term**
>
> **Recognition sequence**
> The specific, short palindromic DNA base sequence to which a restriction endonuclease binds, catalysing hydrolysis of the phosphodiester bonds at that point in the DNA.

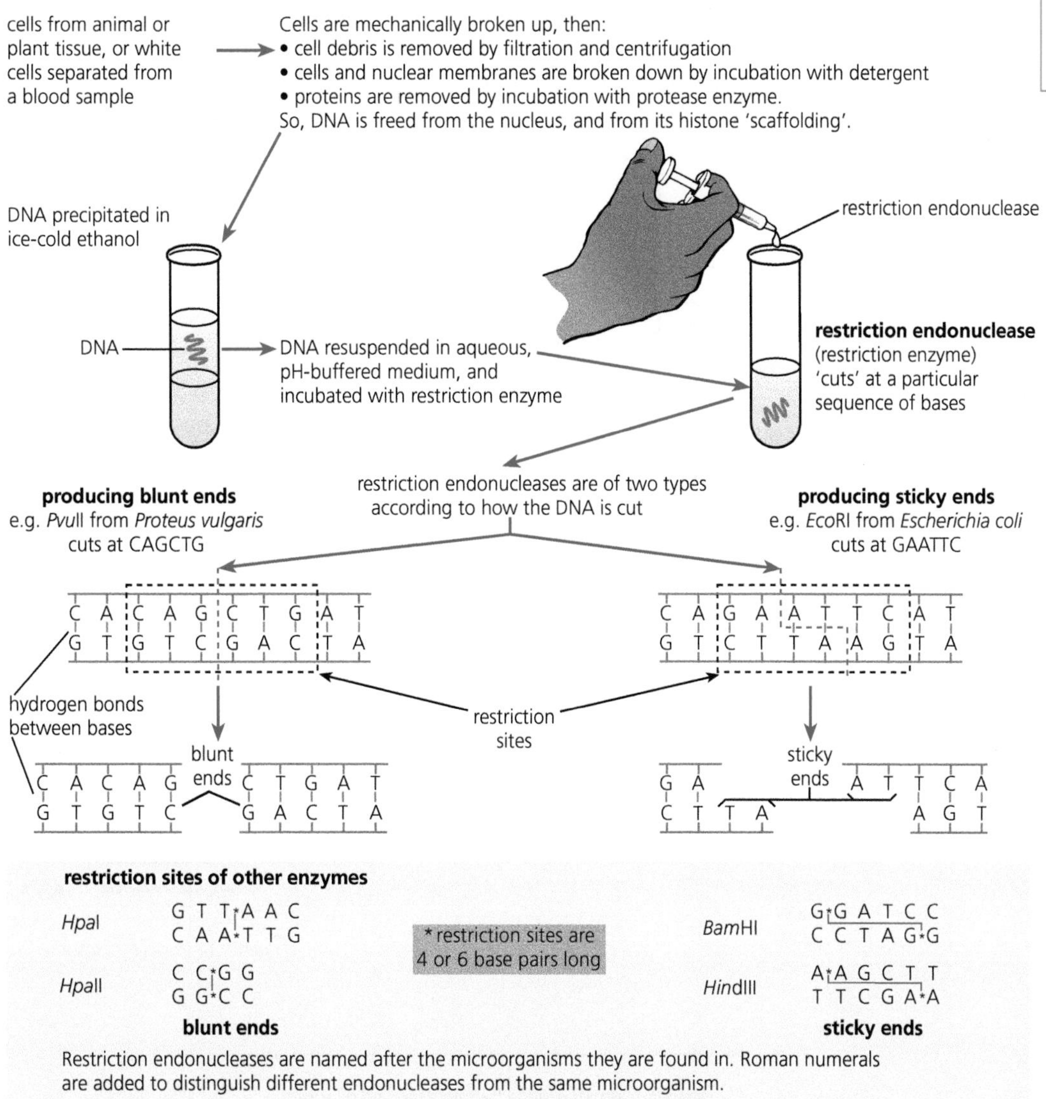

Figure 7.13 Using restriction endonucleases to produce fragments of DNA

Figure 7.13 shows a further property of restriction endonucleases – they produce one of two types of 'cut'. *Pvu*II, *Hpa*I and *Hpa*II make straight cuts, producing so-called **'blunt ends'**. *Eco*RI, *Bam*HI and *Hin*dIII make staggered cuts, producing fragments with overhanging ends, so-called **'sticky ends'**. DNA fragments with 'sticky ends' bind more readily to other DNA molecules. As you will see in the next section, DNA fragments with 'blunt ends' are made into 'sticky ends' when used during *in vivo* methods of DNA amplification.

Producing recombinant DNA

Before a DNA fragment can be cloned using *in vivo* technology, it must be added to a **vector**, which will be used to transfer it into living cells. When considering the transmission of diseases in Chapter 5, we used the term 'vector' to describe an agent that transports between one organism and another. In recombinant DNA technology, a vector is carrier DNA, used to carry a target fragment of DNA into a new organism. One such vector is commonly found in bacteria.

Figure 7.14 reminds you that a bacterium, such as *Escherichia coli*, contains two types of genetic material. One is the long, double-stranded DNA in the form of a ring, called the nucleoid. The other type is the more numerous, and much smaller rings of DNA, called plasmids.

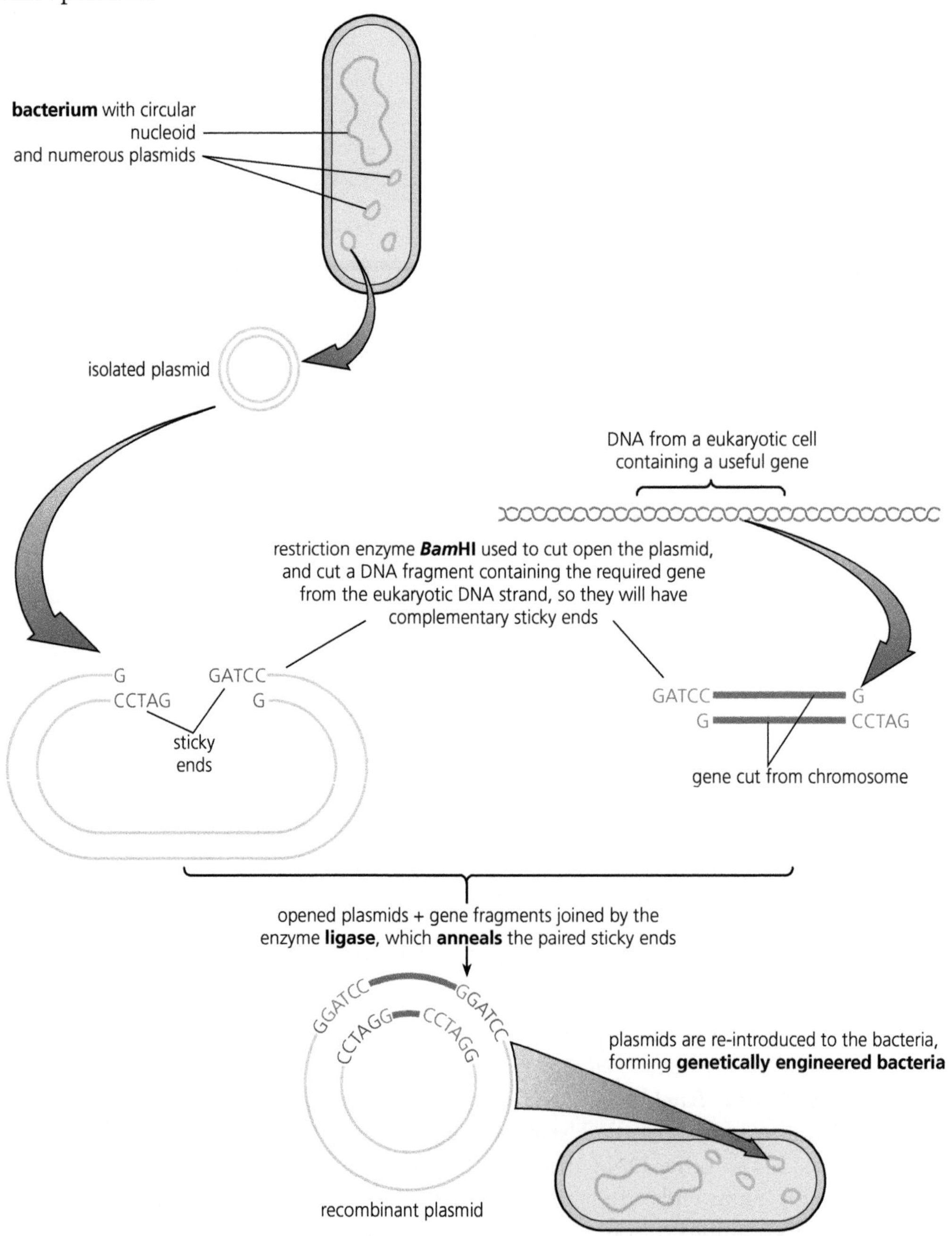

Figure 7.14 Using bacterial plasmids as DNA vectors

Plasmids are relatively easy to isolate from a bacterial cell and to reintroduce into a bacterial cell. They are, therefore, extremely useful as DNA vectors. In a bacterial cell, the plasmids replicate themselves independently of the nucleoid, so any gene added to a plasmid will be copied many times and will be passed on to the daughter cells when the bacterium divides by binary fission.

As you can see in Figure 7.14, a plasmid is cut open using a restriction endonuclease. If it produces 'sticky ends', the same enzyme is also used to cut the DNA fragment to be inserted. If not, complementary 'sticky ends' are added to the blunt ends of both the plasmid and the DNA fragment. The open plasmids and DNA fragments are then mixed with a **DNA ligase** – an enzyme that catalyses the formation of phosphodiester bonds between the ends of the plasmids and the DNA fragments. Ligase occurs naturally in the nuclei of eukaryotic organisms, where it repairs DNA that has been damaged during DNA replication. Figure 7.15 shows in greater detail how, following the alignment of their complementary base pairs, a ligase joins up the DNA of the plasmid and of the fragment – a process called **annealing**.

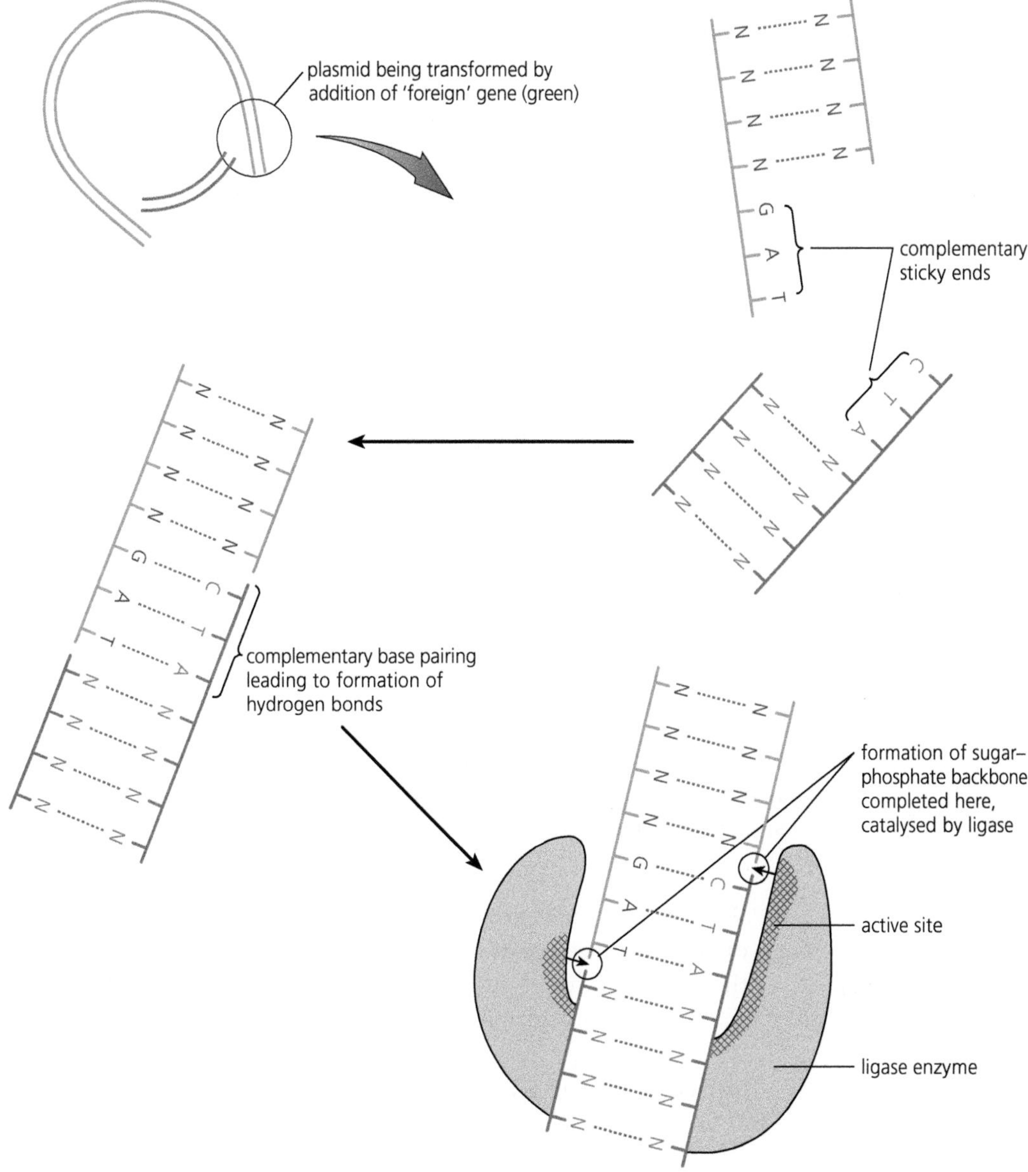

Figure 7.15 Annealing of a cut plasmid and a DNA fragment is catalysed by ligase

Transferring the recombinant DNA into a host bacterium

Most species of bacteria are able to take up plasmids. There are several methods, mostly found by trial and error, which can be used to encourage this. One involves soaking bacterial cells, together with plasmids, in an ice-cold solution of calcium chloride, followed by a brief heat shock at 42 °C for 2 minutes. Bacteria that have taken up these foreign plasmids are said to be **transformed**. If the DNA fragment within the plasmid contains one or more entire genes, they are also described as **transgenic**.

Identifying transformed bacteria

A diagram such as Figure 7.14 makes it easy to see that a bacterium has taken up a plasmid and been transformed. In reality, biologists can see neither the bacteria nor the plasmids. In effect, they carry out the above procedures not knowing whether:

- any plasmids contain recombinant DNA or whether they joined back on themselves (**self-ligated** plasmids)
- any bacteria have taken up plasmids
- any bacterium that has taken up a plasmid has taken up a self-ligated plasmid or a plasmid containing recombinant DNA.

They need methods by which they can identify transformed bacteria so that they can grow these, and only these, in a culture medium. There are several ways in which they can identify transformed bacteria. Most involve the use of marker genes on the plasmids. You are required to recall and understand only one method – the use of antibiotic-resistance marker genes and replica plating.

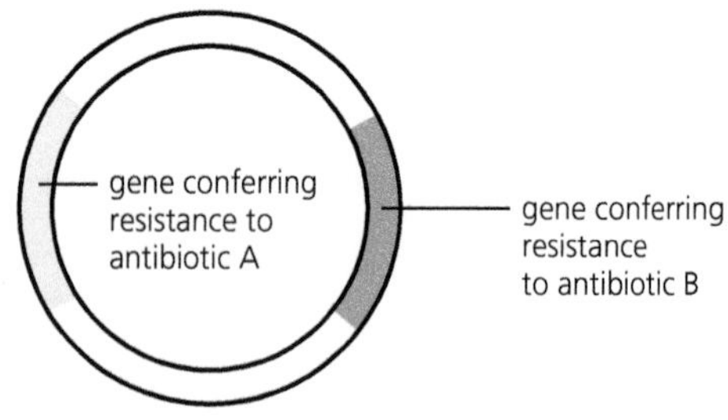

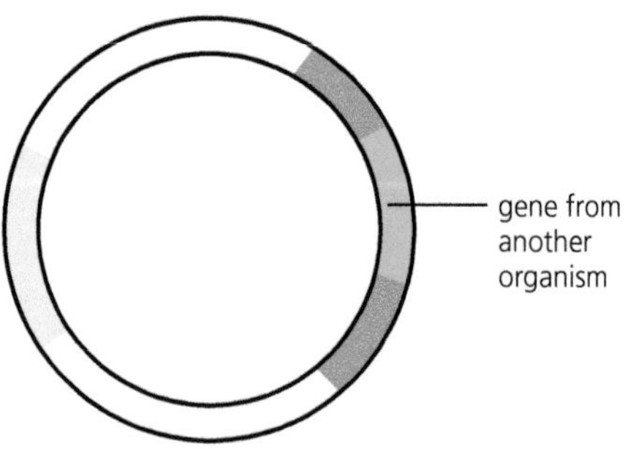

Figure 7.16 Using genes for antibiotic resistance as marker genes

Figure 7.16(a) shows a plasmid that contains genes conferring resistance to two antibiotics, A and B. This plasmid is used as a vector for transferring a gene from another organism. The recognition sequence for the restriction enzyme used to cut this plasmid is within the gene conferring resistance to antibiotic B.

Figure 7.16(b) shows the appearance of a plasmid into which the 'foreign' gene has been successfully annealed. Using the heat-shock treatment, some bacterial cells will take up these plasmids. But how will we be able to tell which bacteria have taken up the plasmid with the 'foreign' gene?

Initially, we would not know. So we culture all the bacteria on agar plates. Figure 7.17 shows the appearance of one agar plate. Six colonies have grown on it; each colony being a clone of a single cell.

- We now take an imprint of this plate by gently pressing down on it with a piece of sterile velvet (or sterile filter paper). We then press the velvet onto the surface of an agar plate in which the agar contains antibiotic A.
- We repeat this process, this time pressing the velvet onto the surface of an agar plate in which the agar contains antibiotic B.

The upper plate in Figure 7.17, representing normal agar, has six colonies. The plate containing antibiotic A has only four of these colonies – 1, 2, 4 and 6. Since the bacteria that gave rise to these colonies were able to survive, they must be resistant to antibiotic A. We can, therefore, conclude that they successfully took up the plasmids and that the bacteria that gave rise to colonies labelled 3 and 5 did not.

Of the four colonies that are resistant to antibiotic A, only two showed resistance to antibiotic B – those labelled 1 and 6. Since these bacteria are resistant to antibiotic B, we know that the gene conferring this resistance must be intact. The bacteria forming colonies labelled 2 and 4, however, were not resistant to antibiotic B. The reason for this must be that they contained the 'foreign' gene and it has disrupted their gene for resistance to antibiotic B. The bacteria in these colonies will have plasmids with the structure represented in Figure 7.16(b). We can, therefore, conclude that the bacteria in colonies 2 and 4 are transgenic bacteria – they carry the 'foreign' gene. We would now transfer samples of those colonies from the original agar plate to sterile nutrient medium and allow them to increase in number.

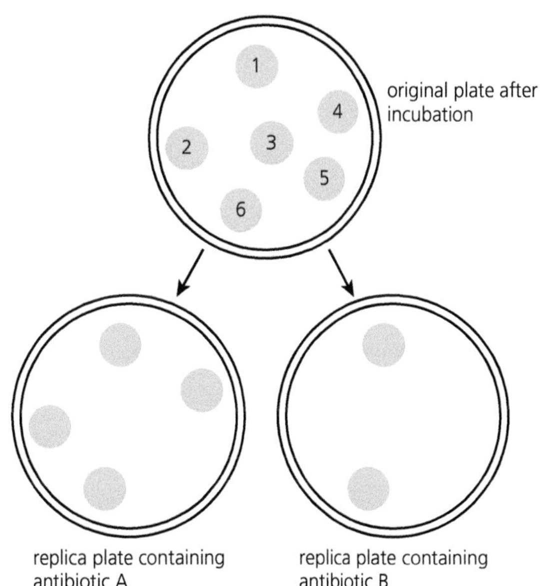

Figure 7.17 The results of the replica plating of six bacterial cultures onto agar plates contain antibiotic A or antibiotic B

Other methods used to insert DNA into cells

In the above account, you have seen how bacteria can be transformed using recombinant plasmids. This technique can be useful if you wish to amplify the target DNA – the transformed bacteria multiply rapidly and replicate the target DNA at each division.

If scientists wish to transform cells other than bacteria, they can use different vectors to transfer the target DNA. The following four methods are commonly used.

- **Viruses** affect the cells they infect by inserting their own nucleic acid into a host cell. You can remind yourself of this by looking back to Figure 4.14 on page 69. This ability can be exploited by using viruses to transfer recombinant DNA into host cells. Used in this way, however, the virus must be weakened so that it does not cause the death of host cells.
- **Liposomes** are small lipid droplets that can be used to coat plasmids that contain recombinant DNA. The lipid coating enables the droplet, with its recombinant DNA, to cross cell surface membranes. Figure 7.19 on the next page, shows how this technique has been used, with mixed success, to transform epithelial cells in the lungs of human sufferers of cystic fibrosis.
- **Gene guns** use air pressure to enable particles of heavy metals, such as gold or titanium, that are coated with recombinant DNA to be 'fired' through the surface of cells. This technique is often called bioballistics, abbreviated to biolistics. The particles coated with DNA can be fired directly into an organ of the target organism or into a tissue culture containing its cells. Figure 7.18 shows a gene gun being used on a plant organ.
- **Electroporation** uses electricity to create pores in the cell surface membranes of the target organism. The affected cells are then able to take up naked recombinant DNA and incorporate it into their own genomes.

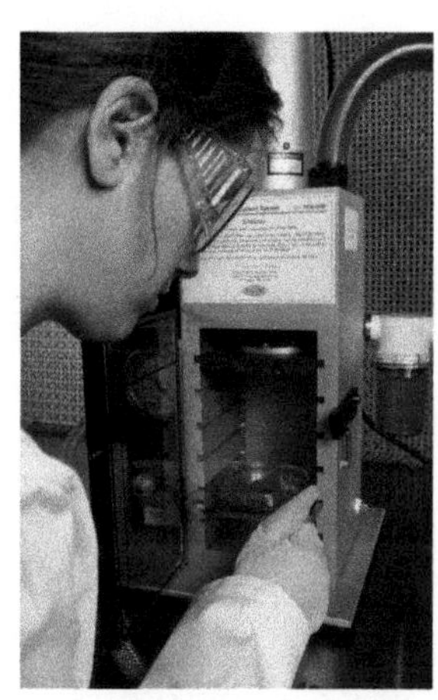

Figure 7.18 A gene gun in use

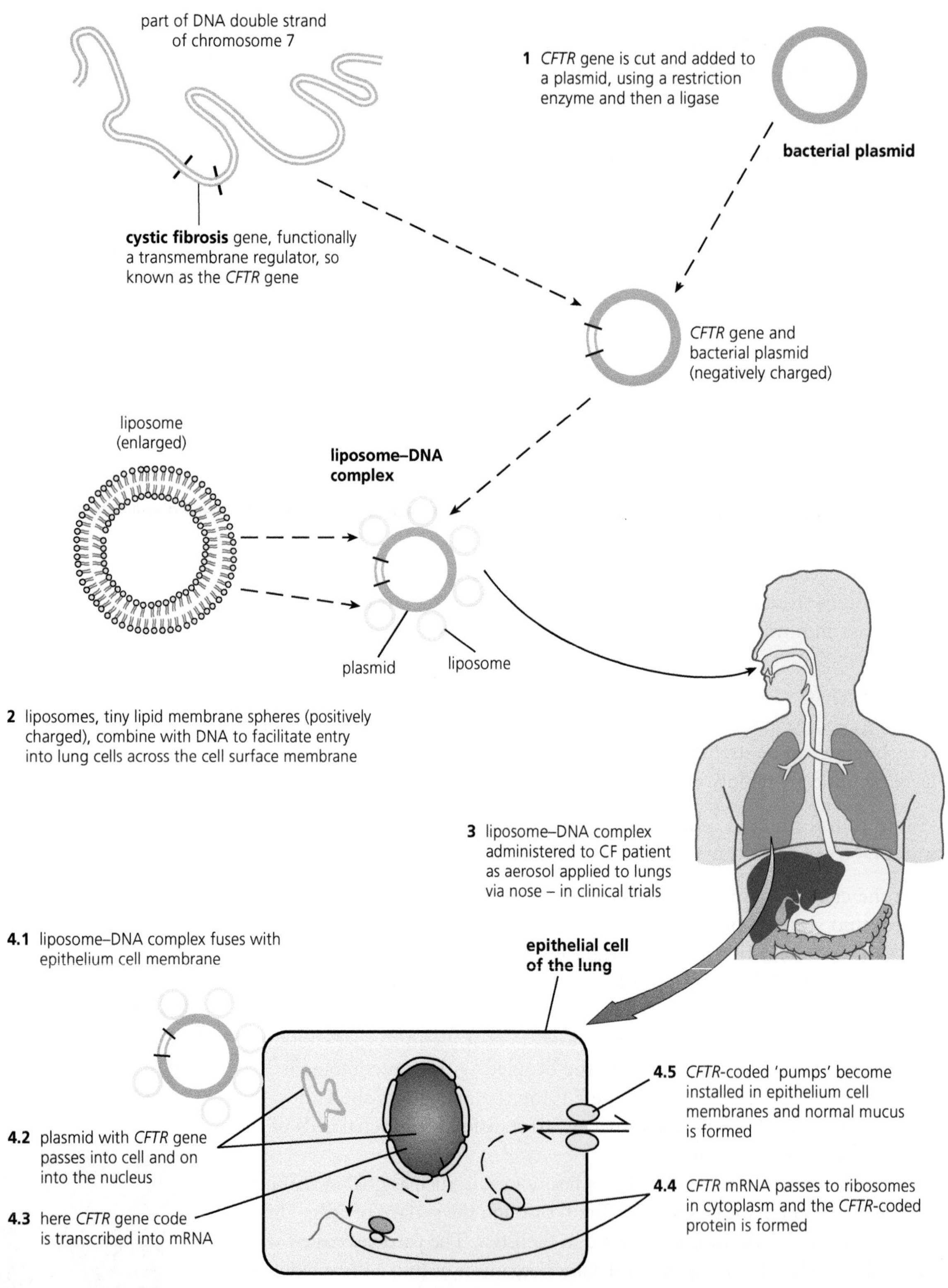

In recent clinical trials some 20% of epithelium cells of cystic fibrosis patients were *temporarily* modified (i.e. accepted the *CFTR* gene), but the effects were relatively short-lived. This is because our epithelium cells are continually replaced at a steady rate, and in cystic fibrosis patients the genetically engineered cells are replaced with cells without *CFTR*-coded pumps. Patients would require periodic treatment with the liposome–DNA complex aerosol to maintain the effect permanently.

Figure 7.19 Using liposomes to transfer recombinant DNA into cells lining the lungs of cystic fibrosis sufferers

Two uses of recombinant DNA technology

Recombinant DNA technology can be used in several ways. Food shortage is a worldwide problem. Using traditional breeding techniques, farmers have selectively bred plants and animals to increase the production of crops, milk and meat for centuries. Using recombinant DNA technology, genetic changes can be made to occur much more quickly than these traditional breeding programmes. The transgenic organisms so formed are referred to as genetically modified (GM) organisms. The production of GM organisms is now a major industry, dominated by a small number of manufacturers. Recombinant DNA technology is also used in pure academic research, helping biologists to understand gene function. We will look at one example of each of these uses of recombinant DNA technology.

The use of recombinant DNA technology in the production of genetically modified soya

Beans of the soya plant (*Glycine max*) are a staple food source for humans and livestock and are used worldwide as a source of protein and oils. Although the growth of genetically modified plants (GM plants) within the European Union is banned under EU regulations, well over half the world's soya bean crop is from GM soya plants.
Each modification has involved the introduction of recombinant DNA containing a gene that changes the properties of the soya bean plants. In some cases, the recombinant DNA includes a gene that leads to an increase in crop yield, for example by conferring:

- resistance to glyphosate (a commonly used weed killer that would also affect the soya plant)
- resistance to pests, including fungi, nematode worms and insects
- tolerance to drought and to soil salinity.

In other cases, the recombinant DNA includes a gene that changes the biochemical components of the soya beans, for example:

- the balance of the fatty acids, increasing the percentage of oleic acid and reducing that of linoleic acid. The benefit of this is that, during cooking, linoleic acid is more easily oxidised to produce trans fats than is oleic acid. Trans fats are harmful to human health.
- production of active pharmacological ingredients (so-called molecular pharming).

Gene guns and electroporation have been used to insert the recombinant DNA into soya plants. In addition, a bacterium, *Agrobacterium tumefaciens*, has also been used. When this bacterium infects plants, it transfers its own DNA into cells of the infected plant, creating tumours (see Figure 7.20). Using restriction endonuclease and ligase enzymes, as described above, it has been possible to insert target genes into non-virulent plasmids of *A. tumefaciens*. These transgenic bacteria have then been used to infect soya tissue cultures, resulting in soya plants in which all the cells contain the desired gene.

Figure 7.20 Crown gall disease in a chrysanthemum plant. This tumour is caused by a bacterium, *Agrobacterium tumefaciens*, which can infect over 10 000 plant species

The use of recombinant DNA technology in the production of 'knockout' mice

A 'knockout' mouse is a laboratory mouse in which research scientists have inactivated (or knocked out) one of its genes. The scientists do this by adding into the genome of a mouse, a piece of foreign DNA that disrupts transcription of a gene. To ensure that most of the cells of the adult mouse will have the inactivated gene, the scientists begin with embryonic stem cells from an early mouse embryo. Using a virus, or a linear piece of bacterial DNA, scientists insert the artificial gene into these stem cells. They then culture these stem cells in a laboratory before injecting them into new mouse embryos, which they transplant into the uterus of a female mouse. Some of the tissues of the resulting pups will have the target gene inactivated. By cross-breeding these mice, scientists can produced mice in which all the tissues have the target gene inactivated, that is, for this gene they are homozygous knockout mice.

By knocking out the activity of a gene, research scientists can investigate the function of that gene. Since humans and mice have many genes in common, using mice in this way helps us to understand the function of many of our own genes. In turn, this knowledge has enabled us to understand human diseases such as heart disease, cancer and Parkinson's disease and to develop, and test, drugs or other therapies to control them. Knockout mice are not always helpful in this way, however. One gene associated with over half of all tumours in humans is called *p53*. Although *p53* knockout mice do show an increase in the production of tumours, they occur in different tissues from those affected in humans.

The debate about the use of recombinant DNA technology

Although there are many examples of genetic modifications that are beneficial, geneticists are really producing new organisms when they produce transgenic organisms. Consequently, this work is potentially a source of hazards and it certainly generates concerns. These include the following:

- Will a gene added to a genome function in an unforeseen manner, for example triggering some disease in the recipient?
- Might an introduced gene for resistance to adverse conditions get transferred from a crop plant to a species of weed?
- Is it possible that a relatively harmless organism, such as the human gut bacterium, *Escherichia coli*, might be transformed into a harmful pathogen that escapes the laboratory and infects populations?
- Is there an important overriding principle that humans should not 'change nature' in a deliberate way?
- Recombinant DNA technology is costly so it would mainly benefit the health and life expectancy of people in the developed nations. Wouldn't it be better to use these funds to address the more basic problems of housing, health and nutrition in less developed countries?

A short consideration of the potential benefits and possible harmful effects of genetic modification is given in Table 7.1.

Table 7.1 Some pro and cons of genetic manipulation

Reassurances	Potential dangers
The nutritional quality of foodstuffs can be improved much faster than the traditional methods of animal and plant breeding.	GM foods could have a negative impact on human health. A gene for use in genetic modification might be taken from a plant that causes an allergic response in some people. Inserting that gene into another plant could cause the new host plant to express that allergen. Alternatively, the interaction of genes from two species could result in the production of a completely new allergen.
Genes conferring resistance to adverse environmental conditions will enable the use of previously unsuitable land to grow crops or increase the yield in less hostile environments.	Horizontal gene transfer might result in pest plant species gaining the transferred genes, enabling them also to grow in these adverse conditions and, if as a result they are able to compete more strongly, reduce crop yield.
GM organisms are tested in the laboratory and in field trials before they are grown commercially.	When grown commercially, there might be consequences to the environment that did not appear in trials. For example, a product of GM crops could be toxic to beneficial organisms (see Figure 7.21).
People can choose whether or not they eat GM products.	GM crops will eventually pollinate non-GM crops, so that it will become impossible for people to make a choice about whether or not they eat GM foods.
In humans, inherited conditions that are common in a family can be avoided by the careful use of genetic modification.	Many people consider this to be treating humans as commodities that could have serious socio-cultural consequences. It would also lead to social divides as the technology would be available only to the rich.
Only one gene is involved in the transfer so it will have minimal effect on the genome.	Interactions within the genome are currently poorly understood but we do know that promoter genes can regulate a large number of polypeptide-coding genes. In reality, we do not know the impact a single gene change might have on the genome.

Figure 7.21 There is current scientific controversy concerning whether pollen from GM maize plants landing on its natural food source (milkweed plants) has contributed to the dramatic fall in the population of these monarch butterflies

Test yourself

21 Give **two** features of the recognition sequence of a restriction endonuclease.

22 Explain the importance of 'sticky ends' in DNA technology.

23 What is a 'transformed' bacterium?

24 Suggest why 'transferring' a gene directly into affected cells is unlikely to result in the long-term treatment of inherited diseases in adult humans.

25 Explain the term 'homozygous knockout mice'.

Exam practice questions

1 Which one of the following base sequences is most likely to be a recognition sequence of a restriction endonuclease?

A CCGTAT
GGCATA

B ATCCGC
TAGGCG

C GGGTAT
CCCATA

D GAATTC
CTTAAG *(1)*

2 Replica plating involves:

A making serial dilutions of a bacterial culture

B taking an imprint of bacterial growth on an agar plate and transferring it to a new agar plate

C setting up several repeats of each inoculated agar plate

D other scientists repeating your experiment with bacteria to see if they get the same results *(1)*

3 A transgenic bacterium has:

A gained one or more genes from another organism

B lost one or more genes to another organism

C undergone horizontal gene transfer via conjugation

D many plasmids *(1)*

4 The diagram represents the processes involved in one cycle of the polymerase chain reaction (PCR).

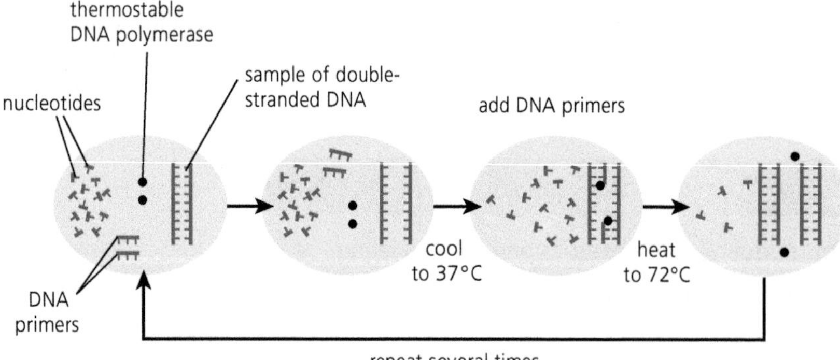

a) Describe the structure of a DNA primer and explain its role in the PCR. *(3)*

b) Explain why the DNA polymerase used in this process must be thermostable. *(2)*

c) How many cycles of the PCR would it take to amplify a single DNA fragment to produce 4000 copies? Explain your answer. *(2)*

5 The diagram shows how enzyme **X** cuts a molecule of DNA.

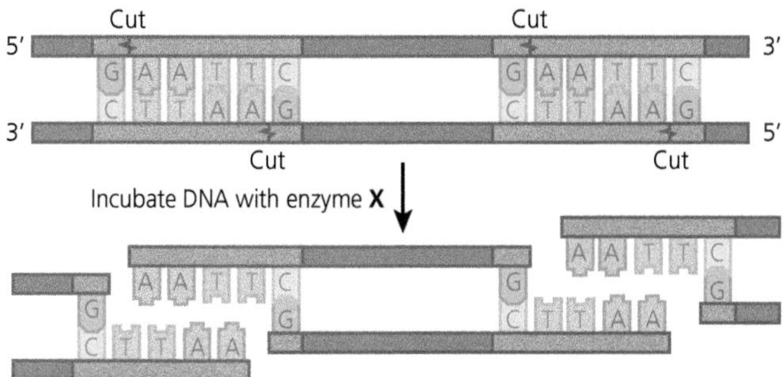

a) What **type** of enzyme is enzyme **X**? *(1)*

b) What is the recognition sequence of this enzyme? *(1)*

c) Explain the importance of the way in which enzyme **X** cuts the DNA. *(3)*

d) Enzyme **X** can be used to cut a gene from a DNA molecule. Suggest why it might be preferable to obtain a gene from mRNA than by cutting it from DNA. *(2)*

6 A scientist used the polymerase chain reaction (PCR) to amplify DNA. She set up four tubes containing the essential ingredients for the PCR to occur. In addition, she added to each tube DNA nucleotides in which the pentose sugar was dideoxyribose. Nucleotides containing this pentose sugar behave in the same way as those in which the pentose sugar is deoxyribose, with one important exception. As soon as a nucleotide containing dideoxyribose is inserted into a developing DNA strand, it terminates further replication.

One of her tubes contained dideoxynucleotides carrying the base adenine (ddNA); the second contained dideoxynucleotides carrying the base cytosine (ddNC); the third contained dideoxynucleotides carrying the base guanine (ddNG); and the fourth contained dideoxynucleotides carrying the base thymine (ddNT).

She then placed samples from each tube into separate wells in an agarose gel and carried out electrophoresis. The diagram shows her results.

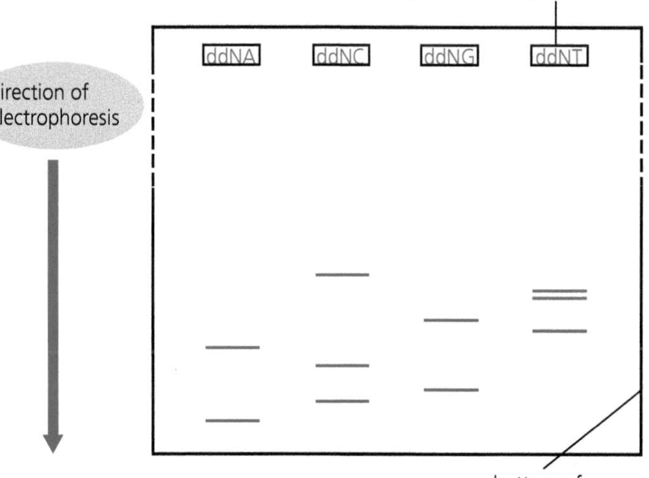

a) Which electrode would she have connected to the gel at the top of the diagram? *(1)*

b) Explain why the fragments of DNA produced the bands you can see in the diagram. *(2)*

c) From the information in the diagram, this scientist was able to deduce the base sequence of the template DNA replicated by the PCR. Explain how she was able to do this and give the base sequence of the template DNA strand. *(4)*

Stretch and challenge

7 Read the following passage and answer the questions that follow it.

A group of scientists wished to insert a gene of interest into the genome of the gut bacterium *Escherichia coli.*

Normal cells of *E. coli* can produce β-galactosidase, an enzyme that hydrolyses the β-glycosidic bond between galactose and another organic residue. This enzyme contains two polypeptide chains, encoded by two genes, *lacZα* and *lacZΩ*. Neither polypeptide is functional alone but, when produced together, the two polypeptides spontaneously form the functional enzyme β-galactosidase.

The scientists chose to work with a mutant form of *E. coli* in which part of the gene encoding β-galactosidase, *lacZα*, is missing.

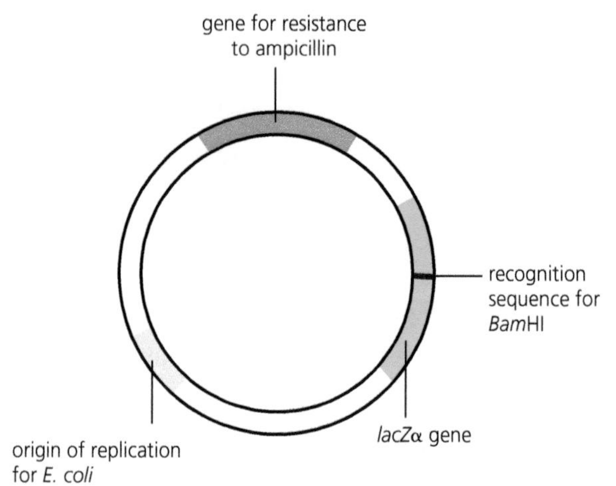

Figure 1 The plasmid pUC used as a DNA vector in this experiment

As a vector, they chose to use a plasmid, called pUC. Figure 1 represents this plasmid, showing some of the genes it carries. The scientists used the restriction enzyme *Bam*HI to insert the gene they were interested in into pUG plasmids. They then mixed these plasmids with heat-shocked mutant *E. coli* and cultured these bacteria on agar plates.

In addition to nutrients and inorganic ions needed for the growth of *E. coli*, the agar contained the antibiotic ampicillin and a lactose analogue, called X-gal. This substance is white but produces a blue-coloured product when hydrolysed by β-galactosidase.

Figure 2 shows the appearance of one of the scientists' plates after incubation for 48 hours.

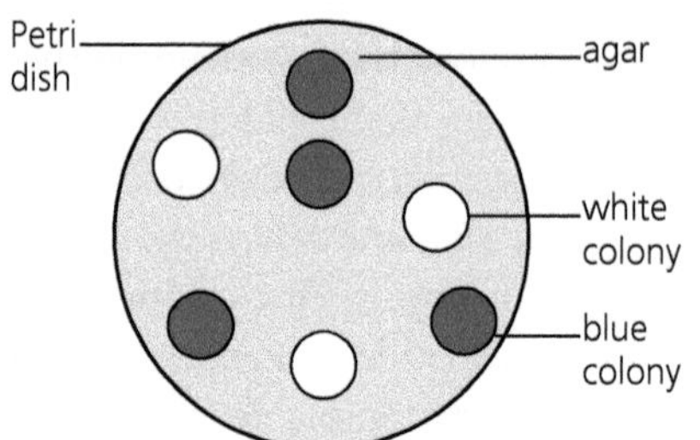

Figure 2 The appearance of one agar plate inoculated with mutant *E. coli* and incubated for 48 hours. Each circle represents a colony of bacteria that had grown from a single cell

a) Explain why β- galactosidase would be able hydrolyse lactose.

b) Suggest the importance of the origin of replication for *E.coli* in the pUC plasmid.

c) Use your knowledge of enzyme activity to suggest the meaning of 'lactose analogue'.

d) Which of the colonies in Figure 2 was formed by cells of *E. coli* that had successfully taken up the gene in which the scientists were interested? Fully explain your answer, using your own knowledge and information within the passage.

e) How would the scientists obtain a pure culture of the transformed bacteria?

8 Developments in molecular genetics have led to the sciences of:

- bioinformatics
- genomics
- pharmacogenomics
- proteomics.

Find out what is involved in each of these branches of biology.

8 Genetics

Prior knowledge

In this chapter you will need to recall that:

- genes encode the amino acid sequences of polypeptides
- a gene mutation changes the base sequence of a gene and might involve the deletion, insertion or substitution of one or more bases. Some mutations lead to a change in the amino acid sequence of the polypeptide encoded by a gene
- alternative base sequences of a single gene are called alleles of that gene
- chromosome mutations include translocation and non-disjunction
- meiosis is a source of genetic variation, through the random segregation of homologous chromosomes during anaphase I and crossing over between non-sister chromatids of homologous chromosomes during prophase I
- the random fusion of gametes during sexual reproduction gives rise to genetic variation in the offspring
- monohybrid inheritance examines the way in which two or more alleles of a single gene controlling a single characteristic are passed from parents to offspring
- genetic diagrams and pedigree diagrams can be used to show patterns of inheritance.

Test yourself on prior knowledge

1 Explain what is meant by a 'point mutation'.

2 Explain why some gene mutations have no effect on the amino acid sequence of the polypeptide encoded by the affected gene.

3 Briefly describe chromosome translocation.

4 Humans have 23 pairs of homologous chromosomes. Assuming that no mutations and no crossing over occurs, how many different chromosome combinations could theoretically be present in:

a) the gametes of each of two parents

b) the offspring of these parents?

5 List each of the stages in a genetic diagram representing a monohybrid cross.

Introduction

This chapter builds on the knowledge and understanding you gained in your GCSE science course and in the first year of your A level Biology course. We will re-examine the origins of genetic variation that you studied in *Edexcel A level Biology 1*, Chapter 3, before building on your study of monohybrid inheritance from GCSE.

Origins of genetic variation – a reminder

Gene mutations and chromosome mutations as sources of genetic variation

Gene mutations are the ultimate source of all new genetic variation. Chromosome mutations are simply rearrangements of existing genetic variation.

Gene mutations

Key term

Point mutation A change affecting the base of a single nucleotide in the base sequence of a gene.

In Chapter 3 of *Edexcel A level Biology 1*, we defined a **gene** as a sequence of nucleotide bases that encodes the amino acid sequence of a polypeptide. You also saw that a mutation involves a change in the number, or sequence, of bases in a particular gene. The three types of gene mutation you examined are exemplified in Table 8.1. Since they involve a change to a single base in the sequence, they are called **point mutations**.

Table 8.1 Three types of gene mutation

Type of mutation	Effect on base sequence	Encoded amino acid sequence
None	UCC CAG GAG CCA	Serine–Glutamine–Glutamic acid–Proline
Deletion	UCC AGG AGC CA	Serine–Arginine–Serine–
Insertion	UCG **C**CA GGA GCC A	Serine–Proline–Glycine–Alanine–
Substitution	UC**U** CAG GAG CCA	Serine–Glutamine–Glutamic acid–Proline

Gene mutations occur randomly at a rate of about 1 in 10^6 base replications. Agents that increase the rate of mutation, including harmful chemicals and radiation, are called **mutagens**. Although cells possess enzyme-catalysed reactions that identify and correct gene mutations, some can remain uncorrected. If they occur in a somatic cell of an adult, gene mutations will only affect that adult. If they occur in cells producing gametes, gene mutations could be inherited from generation to generation.

Chromosome mutations

You looked at chromosome mutations in Chapter 5 of *Edexcel A level Biology 1*, where we reviewed chromosome translocations and chromosome non-disjunction.

Chromosome translocations occur when a chromosome breaks and the resulting fragment joins on to another, non-homologous chromosome. Inheritance of a chromosome that has a translocated fragment can be harmful, for example, about 5 per cent of cases of Down's syndrome are caused by the translocation of part of the long arm of chromosome 21 to chromosome 14.

Chromosome non-disjunction results when members of a homologous pair of chromosomes fail to separate during anaphase I of meiosis. Figure 8.1 shows how this will result in gametes with abnormal chromosome numbers.

- If one of the upper gametes in Figure 8.1 is involved in fertilisation with a normal, haploid gamete, the resulting zygote will have three copies of the long chromosome (known as **polysomy**). Most cases of Down's syndrome result from polysomy involving chromosome 21.

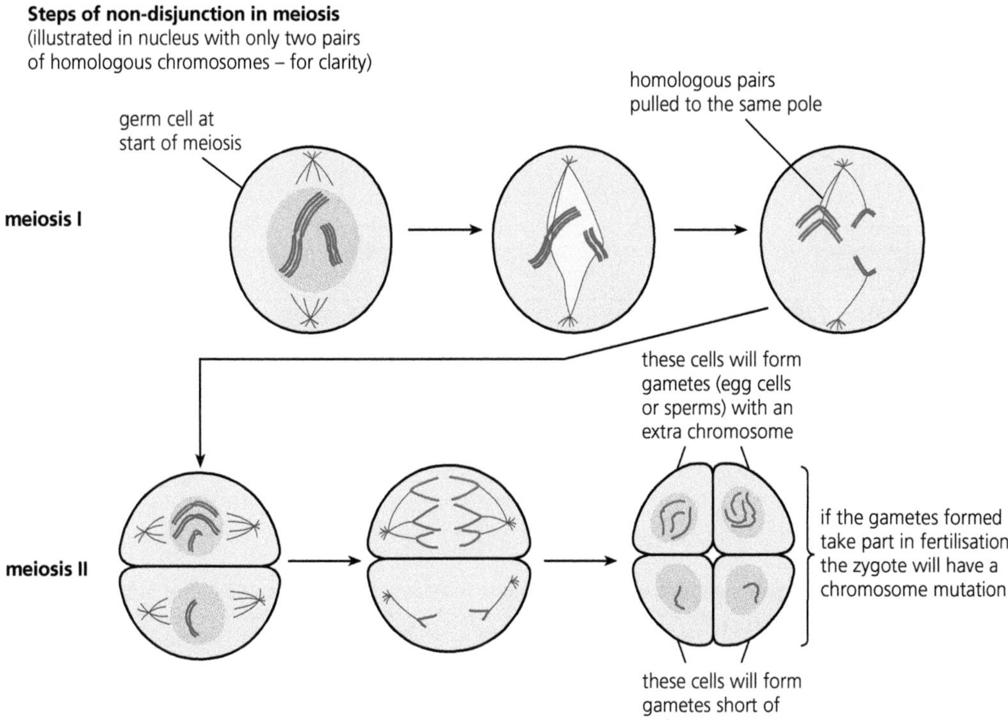

Figure 8.1 Chromosome non-disjunction

- If one of the lower gametes in Figure 8.1 is involved in fertilisation with a normal, haploid gamete, the resulting zygote will have only one copy of the long chromosome (known as **monosomy**). Monosomy is usually lethal. The only example of monosomy in adult humans is Turner's syndrome, in which affected females possess only one X chromosome.

Meiosis and fertilisation as sources of genetic variation

Meiosis was described in Chapter 5 of *Edexcel A level Biology 1*. Two nuclear divisions are involved:

- **Meiosis I** separates the two chromosomes in each homologous pair, one to each end of the dividing cell.
- **Meiosis II** separates the chromatids – the two copies of each chromosome.

The result is haploid daughter cells produced by the division of a diploid parent cell.

Figure 8.2 on the next page, shows two possible outcomes of meiosis. For the sake of simplicity it shows just one homologous pair of chromosomes. In the diagram on the left of Figure 8.2, the homologous chromosomes have remained intact. Because the chromosomes had different alleles of genes A and E, they produce two genetically different haploid gametes, with the combinations of alleles of genes A and E of **AE** and **ae**. The configuration contains only dominant alleles or recessive alleles, in other words **AE** and **ae**. These configurations are known as *cis*. In the diagram on the right of Figure 8.2, crossing over has occurred between two non-sister chromatids. This results in four genetically different haploid gametes, with the combinations of the alleles of genes A and E of **AE**, **ae**, **Ae** and **aE**. The configurations **Ae** and **aE** are known as *trans*. So, we can conclude that, because homologous chromosomes have different alleles at many loci, their random assortment during meiosis produces genetic diversity in the daughter cells and that crossing over between non-sister chromatids further increases this genetic diversity.

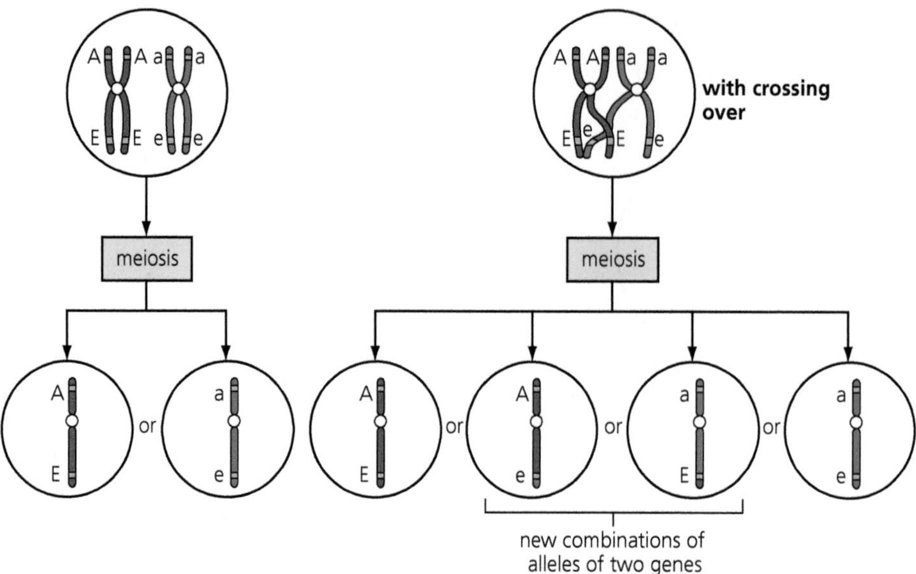

Figure 8.2 Meiosis produces genetic variation amongst the daughter cells. Crossing over further increases the genetic variation

Fertilisation involves the fusion of two haploid gametes to produce a diploid zygote. Meiosis results in genetic diversity in the gametes formed by two adult organisms and which of these different gametes is involved in fertilisation is, again, completely random, further increasing genetic diversity. You saw in Figure 8.2 that even a cell with only one pair of homologous chromosomes involving just one cross-over event can produce four genetically different gametes. The probability of any one of them being involved in fertilisation is 0.25. If the other parent is also able to produce four genetically different gametes, there are 16 possible fusions of any two gametes. The probability of any gamete from one parent fertilising any gamete from the other parent is $0.25 \times 0.25 = 0.0625$ (which we could write as a chance of 1 in 16).

Tip

A probability always has a value between 0 and 1. The expression '1 in 16' refers to chance, not to probability.

Test yourself

1. Define the term 'gene'.
2. What term is used to describe the location of a gene?
3. Identify information in Table 8.1 which shows that the genetic code is degenerate.
4. Which gene mutation, or mutations, in Table 8.1 caused a frame shift? Explain your answer.
5. Describe what is meant by 'polysomy' and explain how it is caused.
6. What is the difference between the *cis* and *trans* combinations of two linked genes?

Key term

Dihybrid inheritance The transfer from generation to generation of two different genes. The two genes might control different features of the phenotype (non-interacting genes) or might both contribute to the control of a single feature of the phenotype (interacting genes). Additionally, they might occur on the same chromosome (linked genes) or on different chromosomes (unlinked genes).

Terminology used in genetics – a reminder

The transfer of genetic information from one generation to the next is known as inheritance. **Genetics** is the study of inheritance. In your GCSE science course, you studied the inheritance of a characteristic controlled by a single gene. This is known as **monohybrid inheritance**. Here, you will study the inheritance of characteristics that are controlled by two different genes, so-called **dihybrid inheritance**.

Some of the expressions used above are rather clumsy. Before going further, we will see how these clumsy expressions can be replaced with technical terms with defined meanings.

Measurable features and combinations of genes

When referring to Figure 8.2 above, we used the expression 'combination of alleles of the genes A and E', which is rather cumbersome. We also used the expression 'inheritance of characteristics' as though every inherited feature could be easily seen. Instead, we use technical terms with exact meanings.

- **Genotype** is the combination of alleles of a particular gene, or genes, present in a haploid gamete or diploid organism. In most of this chapter, we will use the term genotype in the context of two genes considered together.
- **Phenotype** is a measurable feature of an organism. It might be visible, for example, fur colour, or detectable only by chemical or immunological analysis, for example, ABO blood groups. Although the phenotype is affected by the genotype, it is also affected by environmental factors.

  genotype + environmental factors → phenotype

Each member of a pair of homologous chromosomes has the same genes in the same order. If there are two or more alleles of a gene, both homologous chromosomes of a diploid organism might carry the same alleles of that gene or two different alleles of that gene.

- **Homozygous** cells or organisms are diploid and have the **same** allele of a gene under consideration on both copies of a pair of homologous chromosomes. The cell or organism is a **homozygote** for this gene.
- **Heterozygous** cells or organisms are diploid and have **different** alleles of a gene under consideration. The cell or organism is a **heterozygote** for this gene.

The relationship between alleles of the same gene

Although a gene will have an effect on the phenotype, not all alleles have the same ability to affect the phenotypic feature they control.

- **Dominant alleles**, when present in the genotype, always show their effect in the phenotype.
- **Recessive alleles**, when present in the genotype, do not show their effect in the phenotype if a dominant allele of the same gene is also present. Recessive alleles only show their effect in the homozygous condition.
- **Codominant alleles**, when present in the genotype, show their effect in the phenotype regardless of the other allele of the gene.
- **Multiple alleles** occur when there are more than two alleles of a single gene, for example there three alleles of the human ABO blood group ($\mathbf{I^A}$, $\mathbf{I^B}$ and $\mathbf{I^o}$) that you probably studied in your GCSE science course.

Transfer of genetic information

This A level Biology course covers the inheritance of two genes with and without linkage. We will begin by looking at the pattern of inheritance that follows, in the simplest way, from the pattern of monohybrid inheritance you studied during your GCSE science course. Use Figure 8.3 on the next page to refresh your memory of how you can use a genetic diagram to explain the outcome of monohybrid inheritance. The format of Figure 8.3 is that of a genetic diagram and the notes explain the rationale behind its use.

Tip

An organism that is homozygous at a particular locus is sometimes described as **pure-breeding** for that phenotypic feature.

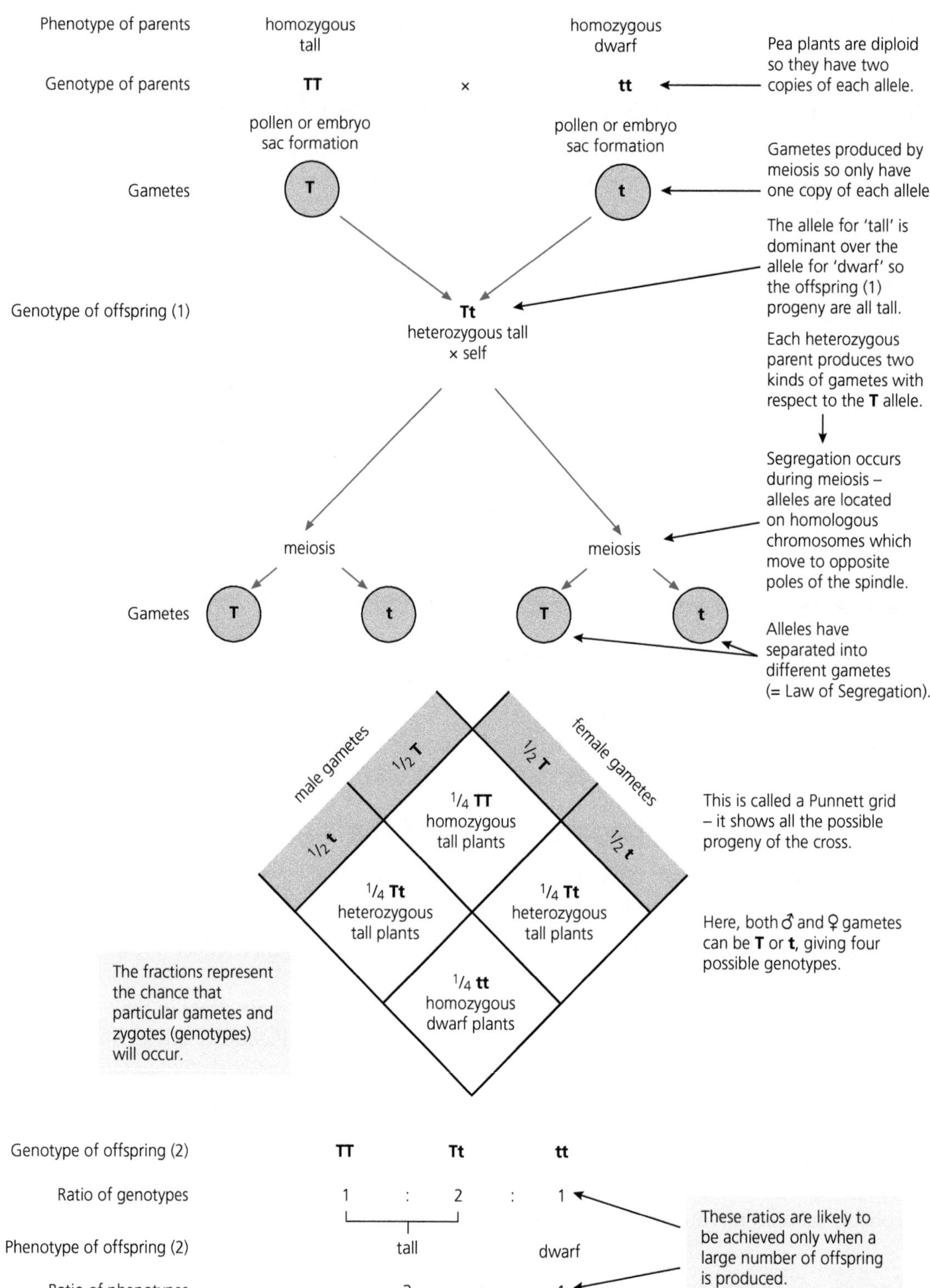

Figure 8.3 Using a genetic diagram to explain the outcome of monohybrid inheritance

Inheritance with sex-linkage

In Chapter 6 we saw in the Extension *X chromosome inactivation in mammals*, page 105, that mammalian females have two X chromosomes (XX) and males have one X chromosome and one Y chromosome (XY). The larger X chromosome has far more genes than does the Y chromosome, many of which are not related to sexual characteristics. One of these is a gene controlling blood clotting. Since this gene is located on the X chromosome, it is said to be **sex-linked**.

> **Key term**
>
> **Sex linkage** A condition in which the locus of a gene is located on a sex chromosome; the gene itself is described as **sex-linked.** Since there are very few genes on a mammalian Y chromosome, sex linkage in mammals almost always refers to a gene with its locus on an X chromosome.

Blood clotting normally occurs following a break or cut in a blood vessel. A series of enzyme-catalysed reactions produces a fibrous network of a protein called fibrin, which traps red blood cells. The fibrin network and trapped red blood cells form a clot, which reduces the likelihood of both the further loss of blood and of the entry of pathogens. Haemophilia is a relatively rare disease in which the blood fails to clot normally. The result is frequent and excessive bleeding that can be fatal.

Haemophilia is the result of the recessive allele of a gene controlling a step in the blood clotting process. As it is recessive, we will represent it with a lower case letter (**h**) and we will represent the dominant allele with an upper case letter (**H**). Furthermore, since we know that this gene is located on the X chromosome, we will show this in our representation of these alleles: $\mathbf{X^h}$ and $\mathbf{X^H}$.

Figure 8.4 on the next page, represents the inheritance of haemophilia in one family. Both parents show normal blood clotting, but the mother is heterozygous for the blood-clotting gene. The eggs and sperm of these parents are haploid, having been formed by meiosis. At fertilisation, a diploid zygote is formed, which develops into a child.

- Since the father has normal blood clotting, his X chromosome must carry the allele for normal blood clotting, **H**. We can represent his genotype as $\mathbf{X^HY}$.
- Since the mother is heterozygous for this gene, one of her X chromosomes carries the allele for normal blood clotting (**H**) and the other carries the allele for haemophilia (**h**). We can represent her genotype as $\mathbf{X^HX^h}$. Since the allele causing haemophilia is recessive, this woman's blood will clot normally: she does not show the disease but is a 'carrier'.

Half the father's sperm are likely to contain a Y chromosome and half an X chromosome. All the mother's eggs will contain an X chromosome but, as a result of the independent segregation of chromosomes during meiosis, half are likely to carry the $\mathbf{X^H}$ allele and half the $\mathbf{X^h}$ allele. Fertilisation is random, so it is a 50 : 50 chance which egg is fertilised and a 50 : 50 chance which sperm fertilises the egg. Figure 8.4 shows how we can represent the formation of gametes, possible fertilisations and the outcome of these fertilisations in the genotypes of the offspring this couple might have. This representation is called a **'genetic cross diagram'** and we will use diagrams like this in most of our explanations of inheritance in the remainder of this chapter.

There are several noteworthy points shown in Figure 8.4.

- The expected ratio of female to male children is 1 : 1.
- Although neither parent is a **haemophiliac** (suffers haemophilia), they could have a haemophiliac child.
- Any haemophiliac child they might have would be male ($\mathbf{X^hY}$).

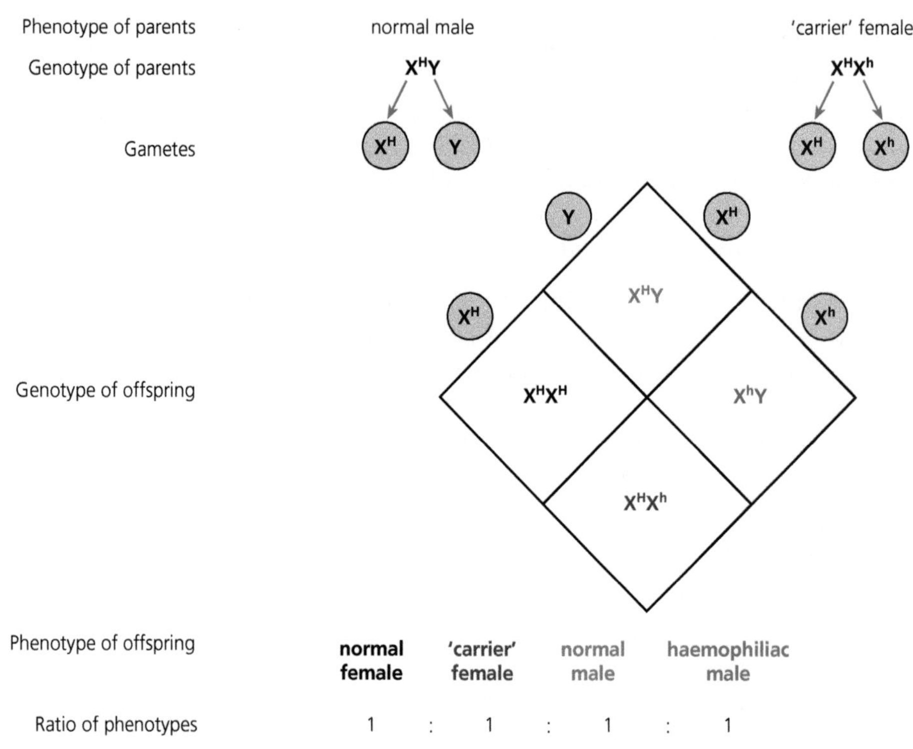

Figure 8.4 Inheritance of haemophilia – an example of sex linkage

Example

Using a pedigree diagram to show the inheritance of haemophilia

Figure 8.5 shows another way in which you can represent the inheritance of a sex-linked gene. It is a **pedigree diagram**. We may investigate the pattern of inheritance of a particular characteristic by researching family pedigrees, where appropriate records of the ancestors exist.

Just as the genetic cross diagram in Figure 8.4 has a fixed layout, so a pedigree diagram involves a set of 'rules'.

- A female is represented as a circle and a male as a square.
- A horizontal line between a circle and a square links two parents.
- A vertical line from the line linking two parents takes us to the offspring from that relationship.
- The offspring from a relationship are shown in birth sequence from left to right of the diagram.

Figure 8.5 also has a key, showing you the phenotype of each individual. Rather than naming them, each individual in Figure 8.5 has been numbered, so you can identify them.

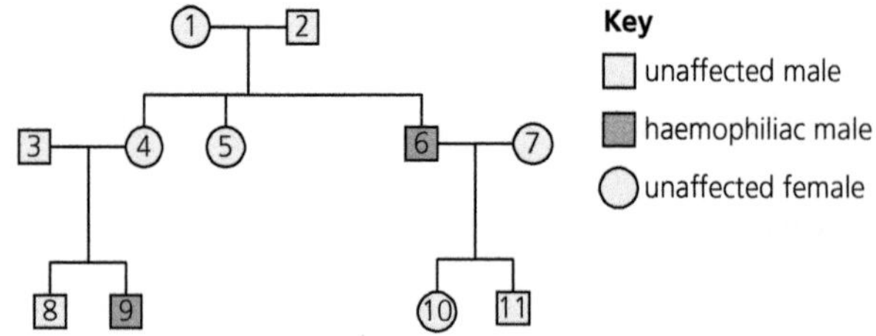

Figure 8.5 A pedigree diagram showing the inheritance of haemophilia in one family

1 How should you interpret a pedigree chart?

2 For which two individuals in Figure 8.5 can you be sure of the genotype?

3 What are the genotypes of individuals 6 and 9?

4 What are the genotypes of individuals 1, 2, 3 and 4?

5 Individual 9 is a haemophiliac. His brother (individual 8) is not. Explain why.

6 What is the genotype of individual 10?

7 Can we be sure of the genotype of individual 7?

8 If individual 7 had been a carrier, what is the probability of her not passing her $\mathbf{X^h}$ chromosome to either of her two children?

Answers

1 When solving an unknown, it is always best to start with something you do know. From your knowledge of the inheritance of haemophilia, you should immediately be able to identify the genotypes of two individuals in Figure 8.5.

2 Since you know that the allele of the gene that causes haemophilia is recessive and you know that females are XX and males XY, you should immediately spot that you can identify the genotypes of individuals 6 and 9 – the haemophiliacs.

3 Individuals 6 and 9 are male (XY). They are both haemophiliacs, so each must have the recessive allele (**h**) on his X chromosome. The genotype of both must be $\mathbf{X^hY}$.

Now you know the genotypes of individuals 6 and 9, you can work backwards from both. Neither of them has a parent who is a haemophiliac. This gives you a big clue to the genotypes of these parents.

4 The fathers (individuals 2 and 3) do not show haemophilia. The genotypes of both must be $\mathbf{X^HY}$. Neither mother (individuals 1 and 4) shows haemophilia but must have passed the haemophilia allele to their sons, in other words they are both carriers with the genotype $\mathbf{X^HX^h}$.

5 Since their father's genotype is $\mathbf{X^HY}$, none of his sperm will carry the $\mathbf{X^h}$ allele. The mother is $\mathbf{X^HX^h}$, so there is a 50 : 50 chance that any one of her eggs will carry the $\mathbf{X^H}$ or $\mathbf{X^h}$ allele. As luck would have it, the egg from which individual 8 was conceived carried the $\mathbf{X^H}$ allele, so his genotype is $\mathbf{X^HY}$.

6 Her father is a haemophiliac, so his X chromosome carries the **h** allele. A girl inherits one of her X chromosomes from her father. As she is not a haemophiliac, her genotype must be $\mathbf{X^HX^h}$.

7 The answer is no. As her children do not show haemophilia, she might be homozygous $\mathbf{X^HX^H}$. On the other hand, she might be heterozygous ($\mathbf{X^HX^h}$) yet, just by chance, neither of her eggs that was fertilised and gave rise to her children carried the $\mathbf{X^h}$ allele.

8 The probability of each child not inheriting the $\mathbf{X^h}$ chromosome from their mother is 0.5, so the probability of neither of them inheriting this chromosome is $0.5 \times 0.5 = 0.25$.

Test yourself

7 Give **two** types of factor that affect the phenotype of an organism.
8 What are 'codominant' alleles?
9 Explain why a son cannot inherit haemophilia from his father.
10 Most haemophiliacs are male. Explain how a female child could inherit haemophilia.
11 What is an 'autosome'?

Figure 8.6 The fruit fly, *Drosophila melanogaster* has been an important animal in genetic research

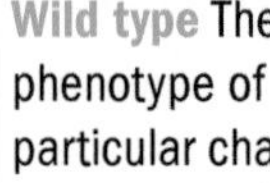

Key term

Wild type The natural phenotype of any particular characteristic, that is, one that is unaffected by any gene mutation.

Dihybrid inheritance with autosomal linkage

An **autosome** is any chromosome other than a sex chromosome, so autosomal linkage describes the condition in which two genes are located on the same chromosome but it is not a sex chromosome. If you look back to Figure 8.2 on page 142, you will see that the alleles of two genes with loci on the same autosome are inherited together, unless crossing over occurs. Bear this in mind as you follow the example of dihybrid inheritance with autosomal linkage given below.

Figure 8.6 shows a fruit fly, *Drosophila melanogaster.* You might have seen flies like this around a bowl of fruit during the summer months. During the early years of the 20th century, an embryologist, named Thomas Hunt Morgan, chose *D. melanogaster* as the organism to use in his research into the mechanisms of inheritance. His work established the science we now call 'genetics' and *D. melanogaster* has continued to be the organism of choice for thousands of geneticists ever since.

D. melanogaster has four pairs of chromosomes – one pair of sex chromosomes and three pairs of autosomes, chromosomes 2, 3 and 4. The fruit fly at the top of Figure 8.7 shows what is called the **wild type** phenotype of *D. melanogaster.* Amongst its features, it has long wings and a yellow body with black stripes. These features are controlled by two genes on chromosome 2: one controlling wing length and the other controlling body colour. The fruit fly at the bottom of Figure 8.7 has vestigial (short) wings and a black body. It is homozygous for a mutant allele of the gene controlling wing length and for another mutant allele of the gene controlling body colour. Table 8.1 shows the effects of these two genes and the two alleles of each gene. It also shows the simple notation we will use in our genetic cross diagram (with the more complex, standard notation used by geneticists to represent the genes in *Drosophila* shown in brackets).

Wild type

Vestigial wings

Figure 8.7 A wild type *Drosophila melanogaster* and a mutant with vestigial wings

Table 8.1 The alleles of two genes on chromosome 2 of *Drosophila melanogaster*

Phenotypic feature	Alleles of gene	Dominant or recessive	Effect on phenotype
Body colour	**G** (vg^+)	Dominant	Results in yellow (wild type) body colour, even in heterozygote
	g (vg)	Recessive	When homozygous, results in black body
Wing length	**L** (b^+)	Dominant	Results in long (wild type) wings, even in heterozygote
	l (b)	Recessive	When homozygous, results in vestigial (short) wings

A dihybrid cross with autosomal linkage in which no crossing over of linked genes occurs

Figure 8.8 represents a cross between two fruit flies. Both are homozygous for the genes controlling body colour and wing length. The female is homozygous dominant (**GGLL**) and the male is homozygous recessive (**ggll**) for these genes. During gamete production, meiosis separates the two copies of chromosome 2 in each diploid cell, so the female adult will produce eggs with the genotype **GL** and the male will produce sperm with the genotype **gl**. All the offspring will be heterozygotes with the genotype **GgLl**.

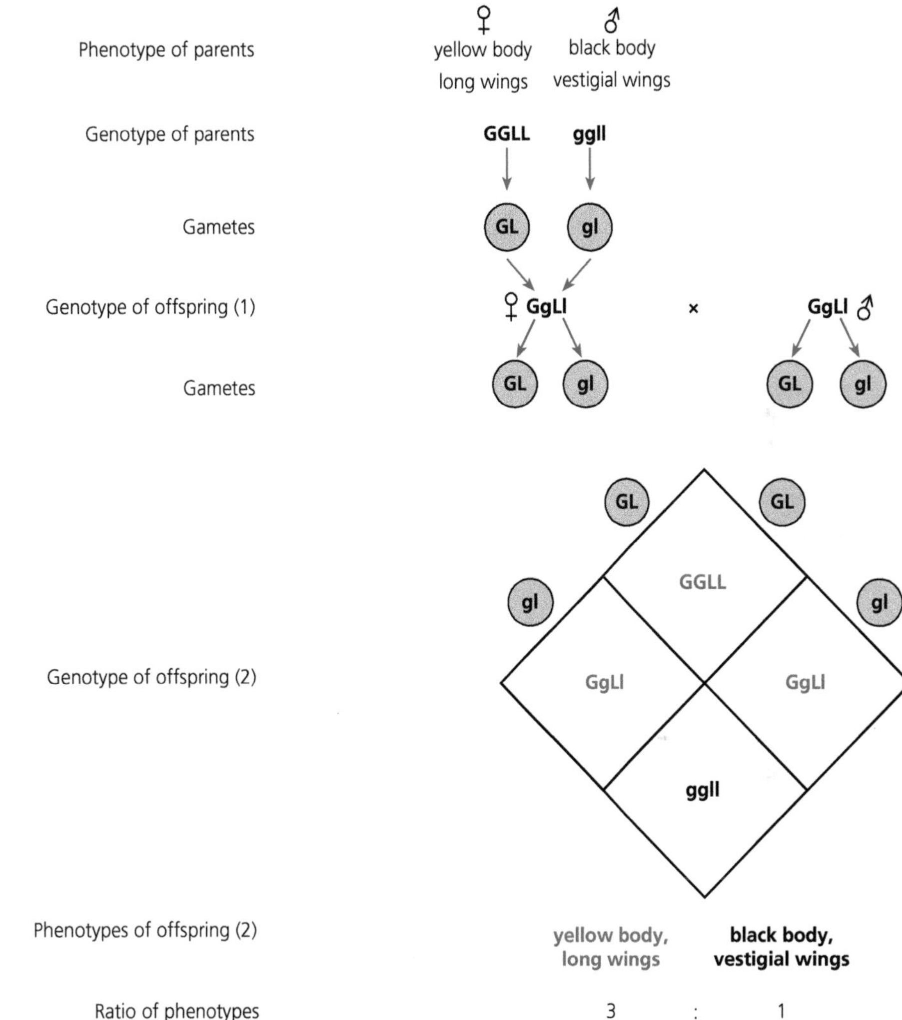

Figure 8.8 A genetic cross between a female fruit fly that is homozygous dominant for the body colour and wing length genes on chromosome 2, with a male that is homozygous recessive for the same genes. One of the female offspring is mated with one of the male offspring to produce a second generation of offspring flies. In this diagram, no crossing over between the two loci has occurred

Tip

Figure 8.8 shows the genotypes of gametes within circles. This makes quite clear to readers, including examiners, how a sequence of letters relates to individual gametes.

It also shows the genotypes of the second offspring generation in the form of a Punnett square. You will find this helpful when constructing your own genetic diagrams.

Since we are dealing with fruit flies, we mate a female and male of this group of brother and sister flies to produce a second offspring generation of flies. You might expect that, since the alleles **G** and **L** are linked on the same chromosome, they will always be inherited together and, similarly, **g** and **l** will always be inherited together. If this were the case, we would only ever get flies that had the same phenotypes as the two parents – wild type and black-bodied with vestigial wings. This is represented in Figure 8.8, where the second offspring generation contains only flies with the original parental phenotypes – yellow bodies with long wings or black bodies with vestigial wings, in an expected ratio of 3 : 1.

Tip

When representing genotypes involving linked genes, some students find it helpful to remind themselves which alleles are linked with which by bracketing them together. In Figure 8.8, this would mean representing the genotypes of the parents as (**GL**)(**GL**) and (**gl**)(**gl**).

Key term

Recombinant An organism with a phenotype that results from the crossing over of linked genes during gamete production in one, or both its parents. The resulting phenotype has a combination of characteristics not shown by either parent.

A dihybrid cross with autosomal linkage in which crossing over of linked genes does occur

Remember, though, that crossing over will sometimes occur. Unusually, crossing over does **not** occur in male *D. melanogaster.* It appears that the males of this species lack the enzyme-catalysed mechanisms that allow close pairing of homologous chromosomes during meiosis, so that chiasmata do not form. Crossing over does, however, occur in female *D. melanogaster.* Sometimes, during meiosis in egg production in a female fruit fly, a crossing over event might occur between the gene for body colour and the gene for wing length, just as is shown for the genes A and E in Figure 8.2 on page 142. If so, we might expect a small number of her gametes to have genotypes of **Gl** and **gL**. You can see from Figure 8.9 that this does occur. As a result, we find a small number of flies in the second offspring generation with new phenotypes – yellow bodies with vestigial wings and black bodies with long wings. These individuals are called **recombinants** and their appearance in the offspring (2) generation shows us that crossing over must have occurred. The number of recombinants will depend on the frequency with which crossing over occurs between the two genes during egg production by the females. This frequency is greater the further apart the two loci are.

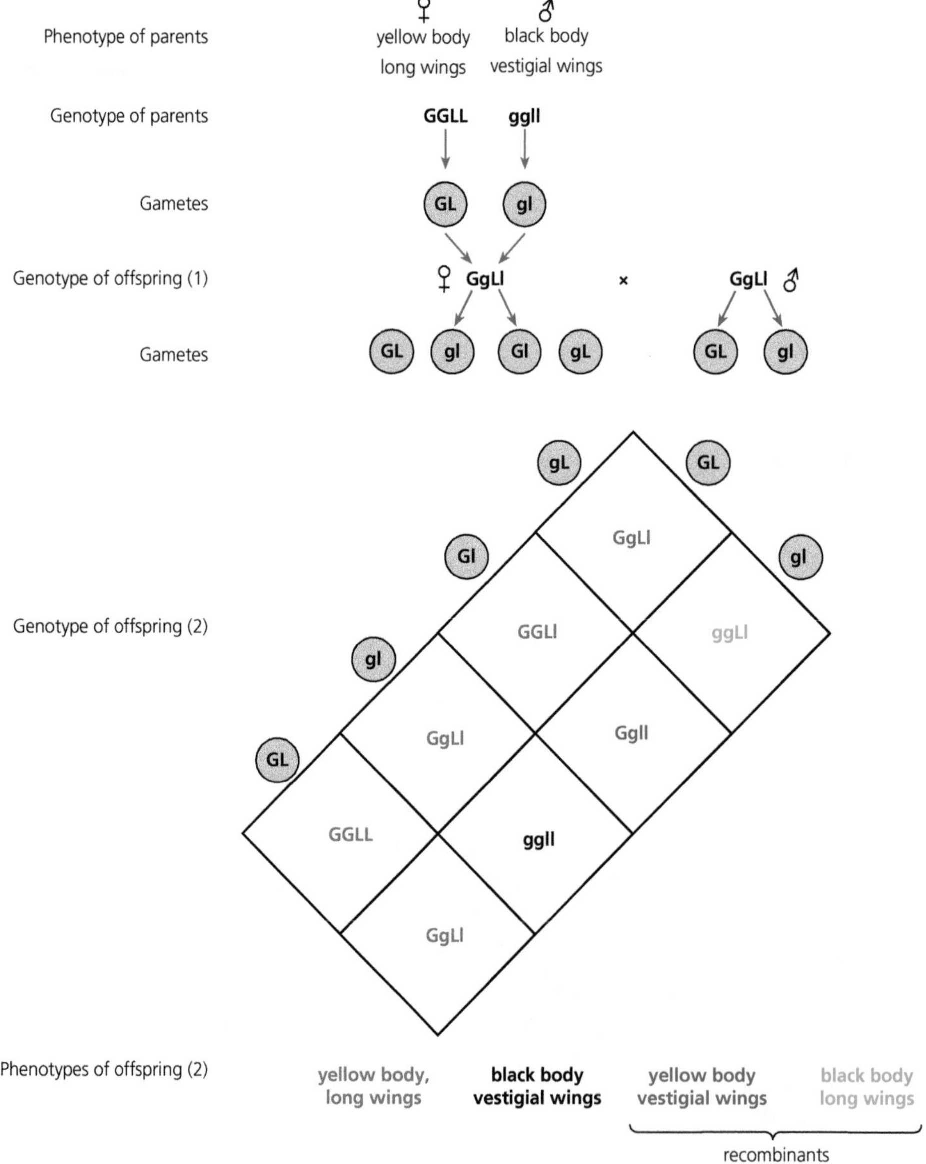

Figure 8.9 The same crosses are shown in Figure 8.8. In this diagram, however, crossing over between the two loci has occurred in the females of the offspring (1) generation

Dihybrid inheritance with non-interacting unlinked genes

Two genes are said to be **unlinked** if they occur on different chromosomes. They are **non-interacting** if they control two different phenotypic features.

Having used humans and fruit flies in the examples above, we will balance the biological examples by using plants here. Gregor Mendel's experiments with pea plants were important forerunners of the science of genetics. We will look at an example of dihybrid inheritance with non-interacting genes using pea plants, *Pisum savitum*, as an example (see Figure 8.10). Table 8.2 shows these two phenotypic features, the notation we will use to represent the two alleles of each of the genes controlling these features and their relationship.

Table 8.2 The alleles of two unlinked genes of *Pisum savitum*

Phenotypic feature	Alleles of gene	Dominant or recessive	Effect on phenotype
Flower colour	**F**	Dominant	Results in violet flowers, even in a heterozygote
	f	Recessive	When homozygous, results in white flowers
Stem length	**T**	Dominant	Results in tall stems, even in a heterozygote
	t	Recessive	When homozygous, results in short stems

Figure 8.11 on the next page represents a cross between two pea plants. Both are homozygous for the genes controlling flower colour and stem length. One plant is homozygous dominant (**FFTT**) and the other is homozygous recessive (**fftt**) for these genes. Since gametes are haploid, one adult will produce gametes with the genotype **FT** and the other will produce gametes with the genotype **ft**. All the offspring will be heterozygotes with the genotype **FfTt** and will be tall plants with violet flowers.

Pea plants are able to self-fertilise – the pollen will germinate if it lands on the stigma within the same flower and fertilise the nuclei within its embryo sac (see *Edexcel A level Biology 1*, Chapter 6). If the offspring (1) plants are allowed to self-fertilise, they will produce an offspring (2) generation. Remember that the two loci are not on the same chromosome, so random segregation of chromosomes will result in all possible combinations of the two sets of alleles – **FT**, **Ft**, **fT** and **ft** – in equal frequencies in female and male gametes. The random fertilisation of these gametes will produce 16 possible fertilisations, all of which are shown in Figure 8.11. The expected ratio of phenotypes of 9 : 3 : 3 : 1 is typical of this type of cross with two, unlinked, non-interacting genes.

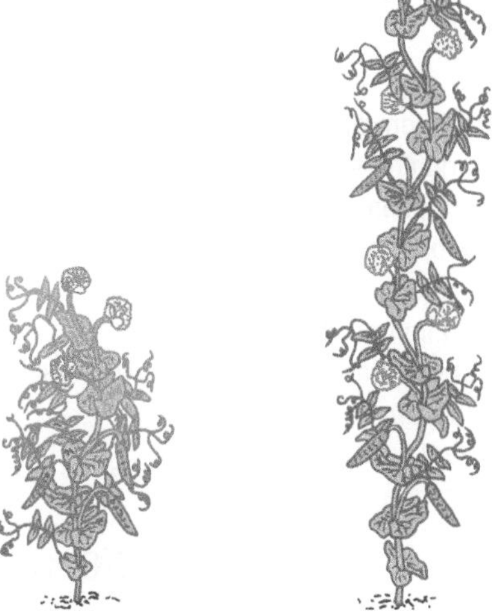

Figure 8.10 Pea plants (*Pisum savitum*) showing one of the features we will use in our example of inheritance - short stems and tall stems

Figure 8.11 A genetic cross between a pea plant that is homozygous dominant for the flower colour and stem length genes with another that is homozygous recessive for the same genes. By allowing the offspring plants to self-fertilise, a second generation of offspring pea plants can be produced

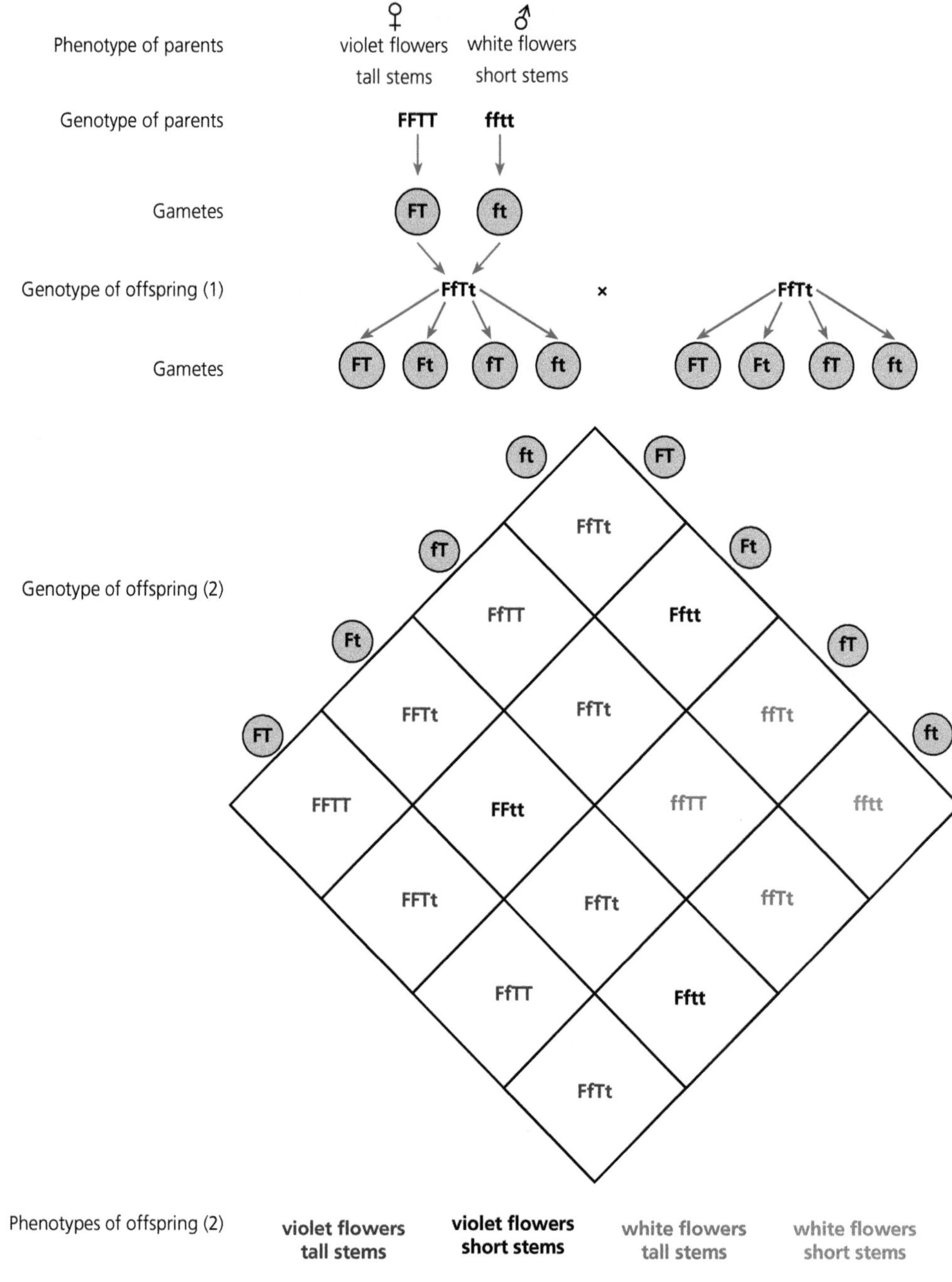

Phenotypes of offspring (2)	violet flowers tall stems	violet flowers short stems	white flowers tall stems	white flowers short stems
Expected ratio of phenotypes	9 :	3 :	3 :	1

Tip

When interpreting dihybrid crosses, the ratio of phenotypes in the offspring (2) generation might give you an immediate clue about whether the two loci are linked or unlinked.

Use of the chi squared (goodness of fit) test in genetics

Once you have a good understanding of the three patterns of inheritance from the previous pages, you will be able to solve genetic puzzles set by your teacher or by examiners. From your genetic diagrams, you will be able to predict the ratio of the different phenotypes resulting from a cross. You will be well aware, though, that random variation is common in biology. In fact, when the results of Gregor Mendel's experiments with pea plants were rediscovered early in the 20th century, the famous statistician R.A. Fisher was highly critical of them – they showed too little random variation from the expected ratios to be believable! However, the underlying mechanisms of these inheritance patterns, hold true.

Table 8.3 The observed results of a genetics experiment with pea plants and the results expected from an understanding of the theory of inheritance

Results	Tall stems with violet flowers	Tall stems with white flowers	Short stems with violet flowers	Short stems with white flowers	Total
Observed	173	55	66	26	320
Expected	180	60	60	20	320

How can you tell whether the results you obtain from a genetics experiment really do match your expectations from theoretical knowledge? Look at the data in Table 8.3. They relate to results from the cross shown in Figure 8.11. From this cross, you would expect the second offspring generation to have the phenotypes: tall with violet flowers, tall with white flowers, short with violet flowers and short with white flowers in a ratio of 9:3:3:1. By dividing the actual number of plants obtained in the second offspring generation into this ratio, you calculate your **expected numbers** of each phenotype. You can compare these with the actual numbers you found – the **observed numbers**.

Since the data shown in Table 8.3 are categoric, you use the chi squared χ^2 test, described in Chapter 15, to find whether your observations are close enough to your expectations for your expectations to be valid. You calculate the value of χ^2 as follows.

$$\chi^2 = \sum \frac{(O - E)^2}{E}$$

Table 8.4 expands on Table 8.3, showing the working to calculate the value of χ^2 from these data.

> **Key term**
>
> **Categoric data** Those where only certain values can exist. The values are not continuous, within a range of values, as would be, for example time in an enzyme experiment.

Table 8.4 Calculation of χ^2 from the data in Table 8.3, using calculated values to three significant figures

	Tall with violet flowers	Tall with white flowers	Short with violet flowers	Short with white flowers
O	173	55	66	26
E	180	60	60	20
$(O - E)$	−7	−5	6	6
$(O - E)^2$	49	25	36	36
$\frac{(O - E)^2}{E}$	0.26	0.42	0.60	1.80
χ^2			3.08	

Now that you have a value for χ^2, you need to interpret what it means. You are testing whether the observed results are a 'good fit' to the expected results. Your assumption is that there is no difference between the observed numbers of each phenotype and the expected numbers of each phenotype (your **null hypothesis**). In your example, you have four categories of data (the four phenotypes), so you have three degrees of freedom. When you check this value against a table of critical values for χ^2, you find:

Degrees of freedom	Significance level		
	0.05	0.01	0.001
3	7.81	11.34	16.27

As your calculated value for χ^2 (3.08) is well below the 0.05 significance level, you can say that the difference between your results and the results you expected is not statistically significant (or, put another way, the probability of the difference between the results you obtained and those you expected being due to chance is greater than 0.95).

Consequently, you accept your null hypothesis – there is no difference between your results and the results you expected. This means that your interpretation of the pattern of inheritance on which you based your expectation was valid.

Test yourself

12 One of the male yellow-bodied, long-winged fruit flies from the offspring (1) generation in Figure 8.8 is mated with a black-bodied, vestigial-winged female. What ratio of phenotypes would you expect in their offspring? Explain your answer.

13 How could the cross in Question 12 be used to determine the genotype of a yellow-bodied, long-winged fly? Explain your answer.

14 Would it make any difference to your answer to Question 12 if it had been the female that had been yellow-bodied and long-winged? Explain your answer,

15 Here are the genotypes of four pea plants: **FfTT, FFTt, FfTt, fftt**. Which pea plant would be described as 'pure-breeding'?

16 Explain why the chi squared test is the appropriate statistical test to use when comparing the expected outcome of a genetic cross with the observed results.

Exam practice questions

1 **Pp** represents:

A an allele

B a gene

C a genome

D a genotype *(1)*

2 A cross between two organisms with the genotype **Aa** would produce offspring with genotypes in an expected ratio of:

A 1:1

B 1:2:1

C 2:1

D 3:1 *(1)*

3 An organism with the genotype for two unlinked genes of **AaBb** could produce gametes with alleles of these genes in the expected ratio of:

A 1:1

B 1:1:1

C 1:2:1

D 1:1:1:1

4 In humans, the ability to distinguish red and green as separate colours is controlled by a single gene. Red–green colour blindness results from a recessive allele of this gene, represented as **b**. The dominant allele (**B**) results in the ability to distinguish red and green.

The diagram represents the inheritance of red–green colour blindness in one family

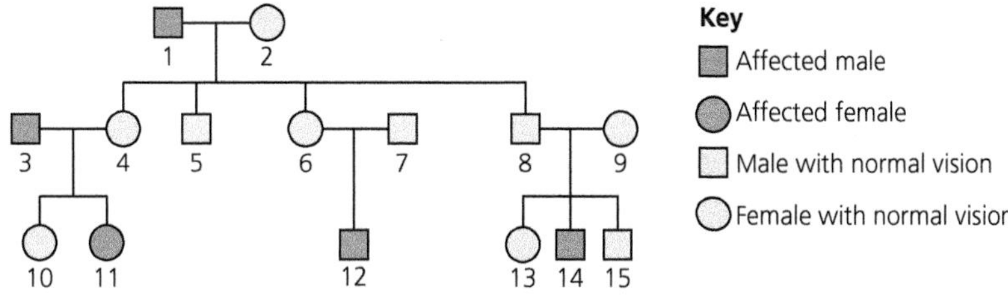

a) Red–green colour blindness is a sex-linked character. Give the evidence from the diagram that supports this statement. *(2)*

b) Give the genotype of person 4. Justify your answer. *(2)*

c) Give the genotype of person 11. Justify your answer. *(1)*

5 Chickens have loose red-coloured skin, called combs, around their heads. The shape of these combs is controlled by two genes, each with two alleles: **P** and **p**, **R** and **r**.

The table shows the phenotypes produced by these genes. The dashes indicate that either of the alleles of a gene could be present.

Genotype	P–R–	P–rr	ppR–	pprr
Phenotype of comb	Walnut	Pea	Rose	Single

The information on the next page shows how a breeder mated a pure-breeding rose-combed chicken with a pure-breeding pea-combed chicken. He then interbred the offspring of these parents to produce an offspring (2) generation.

	♀		♂	
Phenotype of parents	rose comb		pea comb	
Phenotypes of offspring (2)	**walnut comb**	**rose comb**	**pea comb**	**single comb**
Expected ratio of phenotype	**9** :	**3** :	**3** :	**1**

a) Draw a genetic cross diagram to explain the phenotypes produced by this cross. *(4)*

b) What can you conclude about the location of the two genes **P** and **R**? *(2)*

6 The diagram shows two characteristics of the flower of a plant called the sweet pea (*Lathyrus odoratus*).

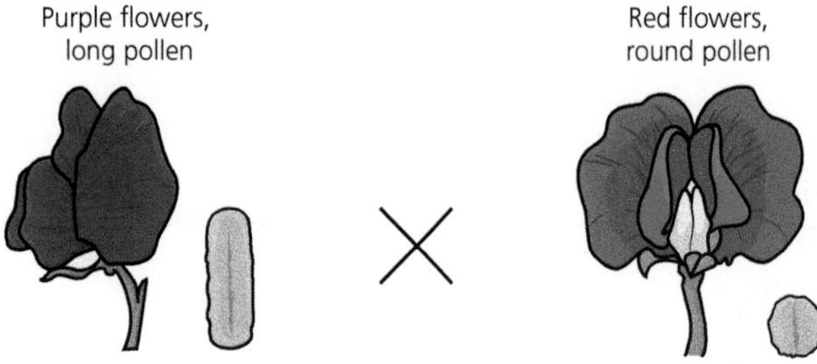

Each of these characteristics is controlled by a gene with two alleles. The gene controlling flower colour has two alleles: the allele for purple flowers (**F**) is dominant and that for red flowers (**f**) is recessive. The gene controlling the shape of the pollen grains also has two alleles: the allele for long pollen grains (**L**) is dominant and that for round pollen grains (**l**) is recessive.

A plant breeder transferred pollen grains from plant 1 that was homozygous dominant for both genes, to plant 2 that was homozygous recessive for both genes.

a) Name the organ of plant 1 from which the plant breeder would take the pollen grains. *(1)*

b) Name the organ of plant 2 onto which the plant breeder would transfer the pollen grains. *(1)*

c) Suggest how the plant breeder would transfer the pollen grains. *(1)*

The plant breeder planted the seed produced by plant 2. After the seeds had germinated and produced adult plants, he ensured these plants self-fertilised. Again, he collected the seeds produced, planted them and recorded the two phenotypes in the offspring (2) generation. He recorded the following number of plants in the offspring (2) generation:

- 284 plants with purple flowers and long pollen grains
- 21 plants with purple flowers and round pollen grains
- 21 plants with red flowers and long purple grains
- 55 plants with red flowers and round pollen grains.

Tip

Question 6 is synoptic, that is, it includes information from another part of your course. It also includes elements of experimental design. Both features are common in Paper 3 questions.

d) Suggest how the plant breeder could ensure that the plants in the offspring (1) generation self-fertilised. *(1)*

e) Explain the results the plant breeder obtained. *(4)*

Stretch and challenge

7 Labrador retrievers are a breed of dog. Their fur can be black, brown or yellow. The colour depends on the inheritance of two unlinked genes.

One gene controls the production of the pigment melanin. This gene has two alleles: a dominant allele, **B**, results in black fur and a recessive allele, **b**, results in brown fur.

A second gene controls whether or not any pigment molecules are deposited in a retriever's fur. This gene has two alleles: a dominant allele, **E**, results in melanin being deposited in the fur (a black or brown dog) and a recessive allele, **e**, results in no melanin being deposited in the fur (a yellow dog).

A dog breeder mated a male dog with the genotype **BBEE** and a female dog with the genotype **bbee**. He then allowed the offspring to mate together to produce an offspring (2) generation.

a) Construct a genetic diagram to show the expected ratio of phenotypes in the offspring (2) generation.

b) Explain why a Labrador retriever might have brown eyes but yellow fur.

c) This cross demonstrates a phenomenon called epistasis. Research the term 'epistasis' and explain what it means.

8 Some human characteristics show continuous variation. Explain the cause of continuous variation in human eye colour.

Gene pools

Prior knowledge

In this chapter you will need to recall that:

- a population is a group of organisms of the same species living in the same habitat at the same time
- genetic variation occurs within populations
- mutation is the ultimate source of genetic variation in all populations. In sexually reproducing populations, random assortment of chromosomes and crossing over during meiosis and the random fertilisation of gametes provide further sources of genetic variation
- genetic variation results in differences in the phenotypes of members of a population
- natural selection is the differential reproductive success of phenotypes within a population
- a gene pool is the sum total of all the alleles of all the genes within a population
- natural selection results in a change in the allele frequencies of genes within a population
- under certain circumstances, natural selection can result in the formation of new species.

Test yourself on prior knowledge

1 Give the main stages in the argument for natural selection.
2 What is meant by the 'struggle for existence'?
3 Give **three** sources of evidence that support the theory of natural selection.
4 Distinguish between allopatric and sympatric speciation.
5 Explain how antibiotic resistance in bacteria is an example of directional selection.

Populations – a reminder

Key term

Population A group of individuals belonging to a single species, living together in one habitat. If the organisms reproduce sexually, members of a population are also able to interbreed.

In Chapter 8, you considered genes and their alleles in the context of inheritance. In this chapter you will consider genes in the context of populations, both of genes within, and gene flow between populations.

Populations can vary in the number of organisms they contain and in the area or volume they occupy. As you saw in Chapter 3, a population containing millions of bacteria could occupy 10 cm^3 of broth culture. A population containing millions of turtles would occupy a much larger volume! The population of garden snails around the compost heap in Figure 9.1 occupies a much smaller area than the population of song thrushes, which might feed on them. In the latter example, the mobility of the animal must be taken into consideration.

In sexually reproducing species, members of a population are able to interbreed. The corollary to this is that if the members of two groups of the same sexually reproducing species are not able to interbreed, they do not belong to the same population. Figure 9.1 illustrates some of the problems in determining whether interbreeding can occur between different groups and, consequently, in defining a population. Garden snails are able to move but only slowly. On this basis, you might consider the garden snails occupying the flower bed on the traffic island in Figure 9.1 as a single population. Snails from this group are unlikely to cross the road safely and mate with the snails living on the compost heap. They probably form a **'closed' population**, that is, one that is completely cut off from neighbouring groups of the same species. Although they move slowly, however, the snails around the compost heap and on the nearby flower bed can probably migrate and interbreed. So these two groups probably form an **'open' population**, that is, a population in which there is considerable movement between members of two or more groups.

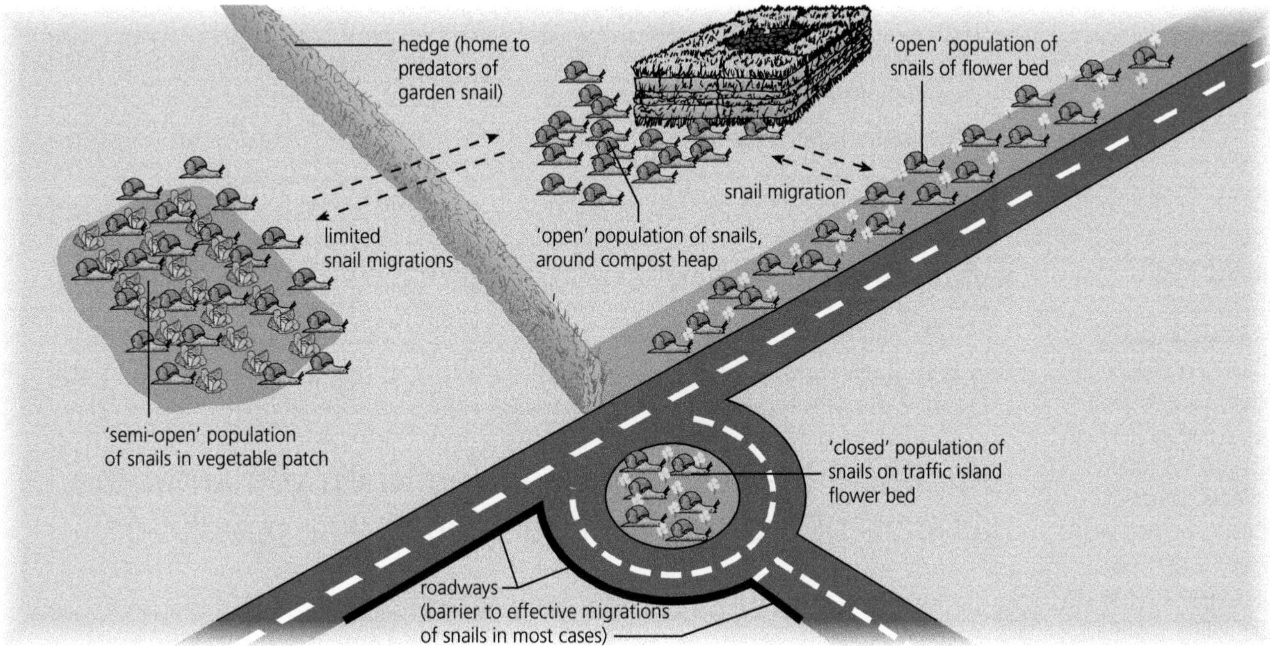

Figure 9.1 Exploring the concept of 'population'

Populations and gene pools

When considering genes within a population, we use the term **gene pool**. As you will see, we normally only consider a single gene at a time. In doing so, we use the term gene pool to refer to the sum of all the alleles of one gene present within a population at a given time. We could, however, use the term to cover the sum of all the alleles of all the genes possessed by members of a particular population at a particular time. Given the complexity of genomes, we would need to use computer software to handle all the data involved.

When breeding occurs between members of a population, a sample of the alleles of each gene in the gene pool will form the genomes (gene sets of individuals) of the next generation – and so on, from generation to generation. In this way, two different populations of the same species might contain different samples of the alleles of the genes forming the genome of this species. Figure 9.2 shows how the gene pools of two populations and the gene pool of the species relate to each other.

Key term

Gene pool The total of all the alleles of all the genes present in a particular population at a given time. Here, we will use the term to refer only to the alleles of a single gene.

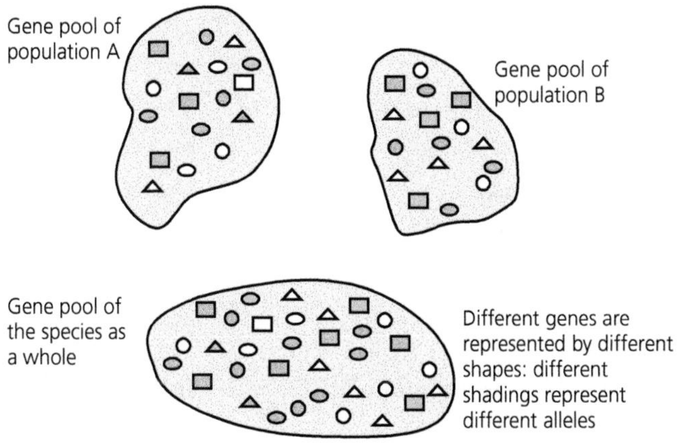

Gene pool of population A

Gene pool of population B

Gene pool of the species as a whole

Different genes are represented by different shapes: different shadings represent different alleles

Figure 9.2 The gene pools of two populations of a single species and the gene pool of the entire species

Test yourself

1 Explain the difference between a biological community and a population.
2 Explain what is meant by 'gene flow'.
3 Will gene flow occur between members of open populations or members of closed populations?
4 Define the term 'gene pool'.
5 Are the gene pools of two populations of the same species identical?

Key term

Allele frequency The number of copies of an allele of a single gene within a population divided by the total number of copies of that gene. Since diploid organisms carry two copies of each gene, the total number of copies in a population is two times the number of organisms.

Allele frequencies within a gene pool

Consider one gene with two alleles. The frequency with which one of these alleles occurs is the proportion it represents of all the alleles of that gene in a given population. It is called the **allele frequency** and is always expressed as a decimal value less than 1.

Estimating the allele frequencies when you know the numbers of the different genotypes in a population

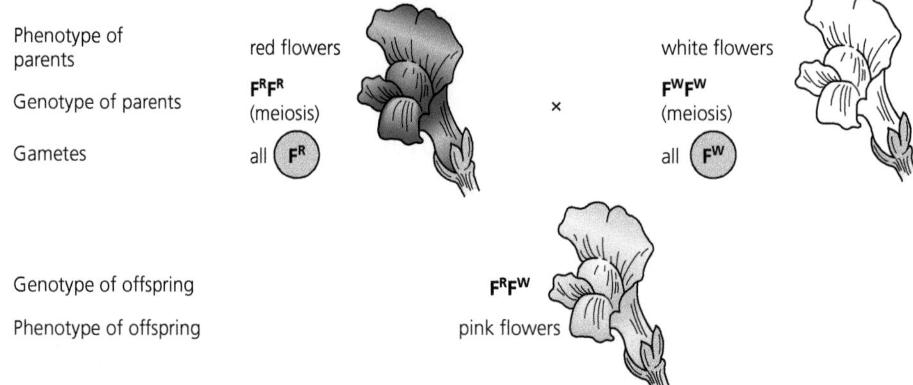

Figure 9.3 Flower colour in snapdragons (*Antirrhinum* sp.) is controlled by two codominant alleles, enabling us to tell all three flower-colour genotypes by observing the phenotypes

Figure 9.3 shows the inheritance of flower colour in snapdragons (*Antirrhinum*). The gene for flower colour has two codominant alleles, $\mathbf{F^R}$ and $\mathbf{F^W}$. Since they are codominant alleles, both show their effect in the phenotype of the heterozygote. As you can see in Figure 9.3, heterozygotes, $\mathbf{F^RF^W}$, have pink flowers. In other words, you can tell the flower-colour genotypes of snapdragon plants just by looking at the colour of their flowers.

Let's suppose that in a particular population of snapdragons:

- there are 1000 diploid plants
- 490 of these individuals are homozygous $\mathbf{F^RF^R}$
- 420 of these individuals are heterozygous $\mathbf{F^RF^W}$
- 90 of these individuals are homozygous $\mathbf{F^WF^W}$.

How can you find the frequencies of allele $\mathbf{F^R}$ and allele $\mathbf{F^W}$?

You start by realising that there are 1000 plants in the population, each with two alleles of this gene, making a total of 2000 alleles.

You then find the frequency of allele $\mathbf{F^R}$:

- 490 plants in the population carry two $\mathbf{F^R}$ alleles, making 980 $\mathbf{F^R}$ alleles
- 420 plants in the population carry one $\mathbf{F^R}$ allele, making a further 420 $\mathbf{F^R}$ alleles
- this gives 1400 $\mathbf{F^R}$ alleles out of the total of 2000 alleles of this gene
- so the frequency of the $\mathbf{F^R}$ allele $= \frac{1400}{2000} = 0.7$

You now find the frequency of allele $\mathbf{F^W}$:

- 90 plants in the population carry two $\mathbf{F^W}$ alleles, making 180 $\mathbf{F^W}$ alleles
- 420 plants in the population carry one $\mathbf{F^W}$ allele, making a further 420 $\mathbf{F^W}$ alleles
- this gives 600 $\mathbf{F^W}$ alleles out of a total of 2000 alleles of this gene
- so the frequency of the $\mathbf{F^W}$ allele $= \frac{600}{2000} = 0.3$

Hopefully you realised that there was an easier way to calculate the frequency of allele $\mathbf{F^W}$. Common sense should tell you that the frequencies of the two alleles added together must equal 1.0. So, once you had calculated the frequency of allele $\mathbf{F^R}$ as 0.7, the frequency of the only other allele available, $\mathbf{F^W}$, must be $1.0 - 0.7 = 0.3$

Estimating the allele frequencies when you do not know the numbers of the different genotypes in a particular population

The above example works fine when you can distinguish between the heterozygotes and the homozygotes. But what happens when you can't, that is, when one of the alleles is dominant. To solve this, you need to use an equation independently discovered by two people and so named after them both – the **Hardy–Weinberg equation**.

To use this equation, you:

- give the dominant allele a frequency of p
- give the recessive allele a frequency of q
- since there are no other alleles, $p + q = 1$
- the frequency of the homozygous dominant individuals is p^2
- the frequency of the heterozygous individuals is $2pq$
- the frequency of the homozygous recessive individuals is q^2
- since this accounts for all the individuals, $\mathbf{p^2 + 2pq + q^2 = 1}$

Key term

Hardy–Weinberg equation
$p^2 + 2pq + q^2 = 1$, where p is the frequency of the dominant allele (**A**) and q is the frequency of the recessive allele (**a**) of a single gene.

Tip

You will not be asked to prove the Hardy–Weinberg equation, only to use it.

Example

Using the Hardy–Weinberg equation

Let's use the same data as our previous calculation, but this time using dominant and recessive alleles of a gene, **A** and **a**.

As in our previous example with snapdragon plants, let's suppose that in a particular population:

- there are 1000 diploid individuals
- 910 of these individuals show the dominant characteristic (in other words are **AA** or **Aa**)
- 90 of these individuals show the recessive characteristic (in other words are **aa**).

As always when finding an unknown, we start with what we do know.

1 What do we know about the genotypes of any of the individuals in this population?

2 What is the frequency of the **aa** individuals in this population?

3 How can we use the frequency of **aa** individuals to find the frequency of the '**a**' allele in this population?

4 How can we use the frequency of the '**a**' allele to find the frequency of the '**A**' allele in this population?

5 What is the frequency of individuals in this population with the genotype **AA**, **Aa** and **aa**?

6 How many individuals in this population had the genotypes **AA**, **Aa** and **aa**?

Answers

1 In this case, we cannot distinguish between the **AA** and **Aa** individuals – they both show the dominant characteristic. We do know, however, that the 90 individuals showing the recessive characteristics must have the genotype **aa**.

2 Since we know that 90 of the 1000 individuals in this population had this genotype, we can calculate their frequency in the population as: $\frac{90}{1000} = 0.09$

3 In the Hardy–Weinberg equation, the frequency of individuals in a population with the genotype **aa** is represented q^2.

So, we find the value of q, the frequency of the **a** allele, as $\sqrt{q^2} = \sqrt{0.09} = 0.3$

4 Since there are only two alleles to consider, the sum of their frequencies must equal 1.

So, if $p + q = 1$, then $p = 1 - q$, which in our case means:

$p = 1 - 0.3 = 0.7$

Since we used the same numbers, you will not be surprised to find that the frequencies of the two alleles in this example are exactly the same as in our previous example in which we could distinguish all three phenotypes.

5 To solve this, we substitute our values of p and q into the Hardy–Weinberg equation:

AA **Aa** **aa**

$p^2 + 2pq + q^2 = 1$

So the frequency of individuals with:

- the **AA** genotype is $0.7^2 = 0.49$
- the **Aa** genotype is $2 \times 0.7 \times 0.3 = 0.42$
- the **aa** genotype is $0.3^2 = 0.09$

6 Since there were 1000 individuals in the population, the number of individuals can be found by multiplying their frequency given above by 1000. So, the number of individuals with the genotype:

- **AA** is $0.49 \times 1000 = 490$
- **Aa** is $0.42 \times 1000 = 420$
- **aa** is $0.09 \times 1000 = 90$

The fact that these values, $490 + 420 + 90 = 1000$, the number of individuals in the population, should reassure us that our calculation is correct.

Tip

When using the Hardy–Weinberg equation to solve problems, always start with the data about homozygous recessive individuals. Their frequency is q^2, enabling you to calculate the value of q ($\sqrt{q^2}$) and of p $(1 - q)$.

Tip

It is always a good idea to check your calculations as you go along. In this case you should expect the sum of the three frequencies to equal 1. Our calculated frequencies are $0.49 + 0.42 + 0.09 = 1.0$, so our calculation is correct.

Test yourself

6 Look back to Figure 9.3 on page 160. If the offspring had been allowed to interbreed, what ratio of phenotypes would you expect in the offspring (2) generation?

7 Continuing the answer to Question 6, which statistical test would you use to see whether any difference between the actual numbers of phenotypes in the offspring (2) generation and the numbers you predicted, was statistically significant? Explain your answer.

8 Explain what is meant by 'statistically significant'.

9 The rhesus blood group in humans is controlled by two alleles of a single gene. The allele for rhesus positive is dominant and the allele for rhesus negative is recessive. In Europe, 16 per cent of the population is rhesus negative. What percentage of the European population is heterozygous for this gene? Explain your answer.

10 What assumption did you make in your calculation in Question 9?

Tip

Test yourself questions 6 to 8 test your understanding of earlier sub-topics.

The Hardy–Weinberg principle (also called the Hardy–Weinberg equilibrium)

G.H. Hardy and Wilhelm Weinberg were independently analysing the behaviour of populations. They both used mathematical models to show that, if certain assumptions are made, the frequency of the alleles of a gene will remain constant from generation to generation in any population.

The seven assumptions underlying the **Hardy–Weinberg principle** are as follows:

- The organisms are diploid.
- The organisms reproduce only by sexual reproduction.
- The generations are discrete, that is, do not overlap.
- Mating is random.
- The population size is infinitely large.
- The allele frequencies are equal in both sexes.
- There is no migration, mutation or selection.

The Hardy–Weinberg principle is of interest because we know that sometimes, in some populations, the composition of the gene pool does change. This can be due to a range of factors, known as **disturbing factors** because they operate to alter the allele frequencies of some genes. Disturbing factors include:

- gene mutation
- migration
- genetic drift
- natural selection.

Key term

Hardy–Weinberg principle Provided several assumptions hold true, the allele frequencies of a gene within a single population will not change from generation to generation.

Tip

Do not confuse the Hardy-Weinberg principle with the Hardy-Weinberg equation. The former proposes an equilibrium within a population of the allele frequencies of a single gene. The latter provides a means to estimate the frequencies of alleles and genotypes in a population.

Gene mutation

Gene mutations are random, rare, spontaneous changes in the base sequence of genes. If they occur during gamete formation in the gonads, they can lead to the possibility of new characteristics in the offspring – for example, the ability to inactivate a pesticide molecule. You examined gene mutations at some length in Chapter 8.

Migration

If you look back to Figure 9.1 on page 159, you can see how two populations could have different frequencies of the alleles of a particular gene. If some individuals leave one population (**emigration**), their alleles will be lost from that population. As a result, the allele frequency of that population will change. Similarly, if those individuals join the second population (**immigration**), their arrival will change the allele frequencies of that population also.

Genetic drift

> **Key term**
>
> **Genetic drift** Random changes in the allele frequency of a particular gene resulting from 'sampling error' during sexual reproduction in a small population.

In much the same way that you are more likely to get a 50 : 50 split in 'heads' or 'tails' if you toss a coin 1000 times than if you toss a coin ten times, the allele frequencies are more likely to be representative at each generation in a very large population than in a very small one. If a population is very small, chance events will play a big part in determining which organisms survive to reproduce successfully. The random changes in the allele frequencies that result from small population size are known as **genetic drift**. Small population size, leading to genetic drift can occur for a number of reasons.

Genetic bottleneck

A sudden hostile physical condition (for example, flooding or drought) could sharply reduce a natural population to a very small number of survivors. On the return of a favourable environment, numbers of the affected population might quickly return to normal (as a result of reduced competition for food sources, for example). The new population, however, would be built from a very small sample of the original population, so would contain a small sample of the original gene pool, possibly with some alleles lost altogether (Figure 9.4).

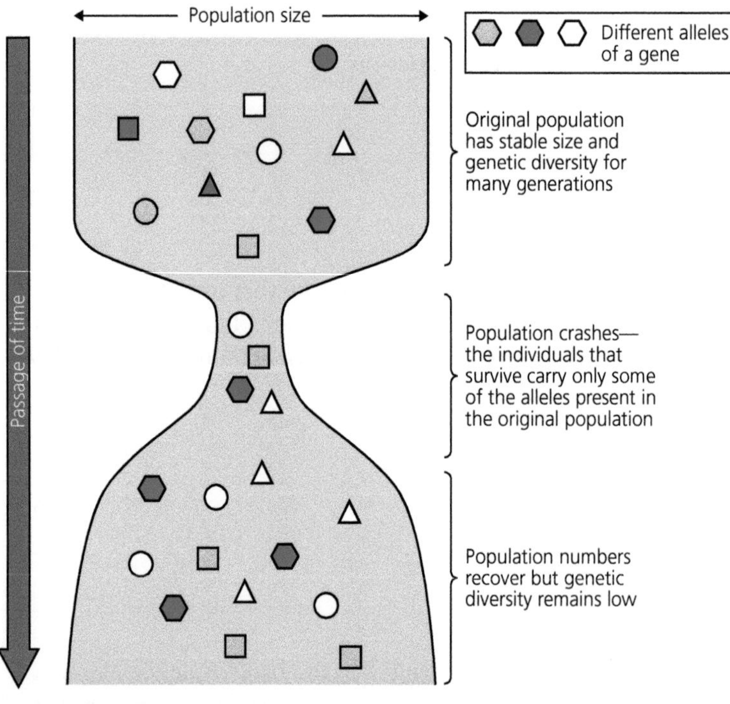

Figure 9.4 A genetic bottleneck results from a sudden reduction in the size of a population and causes a reduction in the genetic diversity of that population

Analysis of the human genome suggests that our own species suffered a genetic bottleneck at some stage in our evolution, as did that of the cheetah (Figure 9.5).

Figure 9.5 There is genetic evidence that the cheetah population went through a genetic bottleneck about 10 000 years ago

Founder effect

A small number of organisms might become isolated in a new environment. As with a genetic bottleneck, the genotypes of these organisms are likely to form a small sample from the original gene pool. When these individuals reproduce to form a larger population, the allele frequencies will remain a small sample of the gene pool of the species. An extreme example of this might occur when a pregnant female mouse is carried on driftwood from the mainland to an unpopulated offshore island (Figure 9.6).

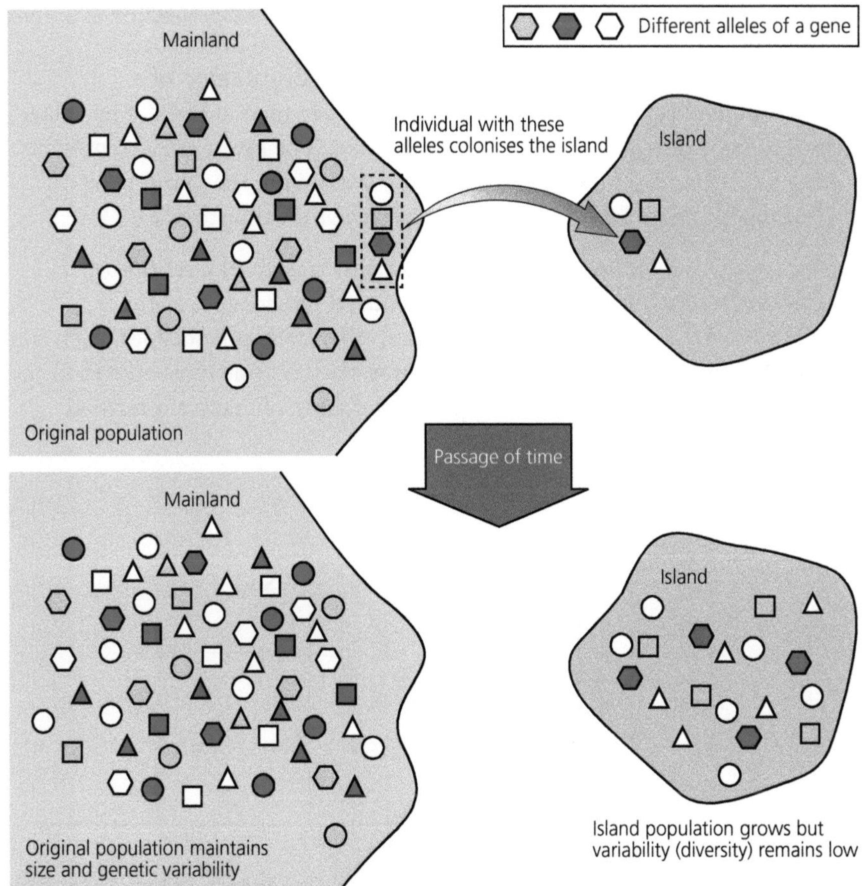

Figure 9.6 The founder effect is shown when a single individual (pregnant or self-fertilising), or small group of individuals, from a mainland colonise an uninhabited offshore island

We can find examples of the founder effect in human populations. On several occasions throughout history, small groups of Europeans emigrated to the USA to escape religious persecution in their own country. On arrival, these people formed self-contained settlements that last to this day. The allele frequencies of many genes in the communities of the Amish and of the Dunkers, for example, are quite different from those of the general population of the USA.

Figure 9.7 The song thrush preys on five-banded snails. It eats more of those that it finds easily, causing pink-shelled snails to be less common in grassland and yellow-shelled snails to be less common in beech woodland

Natural selection

Natural selection acting on a phenotypic feature of members of a population will lead to a change in the allele frequencies of the gene controlling that phenotypic feature. Chapter 8 of *Edexcel A level Biology 1* covered natural selection in some depth. Figure 9.7 provides a reminder of a well-known example – the selective predation by song thrushes (*Turdus philomelos*) on populations of the five-banded snail (*Cepaea nemoralis*) with a particular shell coloration that makes them visible in (say) a woodland habitat, but effectively camouflaged in a grassland habitat.

Natural selection can change the allele frequency in a number of different ways. In Chapter 8 of *Edexcel A level Biology 1*, you learnt about stabilising selection and directional selection. Figure 9.8 serves to remind you of **stabilising selection**. The upper graph shows the distribution of a continuous phenotypic feature, such as human body mass at birth. The values follow a normal distribution with a central mode. The lower curve represents the effect of stabilising selection. As you can see from the lower graph, the extreme birth masses have been eliminated; natural selection has favoured the modal birth mass.

Key term

Polymorphism The existence of two or more body forms (morphs) within a single population.

Figure 9.9 represents a type of selection you did not cover in *Edexcel A level Biology 1* – **disruptive selection**. Here, natural selection has favoured phenotypes at the extremes of the range of values. As a result, the population has two phenotypes, each with a mode that is different from that of the original population. The existence in one population of two or more different phenotypes of the same character is termed **polymorphism**.

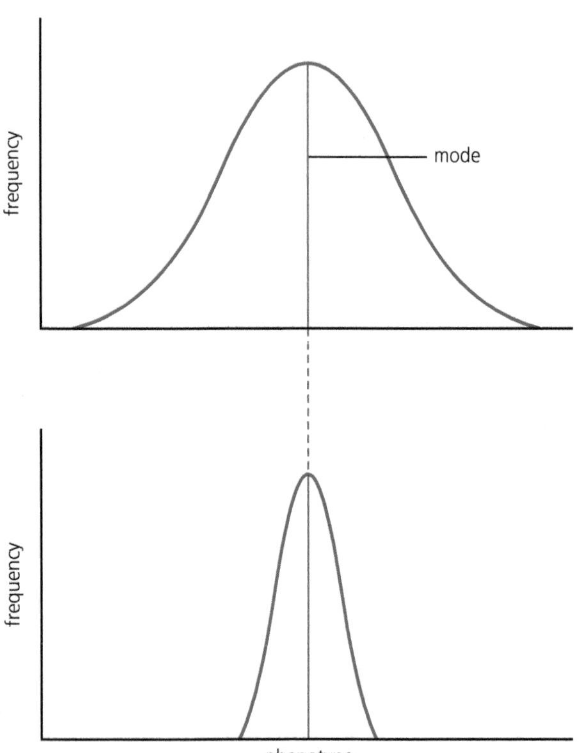

Figure 9.8 Stabilising selection reduces the variation around the mode

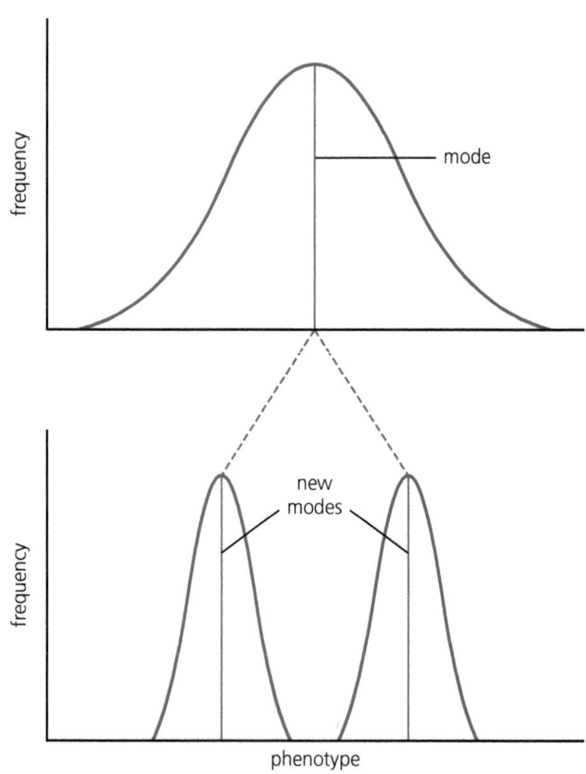

Figure 9.9 Disruptive selection results in two phenotypes (polymorphism) within a population

Test yourself

11 Distinguish between the Hardy–Weinberg principle and the Hardy–Weinberg equation.

12 When asked to calculate allele frequencies in a population, under what circumstances would you **not** use the Hardy–Weinberg equation? Explain your answer.

13 List **four** disturbing factors of the Hardy–Weinberg principle.

14 Analysis of the human genome suggests that our own species suffered a genetic bottleneck at some stage in our evolution. Suggest what this evidence might be.

15 Analysis of several introns in the gene controlling haemophilia showed very little difference between members of the Slavic populations from the European part of Russia and the native ethnic groups of Uzbekistan and Kazakhstan. What can you conclude from this result?

Disruptive selection and polymorphism

The different shell colours of *Cepaea nemoralis* (the five-banded snail) provide an example of a polymorphism. The shell colour that is at a selective advantage is different in different environments.

- In parts of England, where song thrushes prey on these snails, the yellow shell is better camouflaged in grassland than the pink shell but the pink shell is better camouflaged among the beech litter on the floor of a beech woodland than is the yellow shell. Like us, thrushes more readily find the conspicuous shells, so eat more of them. Snails with a yellow shell survive better than snails with a pink shell in grassland but the converse is true in beech woodlands.
- In parts of the Pyrenees, yellow shells reflect more heat than pink shells, so snails with yellow shells are at an advantage in regions with bright sunlight.

In these populations, the polymorphism is stable for many generations, so is called a **balanced polymorphism**.

A **transient polymorphism** is one in which the selective advantage changes over time. An example of this type of polymorphism is industrial melanism. During the industrial revolution in Britain, areas of heavy industry became highly polluted by acidic gases, such as SO_2, which killed the algae and lichens that grew on the surface of trees. The polluting soot also blackened these surfaces.

Like all moths, the peppered moth (*Biston betularia*) rests on surfaces with its wings unfolded. Against the blotchy background caused by lichens, the peppered form shown in Figure 9.10 on the next page had been well camouflaged. In heavily polluted areas, it no longer was and became more susceptible to predation by insectivorous birds. By chance, a gene mutation resulted in a melanic (dark) wing colouration. Moths with this wing colour were much less conspicuous on the blackened, lichen-free trees and so suffered less predation. The result, as you can see in Figure 9.10, was that the lighter, peppered form was more common in unpolluted areas and the melanic form was more common in industrial areas – another example of the effects of disruptive selection.

In the peppered moth (*Biston betularia*), environmental conditions have, at different times, favoured either the pale or the melanic forms.

In these circumstances, the effect of natural selection on the gene pool (in the form of selective predation of moths resting on exposed surfaces by insectivorous birds) has been '**disruptive**'.

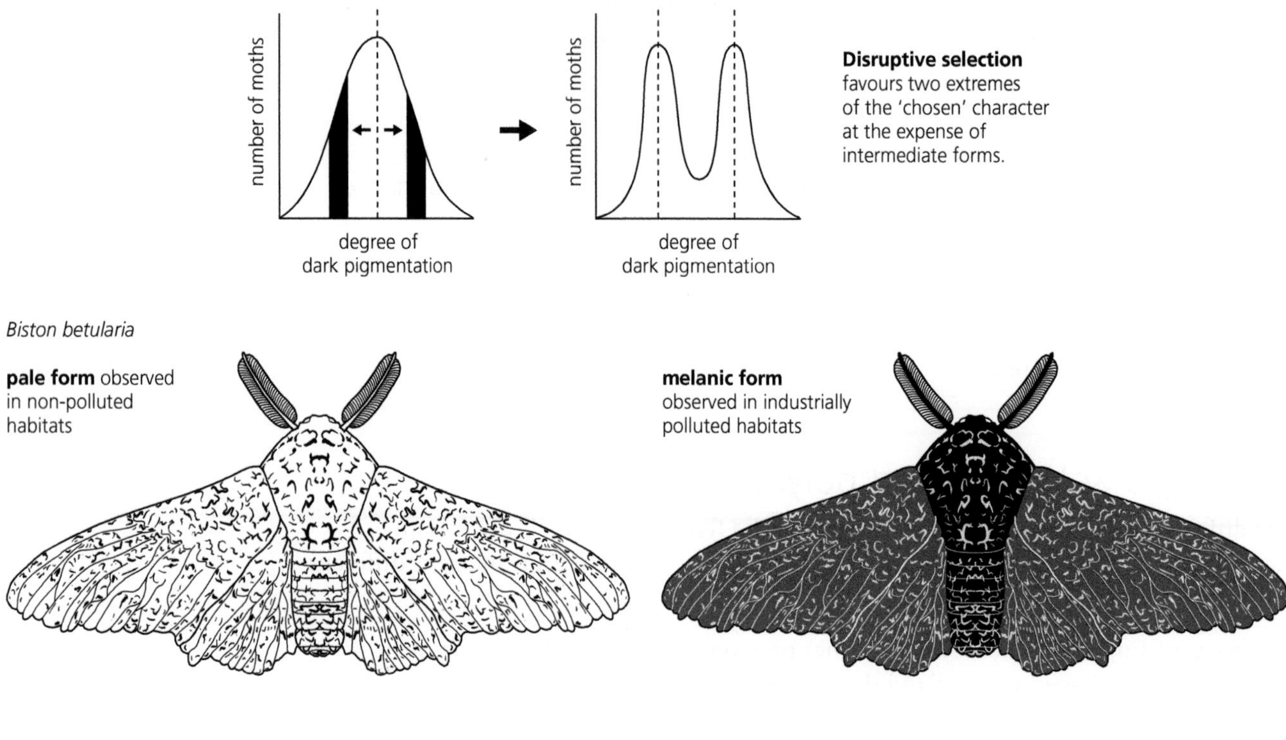

experimental evidence that establishes transient polymorphism

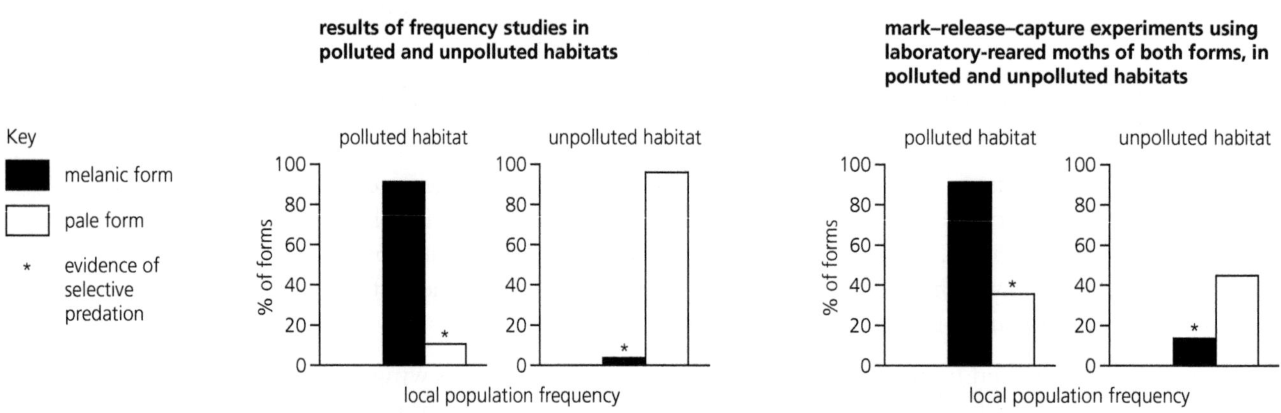

Figure 9.10 The peppered moth (*Biston betularia*) and industrial melanism - a case of a transient polymorphism

Clean Air Acts, passed by the UK government of the 1950s, led to a great reduction in air pollution. As a result, lichens began to grow again in the previously polluted industrial heartlands. As the lichens changed the background on which the peppered and melanic moths rested, insectivorous birds now found the melanic form more conspicuous and ate more of those. Natural selection now favoured the pale, peppered moths and the frequency of these moths increased at the expense of the melanic form.

Balance polymorphisms and speciation

You saw in Chapter 8 of *Edexcel A level Biology 1* how natural selection can lead to speciation. We also differentiated between two models of speciation.

- **Allopatric speciation** might occur following the separation of two populations of the same species by a physical or geographical barrier.
- **Sympatric speciation** occurs without the physical separation of two groups from the same population.

Population geneticists are interested in comparisons between populations of a single species occurring on a mainland and on each of a group of offshore islands. You might recall from Chapter 8 in *Edexcel A level Biology 1* that comparisons between animals on mainland Ecuador and on islands of the Galapagos archipelago were important in the development of Charles Darwin's appreciation of natural selection. Populations on offshore islands could form new species by allopatric speciation.

We can define the term 'species' as a group of organisms that can actually or potentially interbreed to produce fertile offspring. The 'potentially' part of that definition is quite important. We could have two populations of, say, *Cepaea nemoralis*, one in Spain and one in England. They could not naturally interbreed because the distance between them is too great for their migration capability. But if we were to bring them together into the laboratory, they would be able to interbreed and produce fertile offspring. The geographical separation has not led to genetic changes in these two populations that prevent their successful interbreeding; they still belong to the same species.

Balanced polymorphisms are of interest to population geneticists because they might provide opportunities for sympatric speciation to occur. The worked example below provides one example of a balanced polymorphism that might possibly lead to speciation.

Example

Copper tolerance in bent grass

The waste from old copper mines was usually dumped near the mine itself. It formed mounds of soil that have a high concentration of copper ions. Fields around the waste tips of copper mines in North Wales contained populations of the bent grass (*Agrostis tenuis*). But for many years, this plant did not grow on the waste tips, as you can see in Figure 9.11.

Figure 9.11 This hill was formed from waste from a nearby copper mine

1 The seeds from bent grass plants are blown large distances by the wind and some will have landed on the waste tip. Suggest why no bent grass plants grew there.

The waste tip remained plant-free for many years. Eventually, a few plants of bent grass were found growing on the waste tip. Plant biologists from a nearby university found that these plants had become copper-tolerant.

2 Suggest how these plants had become copper tolerant.

3 Did the presence of copper in the waste cause the gene mutation?

4 Although the population of bent grass on the waste tip comprised plants with copper tolerance, the populations in the surrounding areas did not evolve copper tolerance. Suggest why.

5 Explain why the plant biologists were excited that these populations might provide an opportunity for sympatric speciation.

Answers

1 Although copper is a trace element, that is, needed in tiny amounts for healthy plant growth, it is toxic in high concentrations. Consequently, any seeds germinating on the waste tip would be killed by the copper.

2 The copper tolerance resulted from a gene mutation.

3 The answer is no, the gene mutation was a random event. Once it had occurred, however, the mutant plant was able to grow on the copper-contaminated soil and the resulting population contained plants that were copper tolerant.

4 We must assume that a mutation that results in copper tolerance has occurred randomly over the thousands of years that this plant species has existed. In the absence of toxic levels of copper in the soil, this mutation would not confer an advantage and, quite probably, resulted in some selective disadvantage to its possessor. As a result, natural selection would act against the mutation and the allele of this gene would be lost from the population.

5 The copper-tolerant and non-tolerant populations co-exist but any gene flow between them is unlikely to result in successful adults. Plants without copper tolerance will not survive on the waste tips and copper-tolerant plants are likely to be at a selective disadvantage in the unpolluted soil. Effectively, there are closed populations within which changes in allele frequencies could accumulate that would prevent sexual reproduction between them producing fertile offspring.

Test yourself

16 Under what circumstances is genetic drift likely to occur?

17 Of what type of natural selection is the evolution of antibiotic resistance in bacteria an example?

18 Suggest why stabilising selection on human birth mass is of less significance today than in 1950.

19 Following the Clean Air Acts in the UK, the melanic populations of *Biston betularia* became dominated by the lighter, peppered form again. What does this show about the allele of the colour gene causing the melanic body form?

20 What would an unchanging gene pool suggest about a population?

Exam practice questions

1 Warfarin has long been used as a rat poison. The evolution of resistance to warfarin in rat populations is an example of:

A artificial selection **C** disruptive selection

B directional selection **D** stabilising selection *(1)*

2 A gene has two alleles. If the frequency of the homozygous recessive individuals in a population is 0.5, the frequency of the dominant allele:

A is less than that of the recessive allele

B is the same as that of the recessive allele

C is greater than that of the recessive allele

D cannot be estimated *(1)*

3 Humans are genetically less diverse than gorillas as a result of:

A the founder effect **C** stabilising selection

B a genetic bottleneck **D** transient polymorphism *(1)*

4 A population of *Drosophila melanogaster* contains 640 long-winged flies and 360 vestigial-winged flies. What proportion of the long-winged flies are homozygous for this gene? Show your working. *(4)*

5 Rice is grown in paddy fields throughout the world. In many paddy fields, weeds are removed by hand when the rice plants are young. A species of barnyard grass, called *Echinochloa oryzoides*, is a weed that grows in rice paddies.

The structure of the leaves, colour of the leaf bases and time of flowering of *E. oryzoides* is unlike all other species of *Echinochloa*. Instead, *E. oryzoides* is a rice mimic, that is, its appearance and flowering time closely resemble those of rice plants.

a) What is the advantage to *E. oryzoides* of being a rice mimic? *(3)*

b) Suggest how this rice mimicry might have evolved. *(5)*

6 *Mycosphaerella graminicola* is a pathogenic filamentous fungus that causes leaf blotch on wheat. A group of scientists used restriction fragment length polymorphism (RFLP) markers to investigate the DNA of this pathogen from different countries around the world. The table shows their results.

Tip

Question 6 requires you to bring together knowledge and understanding from several chapters and to think about how scientists would carry out an investigation.

Locus	Genetic diversity in samples of fungus from each country						
	Australia	Canada	Denmark	Israel	UK	USA	Uruguay
SS192A	0.77	0.00	0.22	0.02	0.16	0.15	0.22
SS192B	0.00	0.00	0.22	0.48	0.00	0.05	0.00
SS14	0.26	0.53	–	0.36	0.21	0.29	0.22
SS2	0.00	–	0.39	0.50	0.48	0.53	0.50
SL10	0.23	0.61	0.48	0.50	0.56	0.49	0.55
SL53	0.00	0.36	0.47	0.64	0.47	0.70	0.62
SS 43	0.00	0.72	0.65	0.77	0.67	0.49	0.58
SL31	0.24	0.39	0.55	0.74	0.68	0.48	0.79
Mean	0.19	0.37	0.43	0.50	0.40	0.40	0.44

a) Explain the term 'pathogenic'. *(1)*

b) Outline the method by which the scientists would have obtained the data about genetic diversity in the restriction fragments of DNA. *(3)*

c) What causes a restriction fragment length polymorphism (RFLP) at each locus? *(2)*

d) What does a value of 0.00 in the table represent? *(1)*

e) Explain how the data for Australia provide evidence of the founder effect. *(2)*

f) Suggest the origin of the fungus. Use evidence in the table to justify your answer. *(1)*

Stretch and challenge

7 Modelling is an important technique in population genetics. You can model the effects of genetic drift using the diagram. Each of the six individuals in the diagram produces two identical offspring. Each generation, half of these offspring die; the population remains the same size.

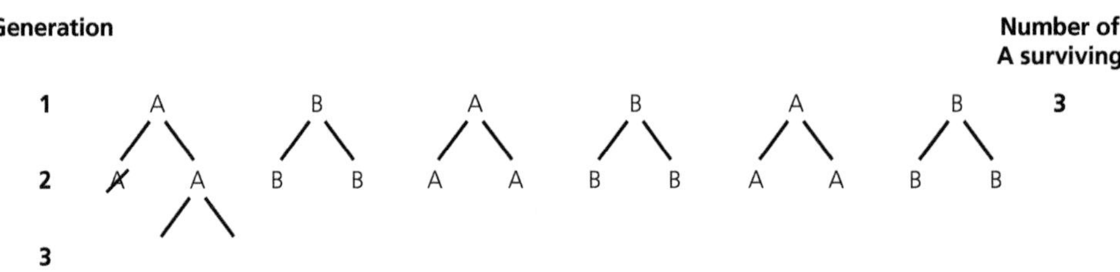

a) Continue the diagram for several generations, deciding which six individuals 'die' by tossing a coin.

b) Plot a graph of the number of surviving **A** offspring against the number of generations.

c) How many generations did it take for either the **A** individuals or the **B** individuals to be lost from the population?

d) Repeat this exercise. Was your answer to question (c) the same? Explain why.

e) Use the internet to find a computer program that will allow you to model genetic drift and use it to explore the effects of different population sizes and starting allele frequencies.

8 How can polyploidy contribute to sympatric speciation in plants?

10 Nervous systems

Prior knowledge

In this chapter you will need to recall that:

- → the nervous system of a mammal has two main parts - the central nervous system (CNS) and the peripheral nervous system
- → nerve cells are known as neurones and have elongated cell bodies called axons surrounded by insulating myelin sheaths
- → there are structural differences between motor and sensory neurones
- → motor neurones carry impulses away from the CNS and sensory neurones carry impulses into the CNS
- → a nerve impulse is called an action potential, which is a wave of depolarisation caused by ion movements in and out of neurones
- → neurones are connected to other neurones and effector organs by synapses
- → there is a microscopic gap between neurone membranes at the synapse
- → transport across membranes can be by diffusion, facilitated diffusion and active transport (see Chapter 9, *Edexcel A level Biology 1*)
- → the simplest connections of the nervous system are reflex arcs
- → the brain has distinct parts with individual functions.

Test yourself on prior knowledge

1. Explain why it is important that neurones have insulating myelin sheaths.
2. What type of compound is myelin?
3. Name the **two** ions that are involved in the formation of action potentials.
4. Describe how action potentials cross synapses.
5. Active transport is needed in the conduction of nerve impulses. Explain why.
6. Name **one** human reflex arc.
7. Name the part of the brain that has areas linked with sensory and motor functions of localised parts of the body.

Introduction

As primitive organisms evolved into larger multicellular forms they developed specialised tissues and organs. In this chapter and in Chapters 11 and 12 you will look at how these specialised units are controlled and coordinated to become an efficient living thing. Control generally refers to how instructions are communicated to individual parts; coordination refers to how all the different parts are made to work together efficiently.

A simple example might be attempting to walk. Obviously instructions must be passed to individual muscles to make them contract and move bones using joints. However, this isn't the whole story. If you simply lift your leg to move it forward you will fall over because you have removed one of your supports. Before you lift your leg you must sway your body so that its centre of gravity is over one leg so that you can lift the other off the ground. This needs careful coordination using many other muscles and sense organs. Too much sway and you will fall one way, too little and you will fall the other. Watching a baby learning to walk by taking its first steps will illustrate all these problems very clearly.

The ability to detect changes and respond appropriately is a life-preserving feature of living things, literally. This characteristic, known as sensitivity, is just as much a property of single cells as it is of whole mammals and flowering plants.

Changes that bring about responses are called stimuli. The stimulus is detected by a receptor, and an effector brings about a response. Since the receptor and effector are often in different places in a multicellular organism, mechanisms of internal communication are essential. In animals, internal communication involves both the nervous system and endocrine system, which you will look at in Chapter 11. You will start by examining the nervous system, focusing in particular on the human.

The gross structure of the mammalian nervous system

It is most likely that you are already familiar with the gross structure of your nervous system, consisting as it does of the central nervous system (CNS – brain and spinal cord) and all the peripheral nerves (Figure 10.1).

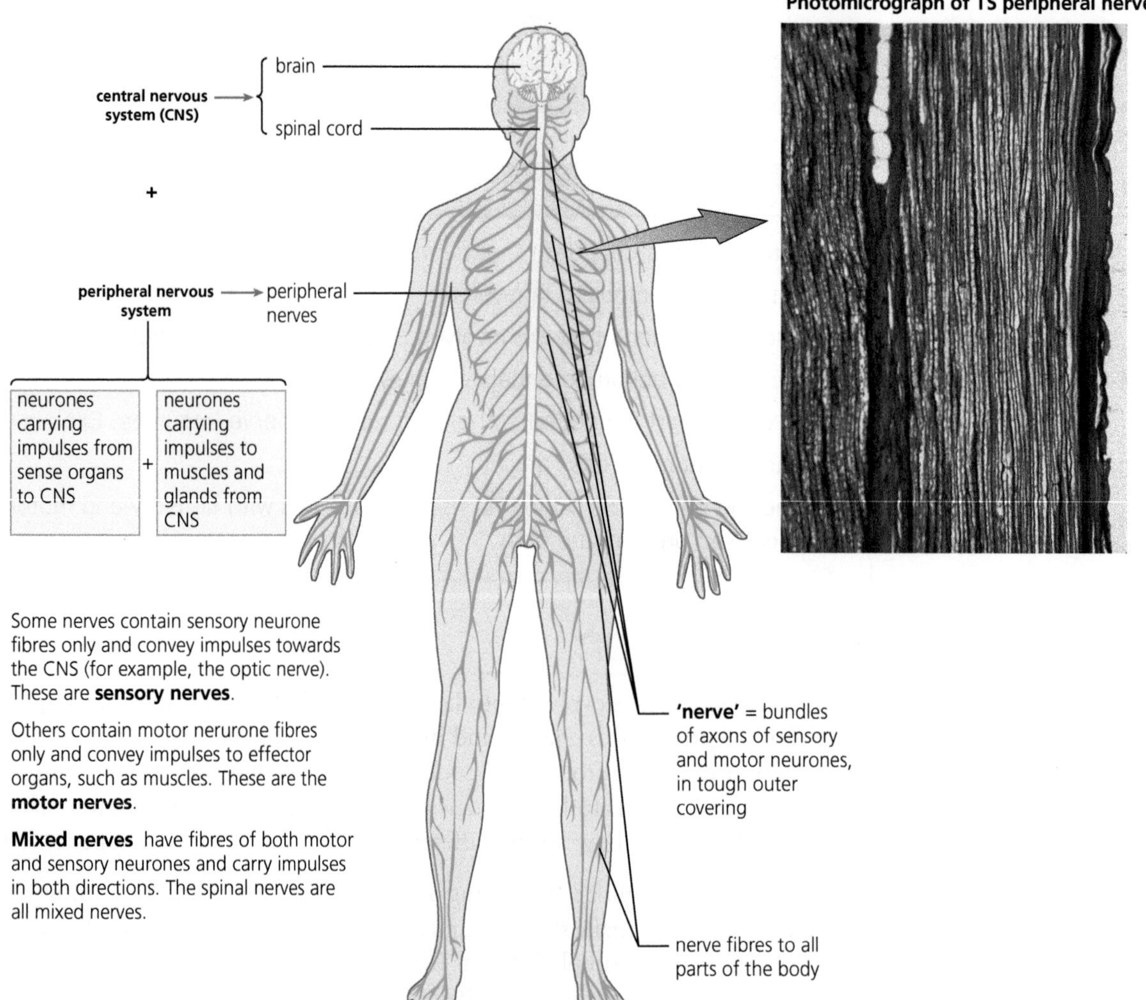

Figure 10.1 The organisation of the mammalian nervous system

The role of the brain in coordination and control of the body's responses is summarised in Figure 10.2.

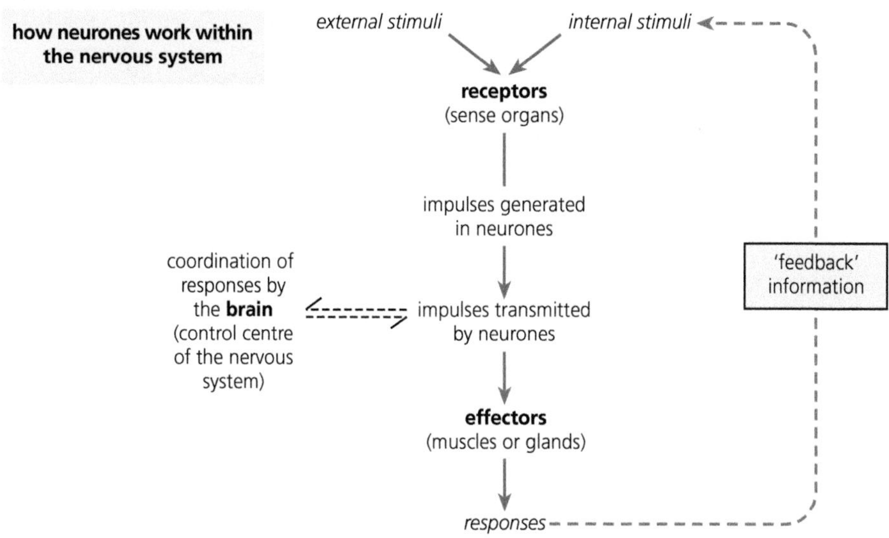

Figure 10.2 Coordination and control by the nervous system

The central nervous system

The spinal cord

The spinal cord is a cylindrical structure with a tiny central canal. The canal contains cerebrospinal fluid and is continuous with the fluid-filled spaces in the centre of the brain. The cord consists of an inner area of **grey matter** (cell bodies and synapses) surrounded by **white matter** (myelinated nerve fibres). You can see a transverse section through the spinal cord in Figure 10.3. The spinal cord is surrounded and protected by the vertebrae of the backbone. In the junction between each pair of vertebrae two spinal nerves leave the cord, one to each side of the body. The role of the spinal cord is to relay action potentials between **receptor organs** and **effector organs** of the body (by reflex action), and between them and the brain. Impulses entering and leaving the spinal cord through the spinal nerves are normally transferred to the brain and back through the neurones it contains.

Key terms

Grey matter Nervous tissue that is made up largely of the cell bodies of neurones.

White matter Nervous tissue made up largely of axons and their fatty myelin sheaths.

Receptor organs Organs that can detect changes caused by a stimulus and initiate action potentials.

Effector organs Organs that can respond to a stimulus to bring about a particular action.

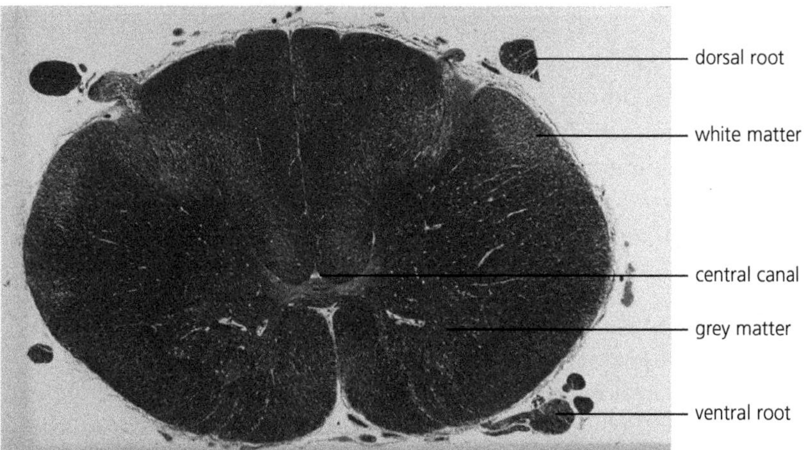

Figure 10.3 Photomicrograph of a transverse section through a spinal cord

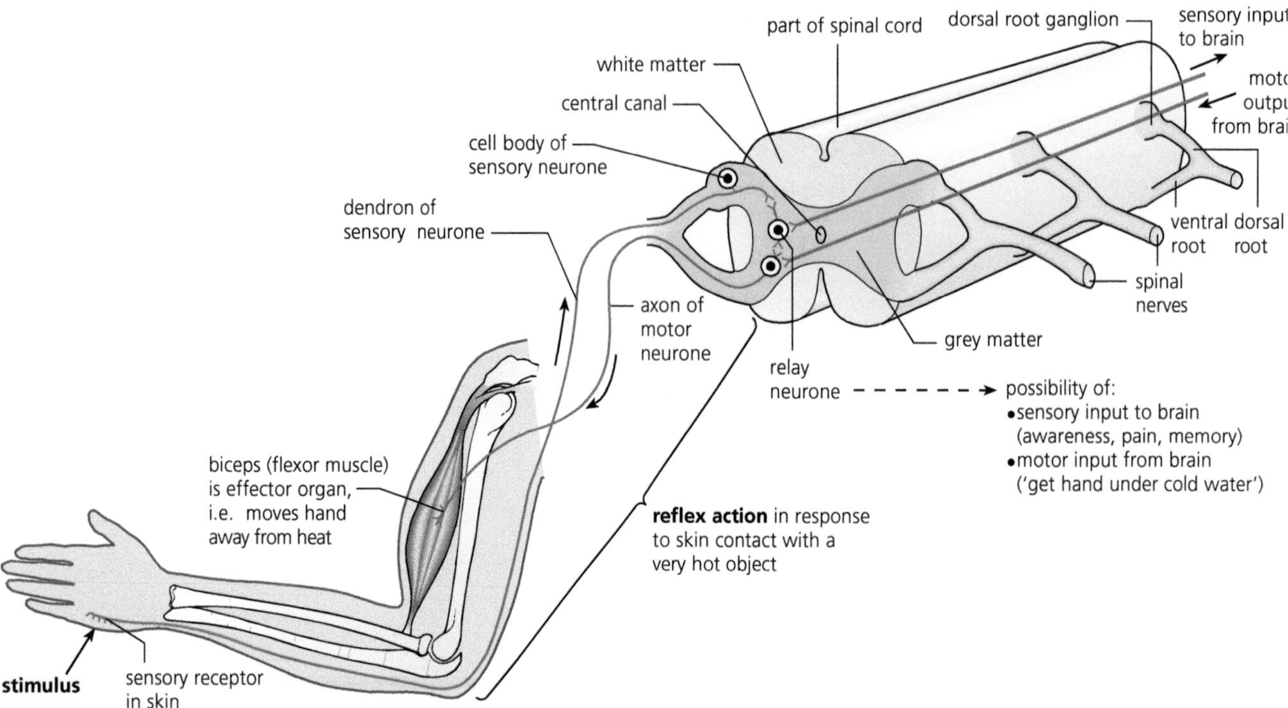

Figure 10.4 The spinal cord with the neurones of a reflex arc, and the associated inputs to, and output from, the brain

> **Key terms**
>
> **Reflex action** A rapid nervous reaction in which there is a direct connection between the sense organ and effector organ, bringing about a similar response to a similar stimulus.
>
> **Interneurones** Short neurones that connect individual neurones to each other.

Very simple **reflex actions** are few in humans but the 'hot hand' reflex shown in Figure 10.4 illustrates how the impulses can be transferred rapidly across the spinal cord. In this reflex action impulses are also sent to the brain, which may override the response, causing the hot object to be tolerated in a dangerous situation or as a test of 'will-power'.

The brain

The brain, a highly organised mass of **interneurones** (see Figure 10.8 on page 182) connected with the rest of the nervous system by numerous motor and sensory neurones, is responsible for complex patterns of behaviour, in addition to many reflex actions. Much activity is initiated by the brain, rather than being mere responses to external stimuli. In summary, the human brain controls all body functions apart from those under the control of simple spinal reflexes. This is achieved by:

- receiving impulses from sensory receptors
- integrating and correlating incoming information in association centres
- sending impulses to effector organs (muscles and glands) causing bodily responses
- storing information and building up an accessible memory bank
- initiating impulses from its own self-contained activities (the brain is also the seat of 'personality' and emotions, and enables you to imagine, create, plan, calculate, predict and reason abstractly).

The vertebrate brain develops in the embryo from the anterior end of a simple tube, the neural tube. This tube enlarges to form three primary structures, known as the forebrain, midbrain and hindbrain (Figure 10.5). The various parts of the mature brain develop from these by selective thickening and folding processes of their walls and roof.

These enlargement processes are most pronounced in mammals, and a striking feature of this group is the enormous development of the cerebral hemispheres, which are an outgrowth of the forebrain.

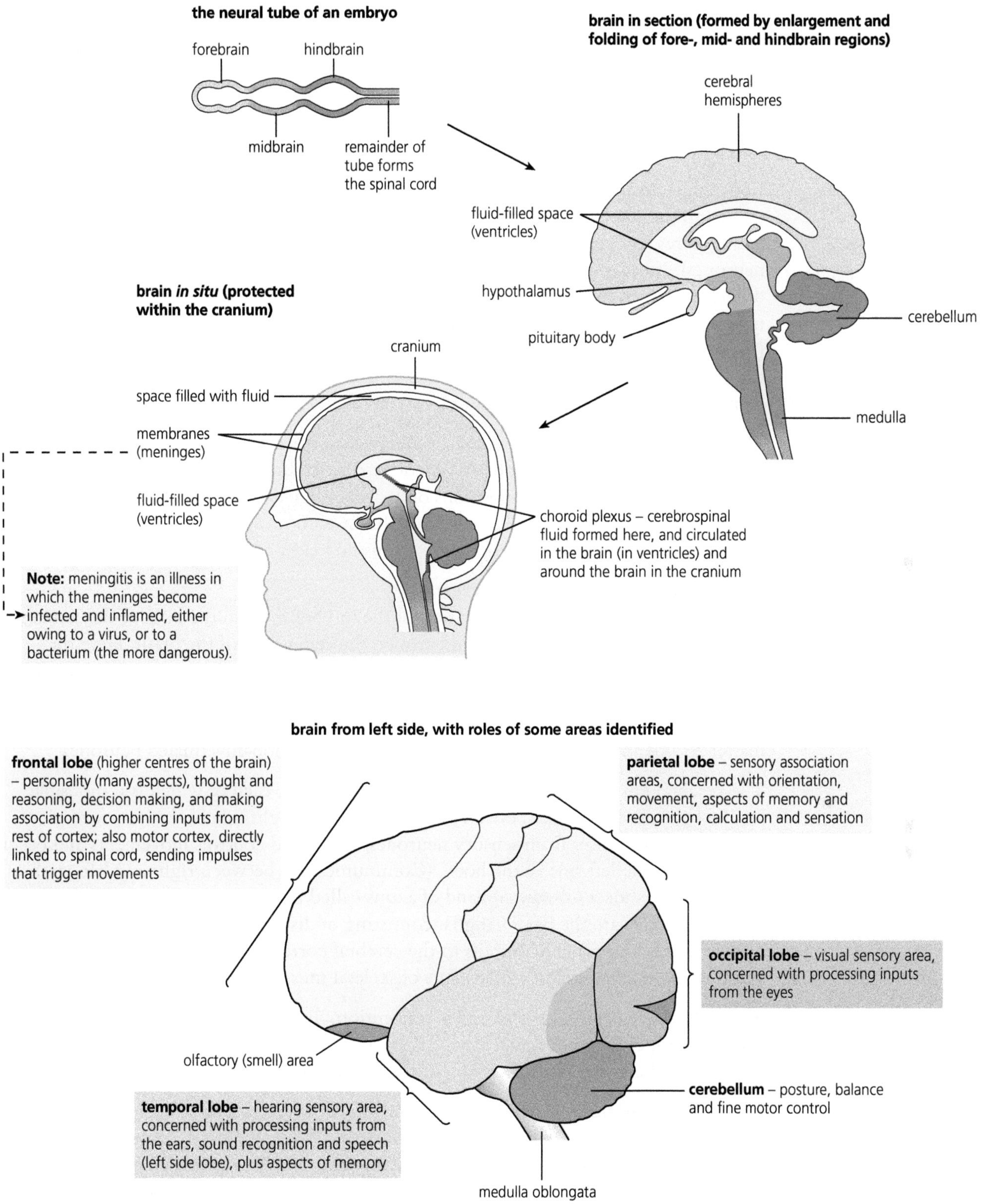

the neural tube of an embryo
forebrain
hindbrain
midbrain
remainder of tube forms the spinal cord
brain in section (formed by enlargement and folding of fore-, mid- and hindbrain regions)
cerebral hemispheres
fluid-filled space (ventricles)
hypothalamus
pituitary body
cerebellum
medulla
brain *in situ* (protected within the cranium)
cranium
space filled with fluid
membranes (meninges)
fluid-filled space (ventricles)
choroid plexus – cerebrospinal fluid formed here, and circulated in the brain (in ventricles) and around the brain in the cranium
Note: meningitis is an illness in which the meninges become infected and inflamed, either owing to a virus, or to a bacterium (the more dangerous).
brain from left side, with roles of some areas identified
frontal lobe (higher centres of the brain) – personality (many aspects), thought and reasoning, decision making, and making association by combining inputs from rest of cortex; also motor cortex, directly linked to spinal cord, sending impulses that trigger movements
parietal lobe – sensory association areas, concerned with orientation, movement, aspects of memory and recognition, calculation and sensation
occipital lobe – visual sensory area, concerned with processing inputs from the eyes
olfactory (smell) area
temporal lobe – hearing sensory area, concerned with processing inputs from the ears, sound recognition and speech (left side lobe), plus aspects of memory
cerebellum – posture, balance and fine motor control
medulla oblongata

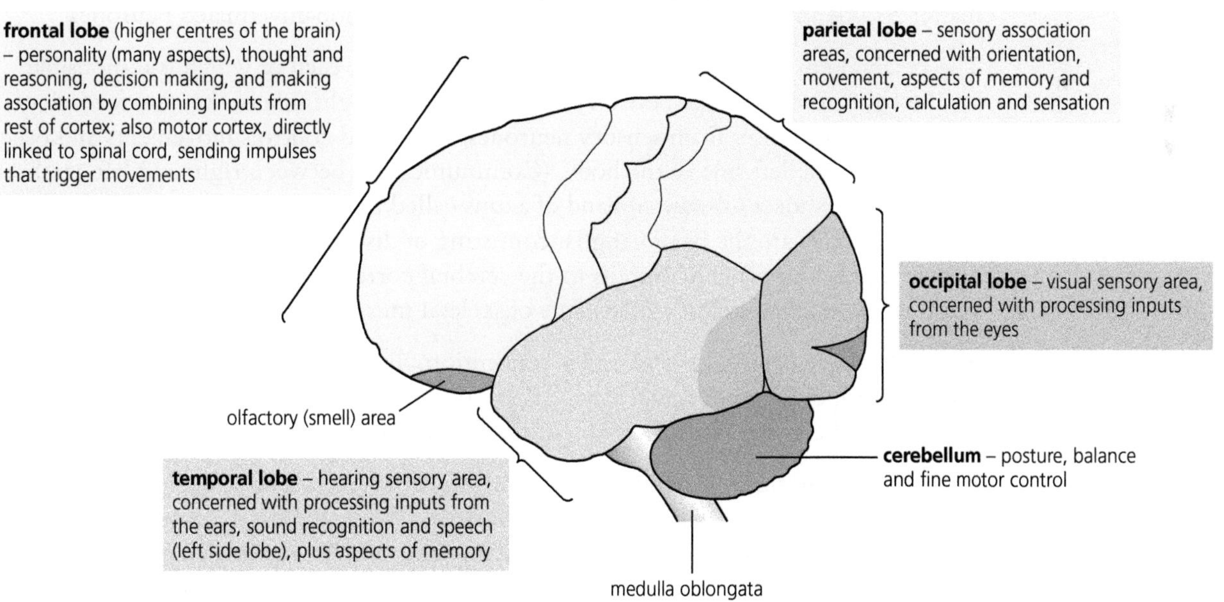

Figure 10.5 The human brain

Key terms

Neuroglia cells Cells found in the brain that are responsible for the support and protection of the neurones of the brain and peripheral nervous system.

Cerebral cortex The thin layer of grey matter covering the cerebral hemispheres.

The human brain contains about 10^{11}–10^{12} interneurones and the same number again of **neuroglia cells**. The majority of these neurones occur in the cerebral hemispheres. There, it is estimated, each interneurone forms synapses with a thousand other neurones. Mammals are the most intelligent of all animals, and their long memory, complexity of behaviour and subtlety of body control are also linked to the development of this brain structure.

Both white and grey matter are present in the brain as in the spinal cord. Grey matter makes up the interior of the brain and white the exterior. However, in the cerebral hemispheres and cerebellum there are additional layers of grey matter (that is, extra neurones). It has long been known, from observation of the effects of brain injury, that different parts of the brain have specific functions. Within the brain as a whole, certain tasks and roles are localised. For example, near the hypothalamus are the thalamus (the 'relay station' for impulses to the cerebral cortex from the rest of the brain and the spinal cord) and the hippocampus (responsible, together with parts of the cerebral hemispheres, for long-term memory).

More recently new techniques have been developed that have dramatically increased our knowledge of brain activity. Magnetic resonance imaging (MRI) uses powerful magnetic fields to detect the positions of hydrogen nuclei, hence it is able to show changes in activity in different parts of the brain without affecting the patient. The functions of some parts of the brain are discussed below.

Cerebrum (cerebral hemispheres)

The cerebral hemispheres, an extension of the forebrain, form the bulk of the human brain. They are positioned above and around the remainder of the brain. Here the body's voluntary activities are coordinated, together with many involuntary ones. The hemispheres have a vastly extended surface, which is achieved by extensive folding so that it forms deep groves. The surface, called the **cerebral cortex**, is covered by grey matter to a depth of 3 mm, and is densely packed with non-myelinated neurones.

The cerebral cortex is divided into right and left halves, each of which is responsible for the opposite half of the body. This means that the right side of the cortex receives information (impulses from sensory neurones) from, and controls movements and other responses in, the left side of the body. (Communication between right and left cerebral cortices occurs via a substantial band of axons called the corpus callosum.) Within the hemispheres are the basal ganglia, consisting of discrete groups of neurones. They receive inputs and provide outputs to the cerebral cortex, thalamus and hypothalamus, and they control automatic movements of skeletal muscle and muscle tone.

Each side of the cerebral cortex is, by convention, divided into four lobes (frontal, parietal, temporal and occipital lobes). You can see from Figure 10.5 (lower image) that the areas of the cortex with special sensory and motor functions have been mapped out.

It is obvious from this description that not only does the cerebral cortex receive large amounts of sensory information but it also initiates action potentials in motor neurones to bring about complex behaviour. To do this it does not simply respond in a pre-determined way as in a reflex action, it uses a whole host of higher mental activities. For example, memory is used to apply reasoning and experience is drawn on when making decisions on what actions to take. It is this ability that makes mammals the most intelligent life-forms yet known.

Cerebellum

The **cerebellum**, part of the hindbrain, has an external surface layer of grey matter. It is concerned with the control of involuntary muscle movements of posture and balance. Here, the precise, voluntary movements involved in hand manipulations, speech and writing are coordinated. Whilst the cerebellum does not initiate motor activity it plays a vital role in ensuring that such actions are carefully coordinated. Damage to the cerebellum does not result in paralysis. The cerebral hemispheres are still able to direct movement but such movements are clumsy and the fine control needed to manipulate a pen or thread a needle is absent. The description of walking given at the beginning of this chapter is also a good example of coordination of movements brought about by the cerebellum.

Medulla oblongata

The **medulla oblongata**, the base of the hindbrain, is a continuation of the uppermost part of the spinal cord. It houses the regulatory centres concerned with maintaining the rate and force of the heart beat and the diameter of the blood vessels. (We will discuss the detailed role of the medulla in the control of heart rate in Chapter 12.) Also, it is here that a respiratory centre adjusts the basic rate of breathing. And it is in the medulla that the ascending and descending pathways of nerve fibres connecting the spinal column and brain cross over (resulting, as already noted above, in the left side of our body being controlled by the right side of the brain, and vice versa).

Hypothalamus

The **hypothalamus** – part of the floor of the forebrain and exceptionally well supplied with blood vessels – is the control centre for the **autonomic nervous system (ANS)**. Here the body monitors and controls body temperature and the levels of sugars, amino acids and ions in **osmoregulation**. Feeding and drinking reflexes, and aggressive and reproductive behaviour, are also controlled here. The hypothalamus works with a 'master gland' called the pituitary gland, to which it is attached, monitoring hormones in the blood, and controlling the release of hormones. So the hypothalamus is the main link between nervous and **endocrine systems**. We shall look in more detail at the role of the hypothalamus and its close links with the hormonal system when discussing osmoregulation in Chapter 12.

Test yourself

1 Explain how a direct connection between sensory neurone and motor neurone is made in a spinal reflex.
2 State where the cell bodies of motor neurones are found.
3 Describe how spinal nerves leave and enter the vertebral column.
4 Describe how the surface area of the cerebral cortex is adapted to increase its size within the cranium.
5 Suggest the symptoms you would expect to see in a patient with a damaged cerebellum.
6 Explain why white matter is white.
7 Name the major endocrine (hormonal) gland that is linked to the hypothalamus.

Key terms

Cerebellum A part of the hindbrain responsible for the fine coordination of motor activity.

Medulla oblongata A part of the hindbrain containing the regulatory centres for breathing and heart rate.

Autonomic nervous system (ANS) The network of motor neurones carrying impulses to smooth muscle and glands, controlling involuntary (unconscious) actions.

Endocrine system A system of glands, in animals, which secrete hormones into the bloodstream.

Hypothalamus A part of the forebrain concerned with control of the autonomic nervous system and maintaining the constant internal environment. It has close links to the endocrine system.

Osmoregulation The process of controlling the concentration of the body fluids.

The peripheral nervous system

The nerves of the peripheral nervous system (PNS) consist of nerve fibres (**axons** and **dendrons**) arranged in bundles, protected by connective tissue sheaths (Figure 10.6). These nerves consist of:

- sensory neurones carrying impulses to the central nervous system
- motor neurones carrying impulses to muscles and glands.

Many of the motor neurones serve the muscles we use in conscious actions to produce voluntary movements, and they form the **somatic nervous system**.

On the other hand, the **autonomic nervous system** controls activities inside the body that are mostly under unconscious (involuntary) control. It consists of motor neurones running to the smooth muscle of the internal organs and to various glands.

The autonomic nervous system (ANS) acts to maintain the body's internal environment (autonomic means 'self-governing'). The detailed role of the ANS in controlling heart rate and osmoregulation are discussed in Chapter 12.

There is a further complication to the ANS, it is divided into two parts:

- the sympathetic nervous system (SNS)
- the parasympathetic nervous system (PNS).

Key terms

Axons Thin, elongated extensions of neurones that carry impulses away from the cell body.

Dendrons Thin, elongated extensions of neurones that carry impulses towards the cell body.

Autonomic nervous system The network of motor neurones carrying impulses to the heart, smooth muscle and glands controlling involuntary (unconscious) actions.

Somatic nervous system The system of motor neurones carrying impulses to muscles controlling voluntary (conscious) actions.

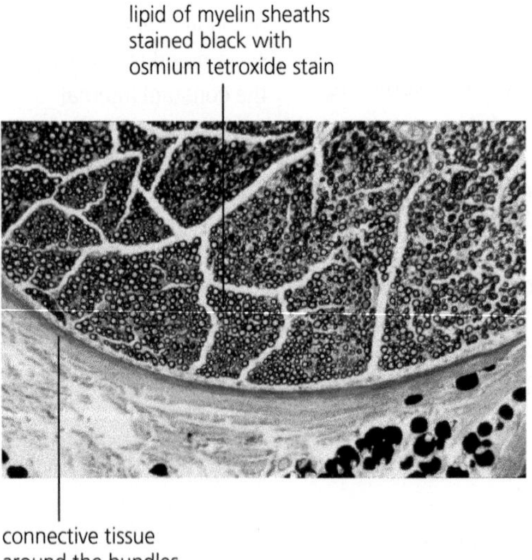

Figure 10.6 A peripheral nerve in TS (×250). The vagus nerve shown here is a mixed nerve having axons of motor neurones and axons and dendrons of sensory neurones

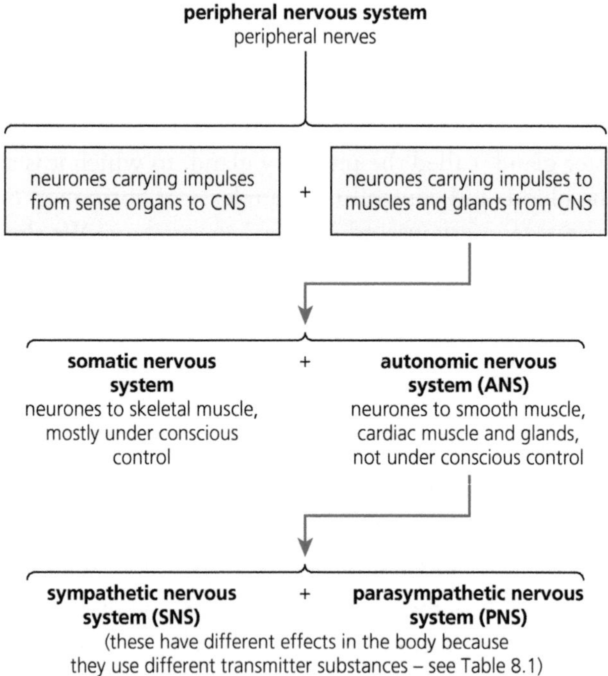

Figure 10.7 Peripheral nerves and conscious/unconscious control

The layout of the peripheral nervous system as a whole is shown in Figure 10.7. In Table 10.1, the key differences between the SNS and the PNS are listed. Note that in some of the functions the two systems are antagonistic in their effects (for example, the SNS causes the heart rate to increase and the PNS causes it to decrease). However, in other cases they may have the same effect on a gland or muscle.

Table 10.1 The autonomic nervous system - roles and responses

Sympathetic nervous system (SNS)	Parasympathetic nervous system (PNS)
More active in times of stress to produce 'fight or flight' responses	Concerned with conservation of energy and the replacement of body reserves
At their junctions with effector tissues (muscles or glands) the neurones release noradrenaline	At their junctions with effector tissues the neurones release acetylcholine
Some of the responses of the two systems	
Increases ventilation rate	Decreases ventilation rate
Causes widening (dilation) of the tissues	Causes constriction (narrowing) of the pupils
Has no effect on the tear glands	Causes the secretion of tears
Has no effect on the salivary glands	Causes the secretion of saliva
Slows peristalsis	Accelerates peristalsis
Constricts bladder sphincter muscles	Causes relaxation of the sphincter muscles of the bladder and contraction of the muscular wall of the bladder (under overall conscious control)

Test yourself

8 Describe the main structural difference between a sensory neurone and an interneurone.

9 Describe the main functional difference between the somatic and autonomic nervous systems.

10 State **two** ways in which the sympathetic and parasympathetic system have antagonistic effects.

Nervous transmission

Neurones – structure and function

You should have already met neurones and their structure in your previous studies but some important details are summarised in this section.

The nervous system is built from specialised cells called neurones. Each neurone has a substantial cell body containing the nucleus and the bulk of the cytoplasm, from which extremely fine cytoplasmic nerve fibres run. The nerve fibres are specialised for the transmission of information in the form of impulses. Most fibres are very long indeed. Impulses are transmitted along these fibres at speeds between 30 and 120 metres per second in mammals, so nervous coordination is extremely fast, and responses are virtually immediate. The three types of neurones are shown in Figure 10.8 on the next page.

- Motor neurones have many fine dendrites, which bring impulses towards the cell body, and a single long axon, which carries impulses away from the cell body.
- Interneurones (also known as relay neurones) have numerous, short fibres.
- Sensory neurones have a single long dendron, which brings impulses towards the cell body, and a single long axon, which carries impulses away.

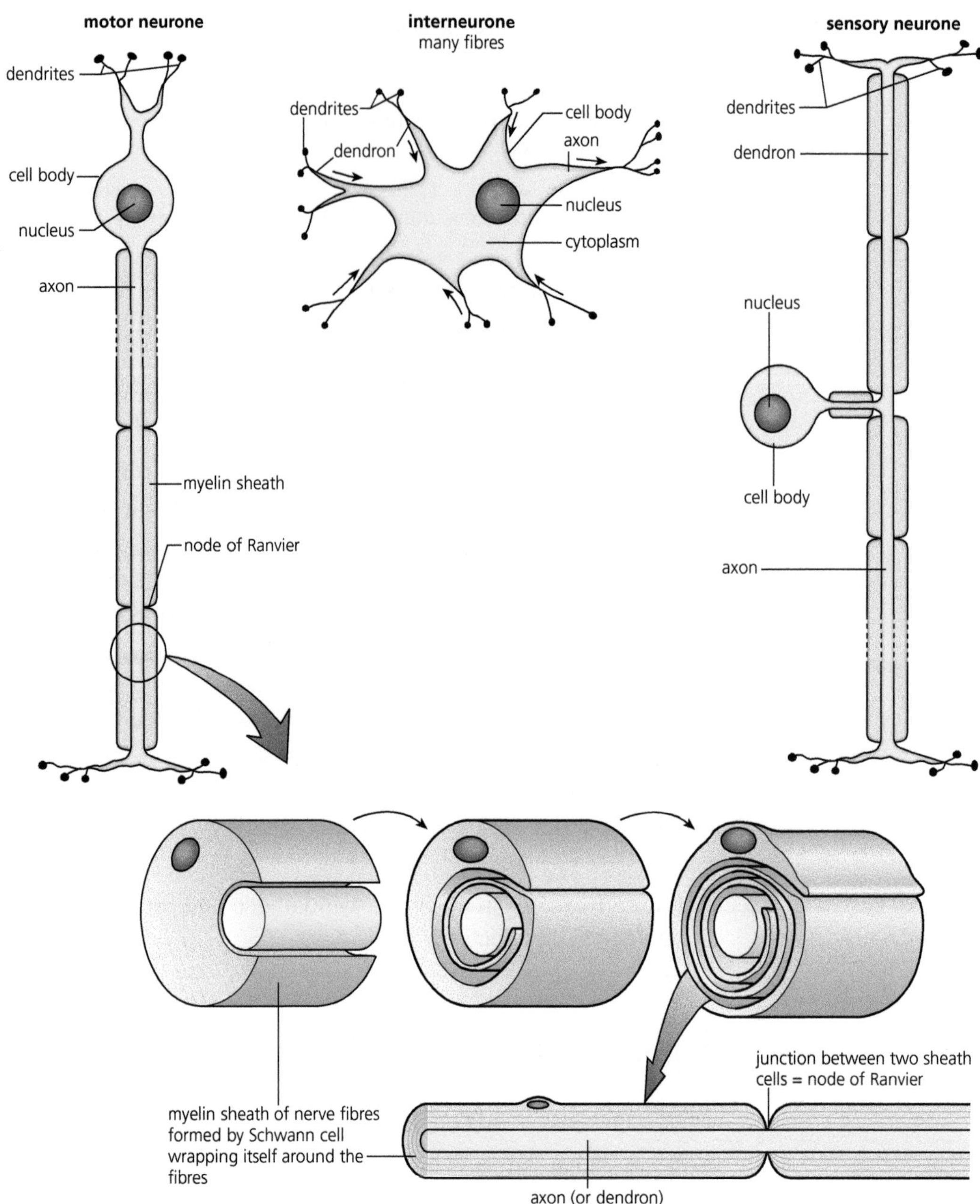

Figure 10.8 Comparing neurones

> **Key terms**
>
> **Schwann cells** Cells found wrapped around dendrons and axons, which produce the myelin sheath.
>
> **Myelin sheath** A layer of fatty tissue secreted by Schwann cells surrounding an axon.
>
> **Node of Ranvier** A small gap in the myelin sheath surrounding an axon.

Surrounding the neurones are different types of supporting cells called neuroglia cells (sometimes shortened to 'glial cells') – also an important part of the nervous system. One type of neuroglia cell is called a Schwann cell. Many of the long fibres (dendrons and axons) are protected by **Schwann cells**. These wrap themselves around the fibres, forming a structure called a **myelin sheath** (Figure 10.8). Between each pair of Schwann cells is a junction in the myelin sheath called a **node of Ranvier**. The myelin sheath and its junctions help increase the speed at which impulses are conducted.

Transmission of an impulse

An impulse is transmitted along nerve fibres, but it is not an electrical current that flows along the 'wires' of the nerves. Rather, the impulse is a momentary reversal in electrical potential difference in the membrane. That is, it is a change in the amounts of positively and negatively charged ions between the inside and outside of the membrane of a nerve fibre (Figure 10.9). This reversal travels from one end of the neurone to the other in a fraction of a second. Between conduction of one impulse and the next, the neurone is said to be resting. Actually, this not the case. During the 'resting' interval between impulses, the membrane of a neurone actively creates and maintains an electrical potential difference between the inside and the outside of the fibre.

How is this done?

The resting potential

Two processes together produce the resting potential difference across the neurone membrane.

- There is **active transport** of potassium (K^+) ions *in* across the membrane, and of sodium (Na^+) ions *out* across the membrane. The ions are transported by a Na^+–K^+ pump, with transfer of energy from ATP. So potassium and sodium ions gradually concentrate on opposite sides of the membrane. However, this in itself makes no change to the potential difference across the membrane.
- There is also facilitated diffusion of K^+ ions *out* and Na^+ ions back *in*. The important point here is that the membrane is far more permeable to K^+ ions flowing out than to Na^+ ions returning. This causes the tissue fluid outside the neurone to contain many more positive ions than are present in the cytoplasm inside. As a result, the inside becomes more and more negatively charged compared with the outside; the resting neurone is said to be **polarised**. The difference in charge, or potential difference, is about −70 mV (the negative sign here is a convention to show the inside is more negative than the outside). This is known as the resting potential. Figure 10.9 summarises how it is set up.

> **Key terms**
>
> **Facilitated diffusion** The movement of molecules by diffusion across a cell surface membrane, which is assisted by the presence of protein molecules in the membrane forming pores or channels through which the molecule can pass.
>
> **Resting potential** The potential difference (approx −70 mV) across the axon membrane when no action potential is formed.

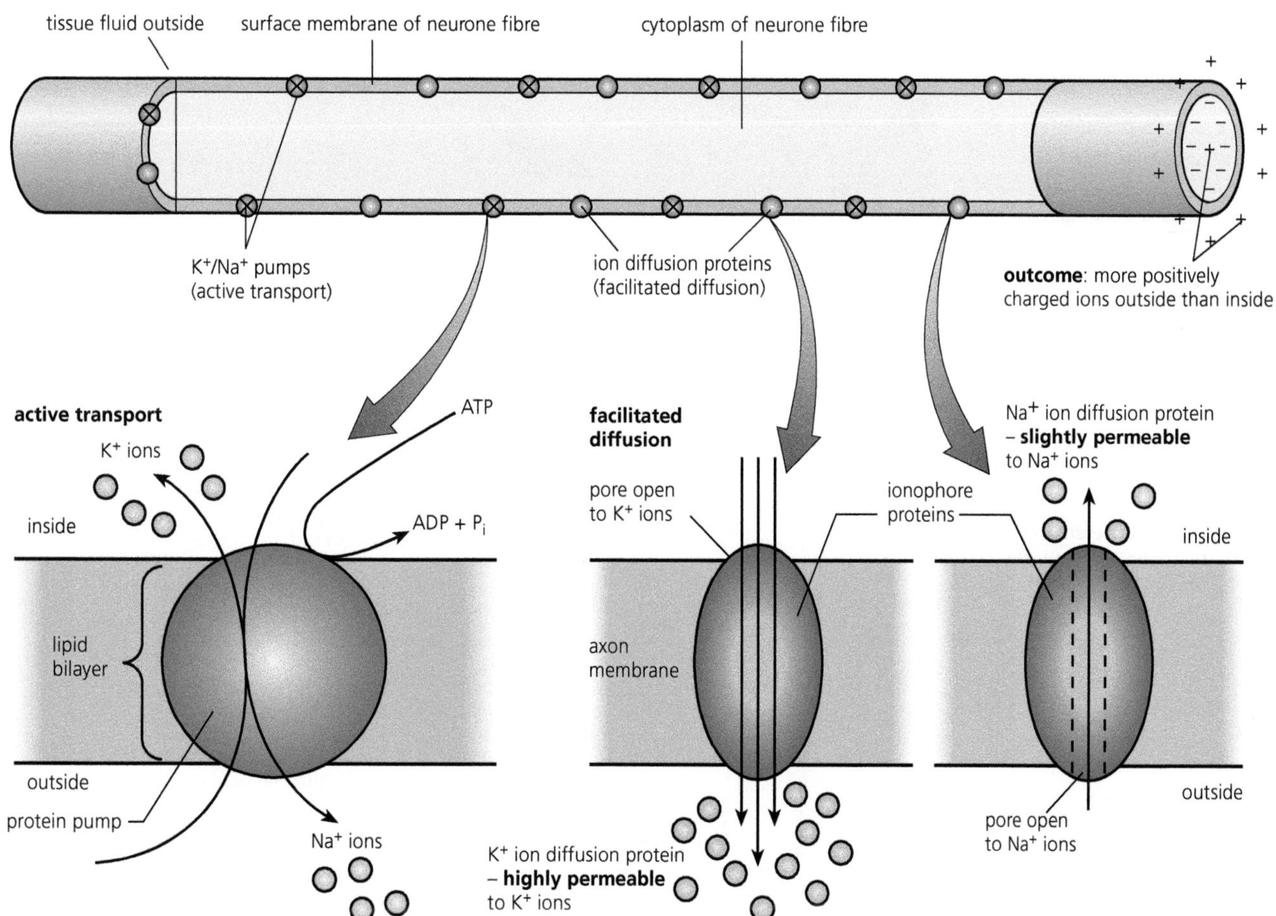

Figure 10.9 The establishment of a resting potential

The action potential

The next event, sooner or later, is the passage of an impulse. An impulse, or action potential, is triggered by a stimulus received at a receptor cell or sensitive nerve ending. The energy transferred by this stimulus causes a temporary and local reversal of the resting potential. The result is that the membrane is briefly depolarised at this point (Figure 10.10).

How does this happen?

The change in potential across the membrane occurs through pores in the membrane, called ion channels because they can allow ions to pass through. One type of channel is permeable to sodium ions, and another to potassium ions. These channels are globular proteins that span the entire width of the membrane. They have a central pore with a gate, which can open and close. During a resting potential, these channels are all closed.

The energy of the stimulus first opens the gates of the sodium channels in the cell surface membrane. This allows sodium ions to diffuse in, down their electrochemical gradient. So the cytoplasm inside the neurone fibre quickly becomes progressively more positive with respect to the outside. This charge reversal continues until the potential difference has altered from −70 mV to +40 mV. At this point, an action potential has been created in the neurone fibre.

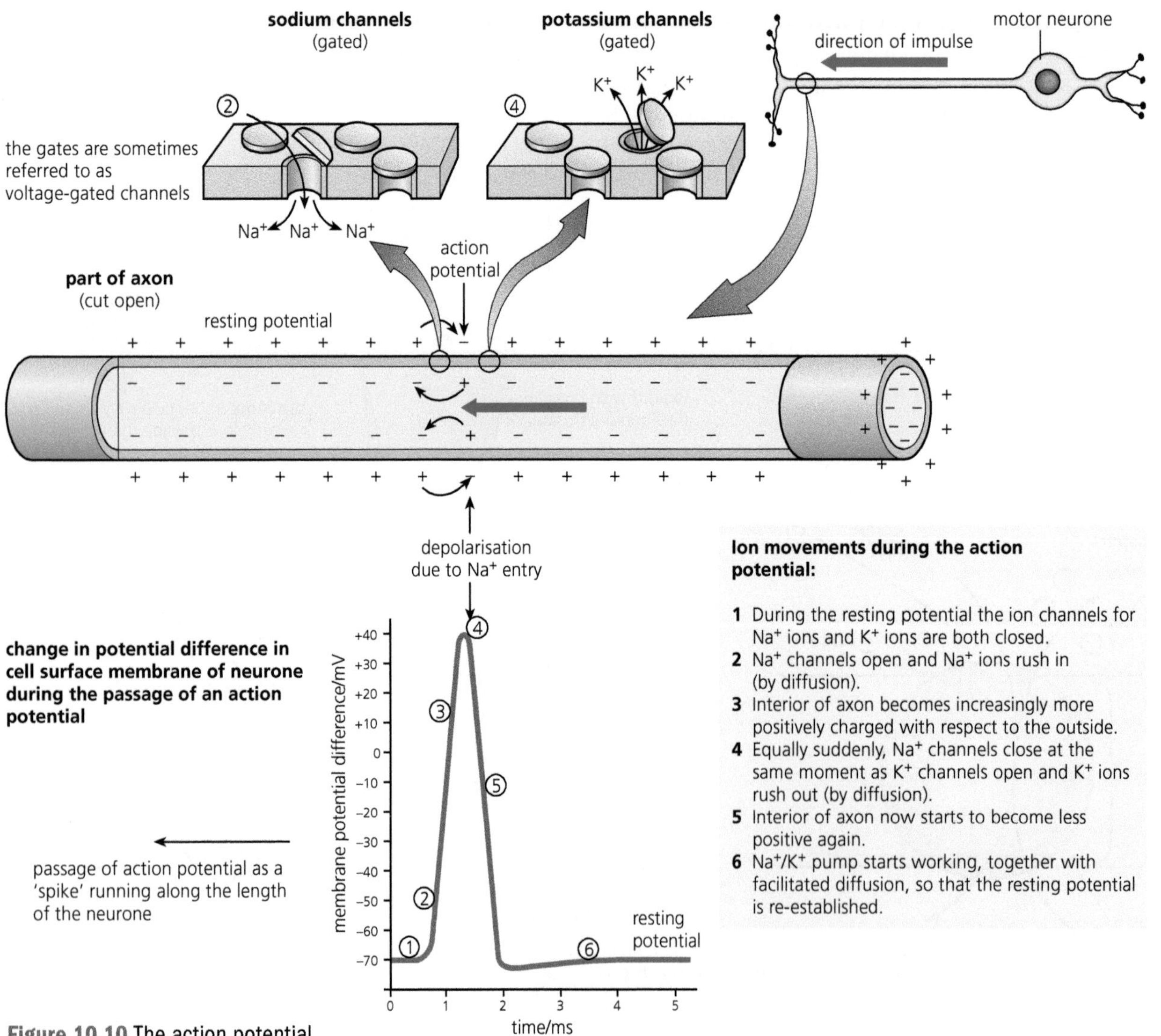

Figure 10.10 The action potential

The action potential then travels along the whole length of the neurone fibre. At any one point it exists for only two thousandths of a second (2 milliseconds) before the membrane starts to re-establish the resting potential. So action potential transmission is exceedingly quick – an example of positive feedback, in fact.

Almost immediately after an action potential has passed, the sodium channels close and potassium channels open. So potassium ions can exit the cell, again down an electrochemical gradient, into the tissue fluid outside. This causes the interior of the neurone to start to become less positive again. Then the potassium channels also close. Finally, the resting potential of −70 mV is re-established by the sodium–potassium pump, and the process of facilitated diffusion.

The refractory period

For a brief period following the passage of an action potential, the neurone fibre is no longer excitable. This is the refractory period. It lasts only 5–10 milliseconds in total. During this time, firstly there is a large excess of sodium ions inside the neurone fibre and further influx is impossible. As the resting potential is progressively restored, however, it becomes increasingly possible for an action potential to be generated again. Because of this refractory period, the maximum frequency of impulses is between 500 and 1000 per second.

The 'all or nothing' principle

Obviously, stimuli are of widely different strengths – for example, the difference between a light touch and the pain of a finger hit by a hammer! A stimulus must be at or above a minimum intensity, known as the threshold of stimulation, in order to initiate an action potential at all. Either a stimulus depolarises the membrane sufficiently to reverse the potential difference (−70 mV to +40 mV), or it does not. If not, no action potential is generated. With all sub-threshold stimuli, the influx of sodium ions is quickly reversed and the resting potential is re-established.

For stimuli above the threshold, as the intensity of the stimulus increases, the frequency at which the action potentials pass along the fibre increases (the individual action potentials are all of standard strength). For example, with a very intense stimulus, action potentials pass along a fibre at an accelerated rate, up to the maximum possible permitted by the refractory period. This means the effector (or the brain) recognises the intensity of a stimulus from the frequency of action potentials (Figure 10.11).

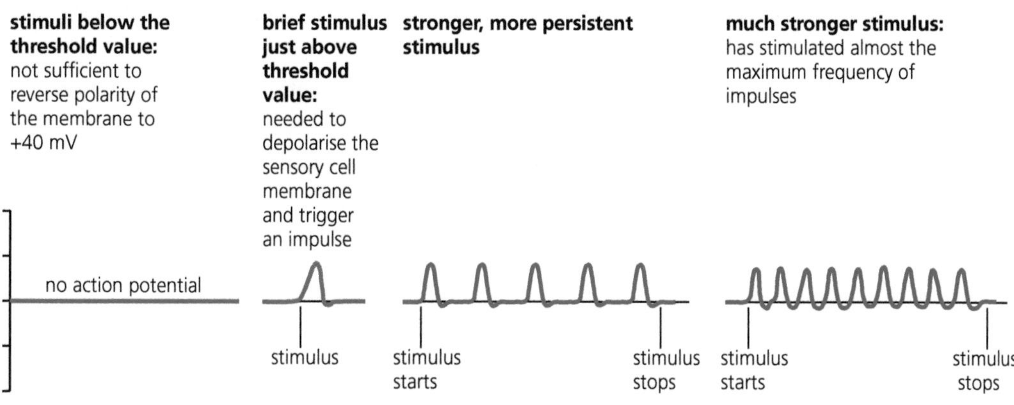

Figure 10.11 Weak and strong stimuli and threshold value

> **Key term**
>
> **Saltatory conduction**
> Conduction of an action potential along a nerve fibre by jumping from node to node, thus increasing the speed.

Speed of conduction of the action potential

The presence of a myelin sheath affects the speed of transmission of the action potential. The junctions in the sheath, known as the nodes of Ranvier, occur at 1–2 mm intervals. Only at these nodes is the axon membrane exposed. Elsewhere along the fibre, the electrical resistance of the myelin sheath prevents depolarisations. Consequently, local depolarisations build up at this point causing them to jump from node to node (Figure 10.12). This is called **saltatory conduction** ('saltation' meaning 'leaping'), and is an advantage, as it greatly speeds up the rate of transmission.

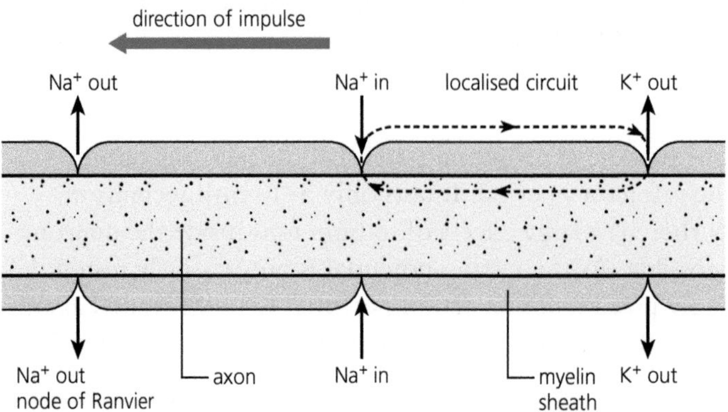

Figure 10.12 Saltatory conduction

Not all neurones have myelinated fibres. In fact, non-myelinated dendrons and axons are common in non-vertebrate animals. Here, transmission is normally much slower because the action potential flows steadily, right along the fibres. However, among non-myelinated fibres it is a fact that large-diameter axons transmit action potentials much more speedily than do narrow ones. Certain non-vertebrates like the squid and the earthworm have giant fibres, which allow fast transmission of action potentials (although not as fast as in myelinated fibres).

> **Test yourself**
>
> 11 Name the passive process that accounts for the transfer of Na^+ and K^+ ions to form the resting potential.
>
> 12 If Na^+ ions move in and K^+ ions move out by the same pumping mechanism, explain why there is a resting potential difference.
>
> 13 State which side of the axon membrane becomes more negatively charged in forming the resting potential.
>
> 14 Describe the change in membrane proteins that causes the onset of an action potential.
>
> 15 State **two** features of nerve fibres that can affect the speed of nervous transmission.

Synapses – the junctions between neurones

Where two neurones meet they do not actually touch. A tiny gap, called a synapse, is the link point between neurones (Figure 10.13). Synapses consist of the swollen tip (synaptic knob) of the axon of one neurone (pre-synaptic neurone) and the dendrite or cell body of another neurone (post-synaptic neurone). Between these is the synaptic cleft, a gap of about 20 nm.

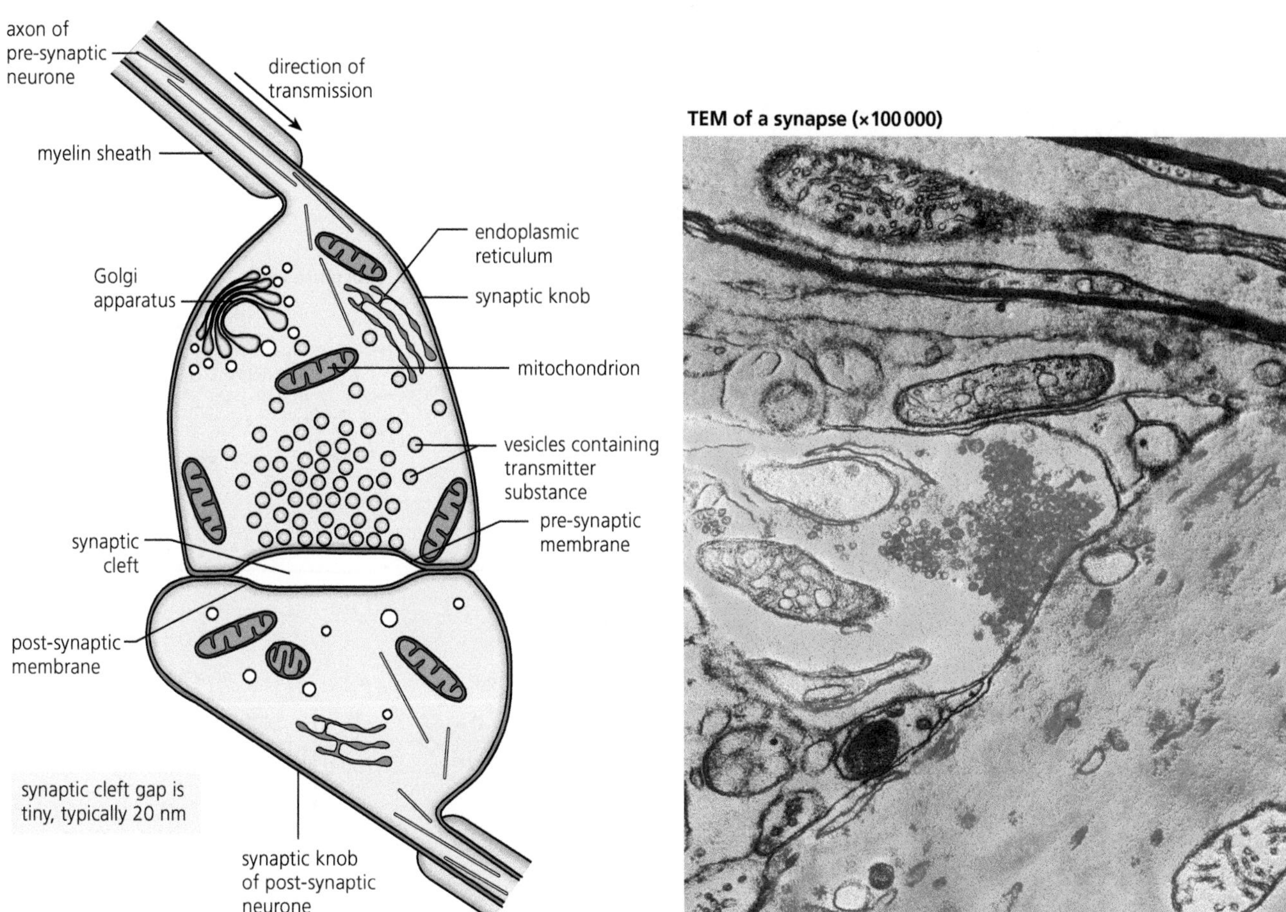

Figure 10.13 A synapse in section

The practical effect of the synaptic cleft is that an action potential cannot cross it. Here, transmission occurs by specific chemicals, known as transmitter substances. These substances are all relatively small, diffusible molecules. They are produced in the Golgi apparatus in the synaptic knob, and held in tiny vesicles prior to use.

Acetylcholine (ACh) is a commonly occurring transmitter substance; the neurones that release acetylcholine are known as **cholinergic neurones**. Another common transmitter substance is **noradrenaline** (released by **adrenergic neurones**). In the brain, the commonly occurring transmitters are glutamic acid and **dopamine**.

Key terms

Acetylcholine (ACh) A commonly occurring transmitter substance at synapses.

Cholinergic neurones Neurones releasing acetylcholine at their synapses.

Noradrenaline A commonly occurring transmitter substance at synapses.

Adrenergic neurones Neurones releasing noradrenaline at their synapses.

Dopamine A neurotransmitter found in the brain.

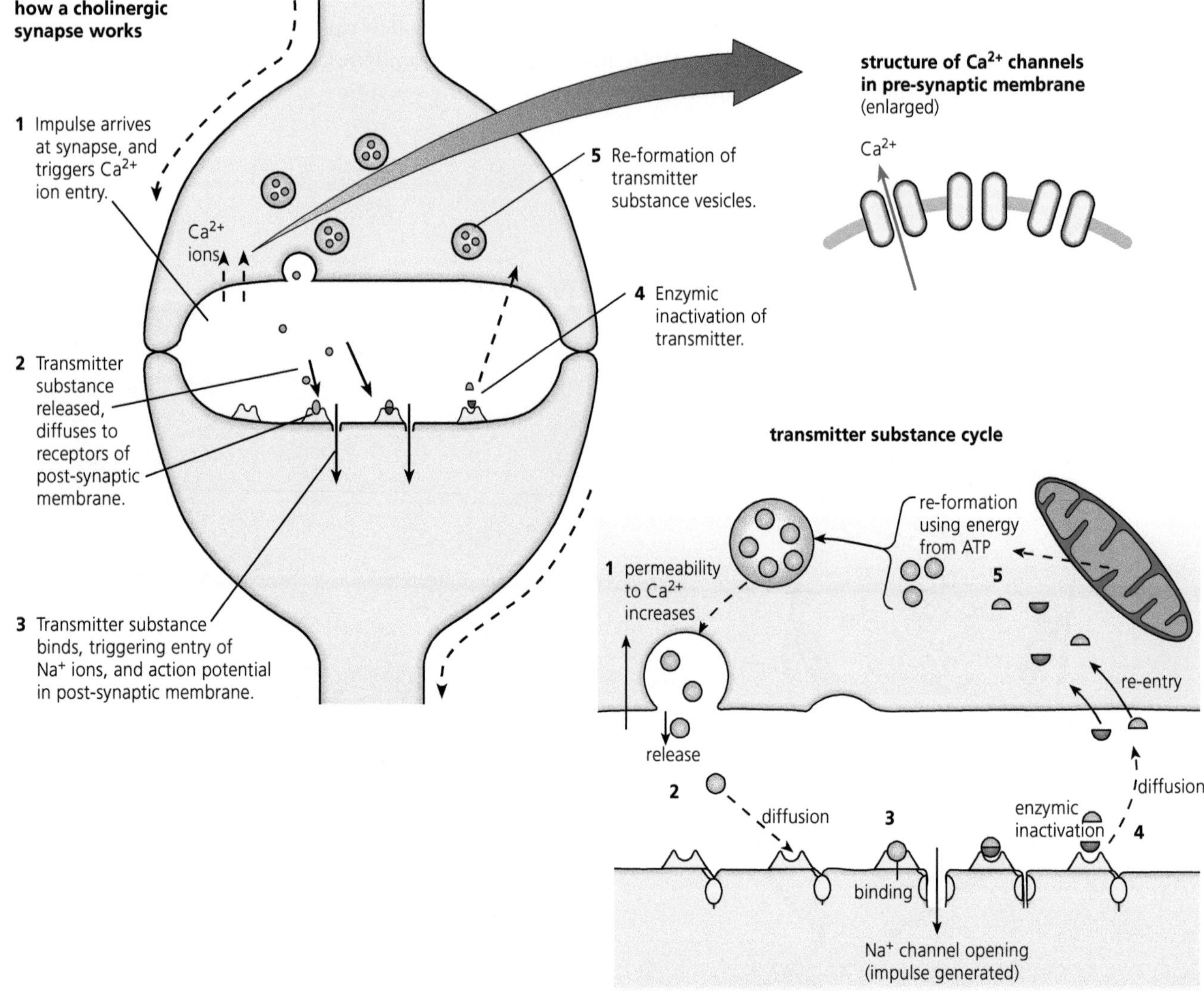

Figure 10.14 Chemical transmission at the synapse

You may find it helpful to follow each step in Figure 10.14.

1 The arrival of an action potential at the synaptic knob opens calcium ion channels in the pre-synaptic membrane. Calcium ions flow in from the synaptic cleft.

2 The calcium ions cause vesicles of transmitter substance to fuse with the pre-synaptic membrane and they release a transmitter substance into the synaptic cleft. The transmitter substance diffuses across the synaptic cleft.

3 The transmitter substance binds with a receptor protein on the post-synaptic membrane.

In the post-synaptic membrane, there are specific receptor sites for each transmitter substance. Each of these receptors also acts as a channel in the membrane that allows a specific ion (such as Na^+ or Cl^-, for example) to pass. The attachment of a transmitter molecule to its receptor instantly opens the ion channel.

When a molecule of ACh attaches to its receptor site, a Na^+ channel opens. As the sodium ions rush into the cytoplasm of the post-synaptic neurone, depolarisation of the post-synaptic membrane occurs. As more and more molecules of ACh bind, it becomes increasingly likely that depolarisation will reach the threshold level. When it does, an action potential is generated in the post-synaptic neurone. This process of build-up to an action potential in post-synaptic membranes is called facilitation.

4 The transmitter substance on the receptors is quickly inactivated. For example, the enzyme cholinesterase hydrolyses ACh to choline and ethanoic acid. These molecules are inactive as transmitters. This reaction causes the ion channel of the receptor protein to close, and so allows the resting potential in the post-synaptic neurone to be re-established.

5 Meanwhile, the inactivated products of the transmitter re-enter the pre-synaptic neurone, are re-synthesised into transmitter substance and packaged for re-use.

Excitation and inhibition at synapses

Although many synapses do function with ACh as their transmitter molecule, the post-synaptic membrane has many receptor sites that will respond to other transmitters. Their effect can be to cause the membrane to be more likely to reach the threshold value (that is to make the potential difference less negative), in which case they will be, like ACh, excitatory. If their effect is to cause the membrane to be less likely to reach the threshold value (that is make the potential difference more negative) then they will be inhibitory, as shown in Figure 10.15.

The transmitter glutamate (an amino acid) attaches to receptor sites and causes the opening of Na^+ channels. The influx of Na^+ ions sets up tiny areas of depolarisation, which are called **excitatory post-synaptic potentials (EPSPs)**. These EPSPs make the membrane less negatively charged and more likely to reach the threshold level to trigger an action potential (which is why they are called excitatory potentials).

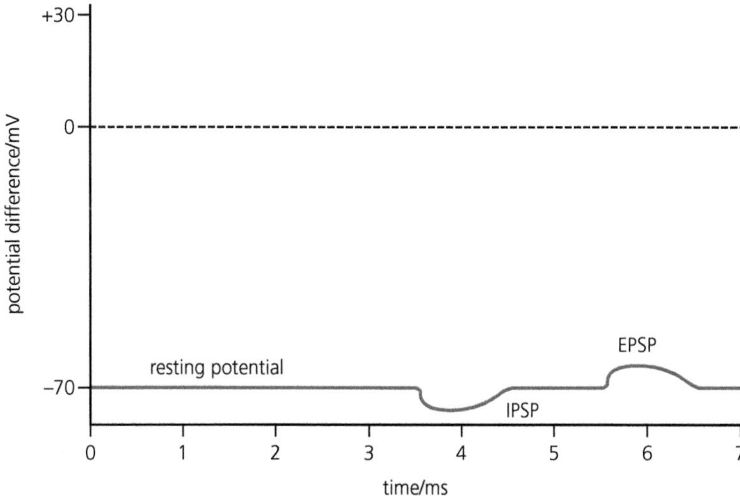

Figure 10.15 The effects of IPSPs and EPSPs on resting potential

The transmitter **gamma-aminobutyric acid (GABA)** is commonly found in the brain. When it attaches to its receptor sites it causes the opening of Cl^- ion channels, which sets up tiny areas of further polarisation that are called **inhibitory post-synaptic potentials (IPSPs)**. These IPSPs make the membrane more negatively charged and less likely to reach the threshold level to trigger an action potential (which is why they are called inhibitory potentials).

Key terms

Excitatory post-synaptic potentials (EPSPs) Small currents making the post synaptic membrane less negative, which makes it more likely to trigger a new action potential.

Gamma-aminobutyric acid (GABA) A neurotransmitter found in the brain.

Inhibitory post-synaptic potentials (IPSPs) Small currents making the post synaptic membrane more negative which make it less likely to trigger a new action potential.

Manipulating synapses using drugs

As synapses use chemical transmitters and are not contained within myelin sheaths, they offer the opportunity for medical intervention, predatory weapons and uncontrolled recreational use. A very common approach is to create artificial molecules that can mimic or block the action of the real transmitters. Some of the most widely prescribed drugs in the world, the benzodiazepines, such as Valium, are used as antidepressants due to their ability to bind to GABA receptors in synapses.

Here are some examples of powerful drugs that act through their effects on synapses.

Lidocaine

Although the name may sound unfamiliar it is very likely that you have experienced its effects. Lidocaine is used as a local anaesthetic, especially by your dentist to 'numb' the nerves of your mouth when working on your teeth. This effect is brought about because lidocaine blocks **voltage-gated Na^+ ion channels**. In the synapse this means the post-synaptic membrane is not able to depolarise, so no action potentials can travel to the brain to record pain. In addition, the pain receptors themselves need to depolarise to initiate pain signals, so they never respond. This is also true of motor neurones in the region affected so your lips also feel 'droopy' and you cannot control them for a while. Fortunately lidocaine is quickly metabolised in the liver so the effect begins to wear off after an hour or so.

> **Key terms**
>
> **Voltage-gated channels** Proteins in cell membranes forming channels, which are opened and closed by changes in potential difference.
>
> **Endorphins** Chemicals released by the brain that produce a feeling of pleasurable relaxation.
>
> **Acetylcholinesterase** The enzyme that breaks down acetylcholine within a synapse.

Nicotine

Nicotine from tobacco smoke has widespread effects on the nervous and hormonal systems. Its main effect is brought about by its ability to bind with acetylcholine (ACh) receptor sites in synapses. This blockage causes more ACh to be produced and a feeling of greater alertness as synapses are excited. At the same time it also causes the brain to release another neurotransmitter called dopamine. High levels of dopamine and ACh stimulate the release of **endorphins** in the brain, which are the chemicals that produce a feeling of relaxed pleasure. They are the brain's 'feel good' compounds.

This gives rise to the double effect of nicotine, an initial increase in alertness followed by a relaxed pleasant sensation. Endorphin release is a very powerful response; the brain quickly associates this pleasurable experience with the action preceding it and this naturally leads to addiction. The effect of nicotine is very enjoyable but very short-lived, a classic formula leading to a craving for more.

Cobra venom

Cobra venom is modified saliva containing a mixture of proteins. It is injected into the bloodstream of prey using fangs (Figure 10.16). One of these proteins binds irreversibly to ACh receptors on the post-synaptic membrane. The enzyme **acetylcholinesterase** has no effect on this venom protein. As a result the Na^+ ion channels remain permanently open and after an initial action potential the membrane is unable to repolarise and no further action potentials can be generated. This causes general paralysis, including respiratory muscles, which results in death due to suffocation.

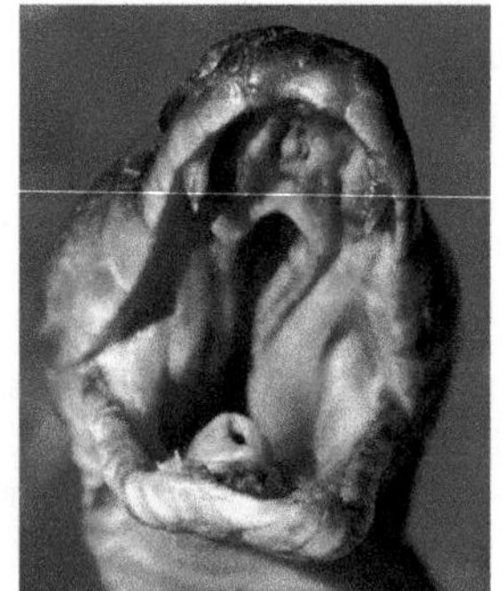

Figure 10.16 Cobra threat posture showing fangs for injection of venom into prey

> **Test yourself**
>
> 16 Name **one** other neurotransmitter apart from acetylcholine.
>
> 17 Name the metal ions that trigger the release of acetylcholine into the synaptic cleft.
>
> 18 The enzyme cholinesterase breaks down acetylcholine. The synapse stops working if this enzyme is not present. Explain why.
>
> 19 Describe the effect of an IPSP on the resting potential.
>
> 20 Suggest why cobra venom victims often suffocate.

Exam practice questions

1 The maximum number of impulses that can be sent down an axon is about $1000\,s^{-1}$. This is because:

A there is a gap between axons at the synapse

B Na^+ gated channels cannot open and close faster than this

C nodes of Ranvier cause delays in conduction

D it takes time for the restoration of a resting potential between each action potential *(1)*

2 The part of the brain that forms the main link between the nervous and endocrine systems is called the:

A cerebrum

B medulla oblongata

C hypothalamus

D cerebellum *(1)*

3 The graph shows the changes in potential difference across an axon membrane as an action potential passes.

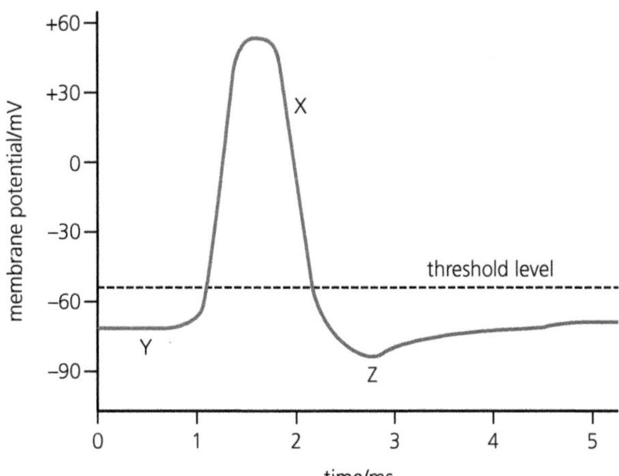

a) What name is given to the potential difference at Y? *(1)*

b) Explain what is meant by the term 'threshold level'. *(2)*

c) Explain how the properties of the axon membrane and the movement of ions account for the changes in membrane potential in region X. *(4)*

d) Explain why the membrane potential at Z becomes lower than that at Y. *(3)*

4 **a)** Explain how the structure of the spinal cord allows the passage of both simple spinal reflexes and voluntary movements in the somatic nervous system. *(5)*

b) Describe how the cerebral cortex and the cerebellum are involved in the initiation and control of voluntary activities. *(5)*

Tip

Question 3 is designed to test your ability to describe processes accurately. It requires some detailed knowledge of action potentials but also tests your understanding of the events in some depth. So, although it is mainly AO1 and straightforward it is not too easy.

Tip

Both sections of Question 4 require a comprehensive explanation in a clear sequence. You will need to know the details involved and also make sure you check exactly what the question is asking. A simple description without emphasis on how the structure and function of the spinal cord and brain are linked will only gain limited credit. Both sections require some application but are mainly AO1.

5 a) The diagram shows a part of a myelinated nerve fibre magnified ×200.

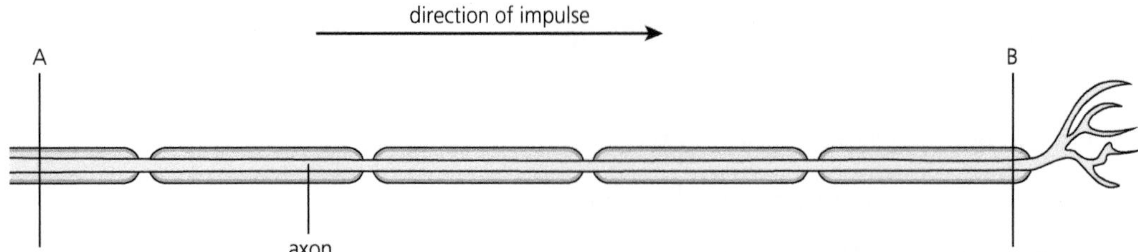

i) The passage of an action potential between the points marked A and B was timed as 0.05 ms. Use the information provided to calculate the speed of the action potential travelling along this fibre. *(3)*

The table below shows the speed of action potentials measured in a number of different nerve fibres.

Nerve fibre	Myelinated/ Non-myelinated	Diameter/μm	Speed of conduction/ms^{-1}
Squid giant axon	NM	500	25
Mammalian type 1	NM	1.5	2
Mammalian type 2	M	17	95
Mammalian type 3	M	9	55
Mammalian type 4	M	3	17

ii) What do these data suggest about the relationship between the myelination and diameter of nerve fibres and the speed of conduction? *(2)*

iii) Comment on the validity of your conclusions based on these data. *(2)*

b) Explain why there is a difference in the speed of conduction in myelinated and non-myelinated nerve fibres. *(4)*

Tip

You will need to revise the magnification formula and your knowledge of standard units such as milliseconds and then apply them to Question 5. In this way it is a synoptic question, a mathematical skills question, a straightforward AO1 question and an AO2 question from this part of the specification. Part (a)(iii) also asks you to apply some judgement in commenting on conclusions so could be classified as AO3. Remember, speed is best expressed as ms^{-1} in your final answer.

Stretch and challenge

6 Research how different neurotransmitters in the brain have enabled pharmacologists to develop treatments for some human disorders.

7 a) i) What is the function of dopamine in the brain and how is it linked to Parkinson's disease?

ii) Why does taking dopamine orally or by intravenous injection have no effect and how is this problem overcome?

b) One function of serotonin is to act as a neurotransmitter in the brain. People suffering from depression are often found to have low levels of serotonin in their brain.

i) One of the world's most prescribed drugs, 'prozac' (fluoxetine), is a SSRI (selective serotonin reuptake inhibitor). Explain how SSRIs are thought to increase serotonin levels in the brain.

ii) What other effects does serotonin have in mammals outside of the brain?

iii) Despite their widespread use, not all doctors accept that they are effective or that depression is simply caused by lack of serotonin. What arguments are used to support their case?

11 Chemical control

Prior knowledge

In this chapter you will need to recall that:

- → mammalian hormones are chemical messengers carried in the bloodstream
- → mammalian hormones affect target organs, often in distant parts of the body
- → hormones are often used to control long-term processes such as growth and reproduction
- → adrenaline produces a 'fight or flight' sequence of actions
- → insulin is used to control blood sugar levels
- → oestrogen and progesterone are important in human reproductive cycles
- → hormones interact with each other to produce a sequence of events
- → many plant hormones are often known as plant growth substances
- → the plant hormone, auxin, is involved in phototropism
- → transcription factors control gene expression (Chapter 6).

Test yourself on prior knowledge

1. State where in the body insulin is produced.
2. Name the target organ for insulin.
3. Describe the effect that auxin has on developing plant cells.
4. State where in the root of a plant auxin is produced.
5. Hormones are carried in the bloodstream but only the target organs respond. Explain why.

You saw in Chapter 10 that the nervous system of mammals is a highly effective and rapid means of controlling and coordinating groups of specialised tissues and organs. The continual need for synthesis of transmitter molecules and the ATP requirements of active transport mean that this is a very energy-demanding means of coordination. This level of energy demand would need to be sustained for the full lifetime of the individual. However, plants do not have nervous systems and they need to exercise control and coordination too.

All of this means that it is an advantage to have other less energy-demanding means of coordination that are better suited to longer and slower processes. Chemical control provides this alternative in both plants and animals. As will be explained further in Chapter 12, both of these means of control need to be carefully coordinated in order to regulate more complex processes in an efficient manner. In this chapter you will look at the details of the principles and individual examples of chemical control in both plants and animals. In Chapter 12 we will explain how these are linked together to bring about control of heart rate, osmoregulation and thermoregulation in mammals.

Homeostasis

Survival of all plants and animals depends upon each cell's metabolism working at its optimum. To achieve this it is essential that the cells' environment, both internal and external is kept at the most favourable level. This process is known as **homeostasis**.

However, the intense activity in all cells means that this environment is constantly changing. The processes of homeostasis therefore need to make adjustments continually, to bring all the different factors back to the correct levels. For this reason there will always be small fluctuations about the optimum level. This is known as a **dynamic equilibrium**. Examples of the many factors that need to be controlled are shown in Figure 11.1.

Mammals are a comparatively recent group in terms of their evolutionary history, yet they have successfully settled in significant numbers in virtually every type of habitat on Earth. This success is directly linked to their ability to control their internal environment by homeostatis.

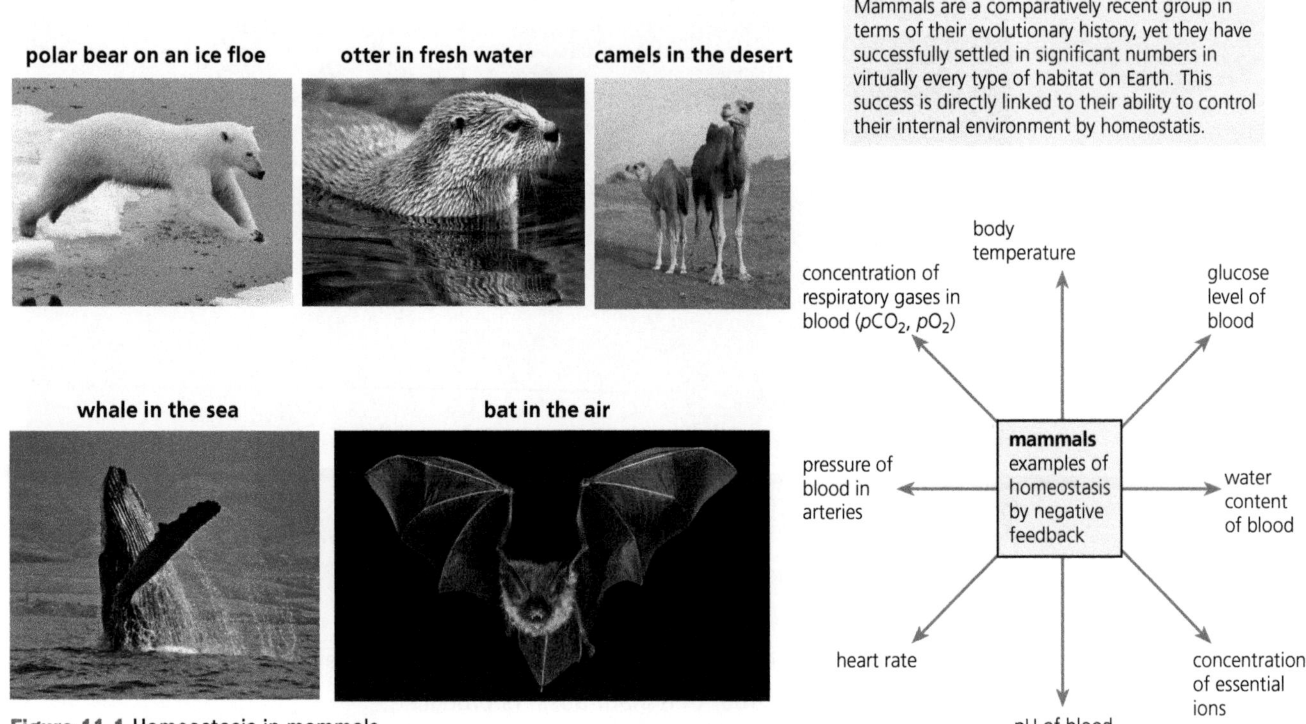

Figure 11.1 Homeostasis in mammals

Negative and positive feedback

An important feature of homeostasis is the use of feedback control. This can be in the form of **negative feedback** or **positive feedback**.

Negative feedback can best be illustrated by taking a simple example of a room, heated by a radiator, whose temperature is controlled by a single thermostat. If the room begins to cool, the thermostat is triggered to switch on the radiator. This causes the temperature to rise and then the thermostat switches off. In other words the more the thermostat is switched on the more it causes an effect that will switch it off. This is therefore **negative feedback**. Negative feedback provides an automatic self-regulating method of control. The process is summarised in Figure 11.2.

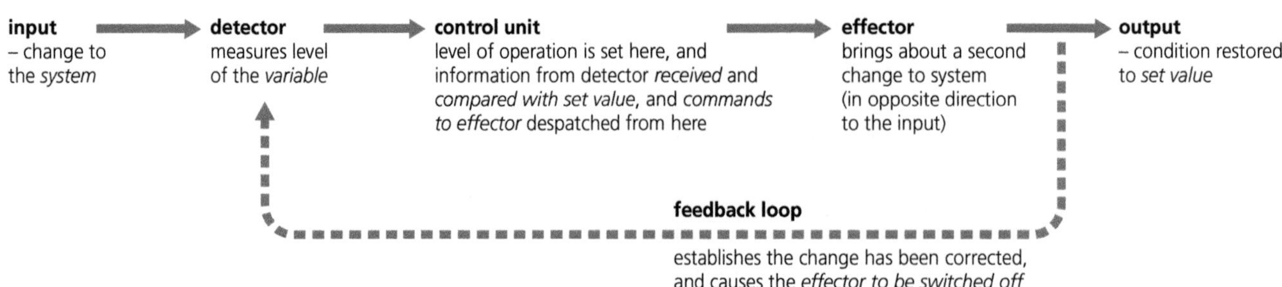

Figure 11.2 The mechanism of negative feedback

This example also illustrates another important principle. With only one heating system and no cooling system then this method will work well in colder weather but in summer the temperature will continue to rise regardless of the thermostat activity. What is needed, of course, is an air-cooling system alongside the radiator heating. As you will see in Chapter 12 the control of body temperature in mammals works in exactly this way.

In mammals the level of blood sugar is partially controlled by the peptide hormone insulin. When the blood sugar levels rise, the β cells of the pancreas (detector) respond by secreting insulin (effector). Insulin stimulates glucose uptake into cells such as muscles and liver by activating glucose transport proteins in the cell membranes. Thus the level of glucose falls and the β cells stop producing insulin (negative feedback). Just as in the room thermostat example there is also a hormone, glucagon, that raises blood sugar levels by a similar negative feedback mechanism. In this way human blood sugar levels are normally maintained between 80 and 110 mg 100 cm^{-3} with short-lived peaks after eating starchy or sugary meals.

Positive feedback is the opposite of this. Again, it is useful to consider an example with which you are probably familiar. Playing an electric guitar, attached to an amplifier and speaker, very close to the speakers will often produce an increasingly loud howling noise. Although generally referred to as just 'feedback' it is an example of **positive feedback**. What happens is that the guitar strings vibrate and this is picked up by the guitar sensors and amplified. If the strings are close to the speaker the vibrations in the air can cause the strings to vibrate more. This is again amplified and the speakers stimulate the strings more until the noise is deafening. In other words vibrations of the strings cause them to produce an effect that makes them vibrate even more. The problem with positive feedback is that unless there is some intervention it will spiral out of control. However, it can be useful to initiate actions in the body provided that there are other feedback loops that will bring it under control. Again, you will meet examples in Chapter 12.

Test yourself

1 Describe the main features of a dynamic equilibrium.
2 The action of insulin is an example of negative feedback. Explain why.
3 Suggest why plants do not have the equivalent of a nervous system.

Chemical control in mammals

Hormones are chemical substances produced and secreted from the cells of the **ductless** or **endocrine glands**. In effect, hormones carry messages around the body – but in a totally different way from the nervous system. Hormones are transported indiscriminately in the bloodstream, but they act only at specific sites, called **target organs**. Although present in small quantities, hormones are extremely effective messengers, helping to control and coordinate body activities. Once released, hormones typically cause changes to specific metabolic actions of their target organs. However, hormones circulate in the bloodstream only briefly. When they reach the liver they are broken down and the breakdown products are excreted via the kidneys. So, long-acting hormones must be secreted continuously to be effective.

Key terms

Endocrine (ductless) glands Glands producing hormones that are delivered directly into the bloodstream without the use of specific tubes (ducts) carrying them to their site of action.

Target organ An organ that will respond to a specific hormone.

The positions of all the endocrine glands of the human body are shown in Figure 11.3.

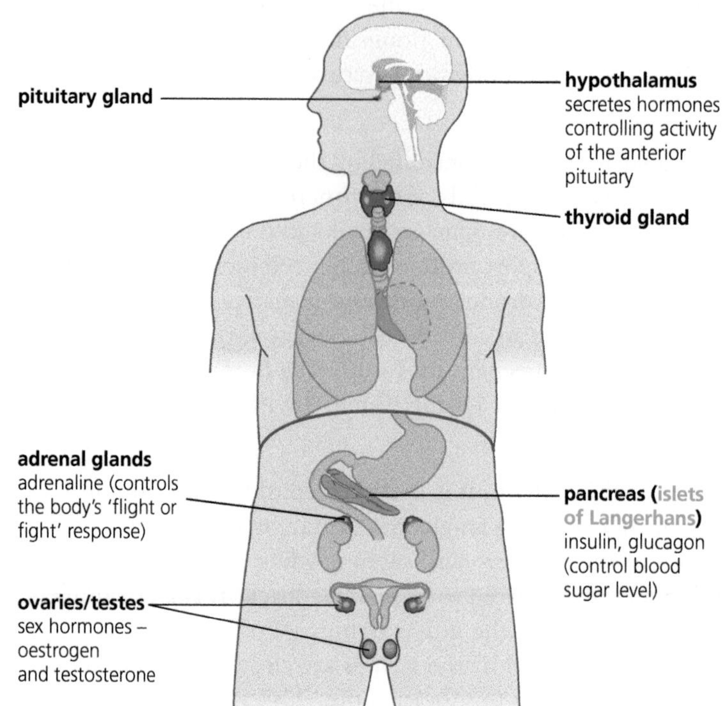

Figure 11.3 The human endocrine system

Key terms

Islets of Langerhans Distinct groups of cells found within the pancreas that produce the hormones insulin and glucagon.

Pituitary gland An important endocrine gland found attached to the hypothalamus at the base of the brain; it coordinates the functions of other endocrine glands.

The endocrine system and the nervous system work in distinctive and different ways when controlling and coordinating body activities. However, the effects of the endocrine system are coordinated by the **pituitary gland**, a 'master gland' working in tandem with the hypothalamus of the brain. The hypothalamus secretes hormones that regulate the functioning of the pituitary. This is an excellent example of how the two systems are very closely linked.

How hormones bring about their effects

Although hormones are distributed to all tissues by the bloodstream, they only affect their target cells. This is because only the target cells have the correct receptors.

There are two types of hormones, which bring about their effects in slightly different ways:

- **Amides** or **peptides** interact with specific receptors on the outside of the cell surface membrane.
- **Steroids** can pass through the cell surface membrane and interact with specific receptors within the cytoplasm.

In both cases a specific enzyme is needed to bring about the final action of altering structure or function. How this is brought about also differs for each type of hormone, as shown in Figure 11.4.

amine/peptide hormone action

peptide hormone

hormone does not enter cell

binds to receptor in cell surface membrane

ATP

cyclicAMP

existing (inactive) protein

activated enzyme

structure/ function of cell altered

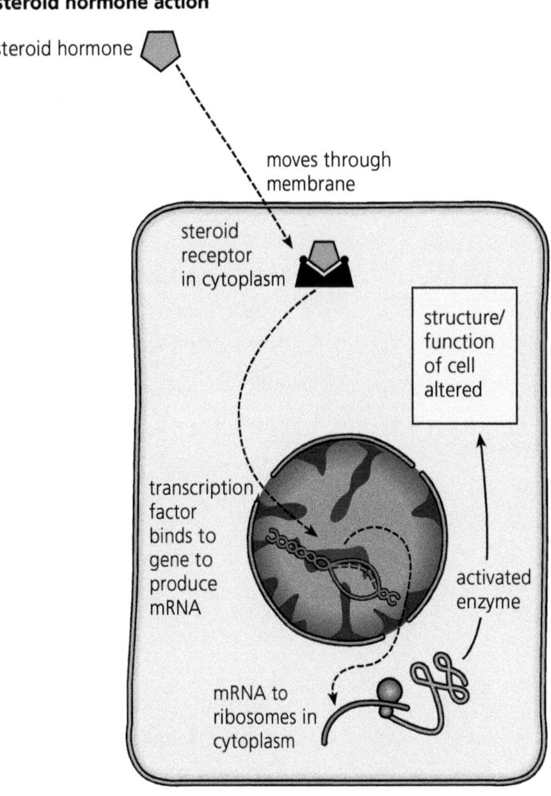

Figure 11.4 How the two types of hormones influence target cells

Peptide/amine hormones such as antidiuretic hormone (ADH) (see Chapter 12) and adrenaline attach to receptors on the cell surface membrane and trigger the release of a **second messenger**, often in the form of a molecule called **cyclic adenosine monophosphate (cAMP)**. This molecule, as its name suggests, is a very close relative of ATP. When the hormone binds to its receptor, cAMP is formed inside the cell and activates existing proteins in the cytoplasm to form enzymes that bring about specific changes. The formation of cAMP continues as long as the hormone binds to the receptor site and in this way the initial signal will be amplified within the cell.

Steroid hormones such as oestrogen and testosterone pass through the membrane before combining with receptors to form **transcription factors**. As we have described in Chapter 6, transcription factors bind to specific genes and as a result protein synthesis is initiated. In this case to produce the enzymes required to bring about specific changes.

Key terms

Second messenger
An additional molecule that is used to trigger a response when a hormone attaches to receptor sites on a target cell surface membrane.

Cyclic adenosine monophosphate (cAMP)
A cellular molecule often used as a second messenger when cells respond to a stimulus from a hormone.

Hormones can have multiple effects

Adrenaline (also known as epinephrine)

One advantage of hormonal control is that a single chemical release can have multiple effects in widespread parts of the body. Given that the hormone is already in the bloodstream, all that is required is that a group of cells has the correct receptor sites on its membrane in order to respond. This provides another level of coordination.

The hormone adrenaline (more correctly called epinephrine) is a good example of multiple coordinated responses (Table 11.1 on the next page). It is known as the 'fight or flight' hormone and although this is rather vague it does infer that it brings about several different responses for one purpose. In simple terms the release of adrenaline from the adrenal medulla is triggered by excitement or fear, so prepares the body for a response to what might be causing these sensations. Hence the ability to make muscles contract a little quicker, mobilise stored food reserves and constrict superficial veins (in case of damage), are all useful actions if we wish to defend ourselves or run away. This is likely to be rather less important in humans than in wild mammals, although it is clear that human sports and athletic performance are enhanced by the excitement of a big occasion.

Table 11.1 The multiple effects of adrenaline

Organ	Effect of adrenaline
Heart	Increases heart rate
Lungs	Increases breathing rate
Circulatory	Vasoconstriction
Liver	Increases breakdown of glycogen in liver to increase blood sugar levels
Muscle	Increases readiness to contract (causing shivering in extreme fear or excitement)

You may find some of Table 11.1 rather familiar. In Chapter 10 (Table 10.1) you looked at the effects of the sympathetic and parasympathetic divisions of the autonomic nervous system. The list for the sympathetic system is almost identical to Table 11.1. The explanation for this is that the sympathetic system uses epinephrine as a neurotransmitter and, in fact, the level of this in the blood comes partly from such secretions. So, not surprisingly, their effects are almost identical. Once again we see a very close link between nervous and hormonal control.

Test yourself

4 Explain the term 'steroids'.

5 Describe how amide and peptide hormones operate in a similar way to steroid hormones.

6 Describe how amide and peptide hormones operate in a different way to steroid hormones.

7 Describe the similarities between cAMP and ATP.

8 Suggest why a 'second messenger' is given this name.

9 Name **one** peptide hormone and **one** steroid hormone.

Chemical control in plants

The fundamental differences between plants and animals are reflected in their levels of control and coordination. Lower metabolic rates, a lack of movement and lower internal temperatures all mean that plants require less rapid control systems. Most plant responses are brought about by changes in growth patterns and growth rates. Therefore our definition of hormones is difficult to apply to plants (Table 11.2) and you will often see the compounds involved described as **plant growth substances**. Work on these substances often uses artificial closely related compounds with similar effects as they are more stable and longer-lasting.

Table 11.2 Differences between plant growth substances and animal hormones

Plant growth substances	Animal hormones
Produced in a region of plant structure, e.g. stem or root tips, in unspecialised cells	Produced in specific glands in specialised cells, e.g. islets of Langerhans in the pancreas (producing insulin)
Not necessarily transported widely or at all, and some are active at sites of production	Transported to all parts of the body by the bloodstream
Not particularly specific – tend to influence different tissues and organs, sometimes in contrasting ways	Effects are mostly highly specific to a particular tissue or organ, and without effects in other parts or on different processes

Auxins

Auxins are the most widely distributed of all plant growth substances. Auxin has been shown to be the compound indoleacetic acid (IAA), whose formula is illustrated in Figure 11.5. Their primary effect is on plant cell walls. As new cells are produced by mitosis in the **meristems** of shoot and root tips, they begin to elongate and differentiate in to specialised tissues. The increase in length of these cells is largely brought about by expansion caused by turgor pressure. Auxin increases the plasticity of the cell walls to allow them to expand further. This effect is the mechanism by which stems and roots respond to gravity and light in growth movements called **tropisms**. However, auxins are also involved in many other aspects of plant physiology as shown in Figure 11.5, both as a separate compound and interacting with other growth substances.

Key terms

Meristem A region of actively dividing cells found in plants, often at the root and shoot tips.

Tropism A plant growth response where the direction of movement is determined by the direction of the stimulus.

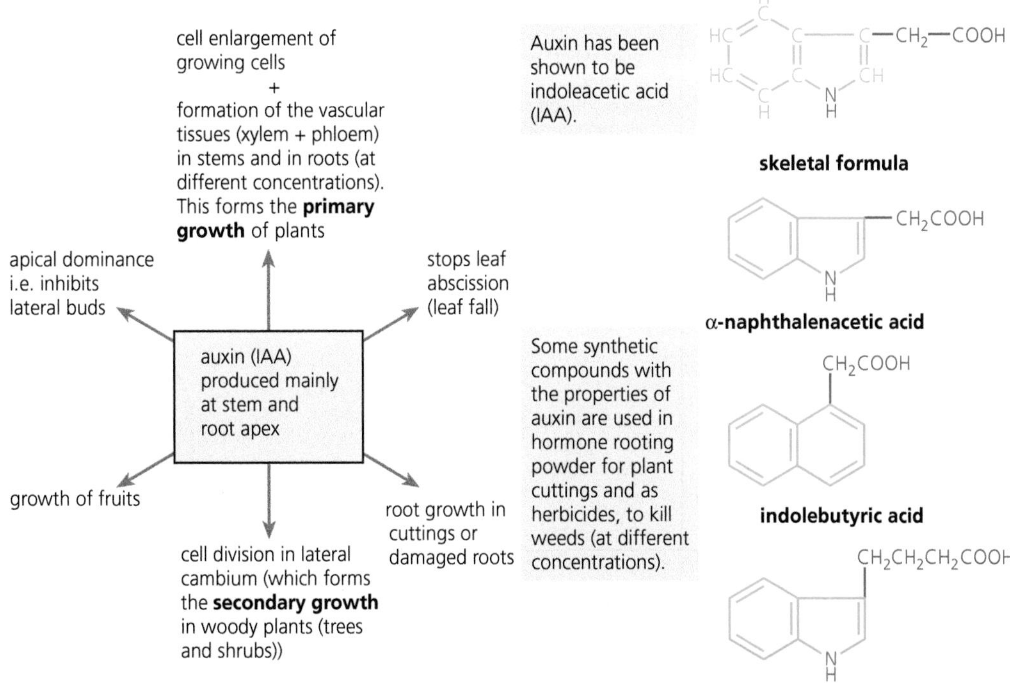

Figure 11.5 The roles of auxin (IAA) in plant growth, and the structure of natural and synthetic auxin

Using coleoptiles to investigate phototropism

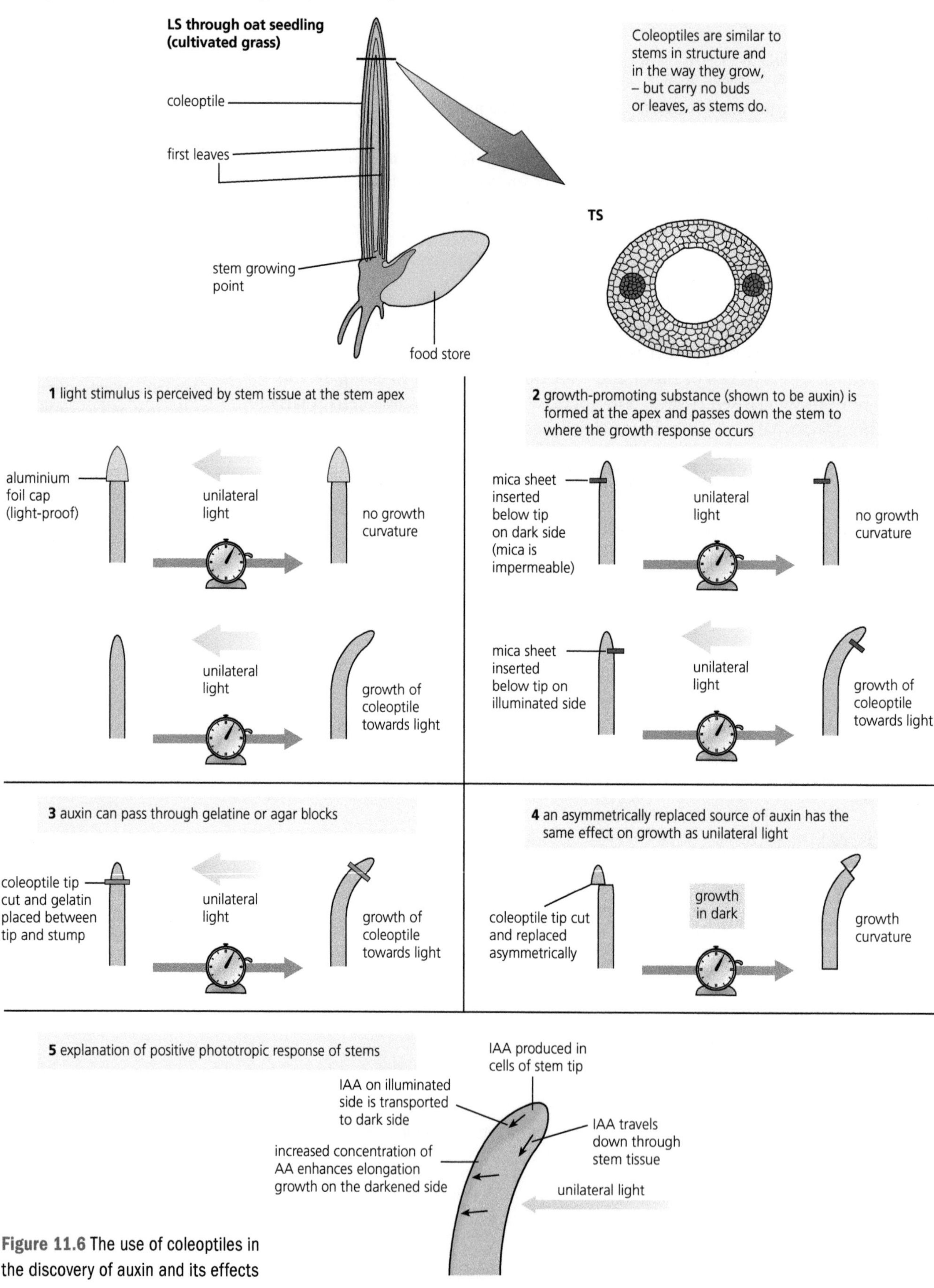

Figure 11.6 The use of coleoptiles in the discovery of auxin and its effects

The **coleoptile** is a sheath of tissue, unique to the grass family, which encloses the shoot of a germinating grass seedling as it grows up through the soil. It grows rather like a stem does, but it is uncluttered by leaves or buds so its growth is easily observed. Experiments have been conducted with oat seedling coleoptiles, the plant organ first used to investigate phototropism (Figure 11.6). In fact, experiments on the responses of oat coleoptiles to unilateral light led to the discovery of 'auxin', later shown to be indoleacetic acid.

Auxin is manufactured by cells undergoing repeated cell division, such as those found at the stem and root tips (and at the tip of coleoptiles). Consequently, the concentration of auxin is highest there. Auxin is then transported to the region of growth behind the tip, where it causes cells to elongate. In the process, the auxin is used up and inactivated.

In one experiment, the tip of the stem or coleoptile is cut off and stood on a gelatine block for a short while. Then the block is placed on a cut stump of stem or coleoptile. Growth in length is found to continue. The explanation is that auxin passes into the gelatine block, so when the gelatine is placed on the stump, the auxin passes down into the tissue and stimulates elongation of the cells. This technique has been used to investigate auxin actions further, as follows.

In stems and coleoptiles exposed to unilateral light, it is the auxin passing down the stem that is redistributed to the darkened side, causing differential growth and the curvature of the stem as shown in part 5 of Figure 11.6.

Cytokinins

Cytokinins are a group of related adenine-based compounds that act as plant growth substances. The basic structure of two of the cytokinins is shown in Figure 11.7. They are produced in meristematic tissues in the roots and shoots. Their principal mode of action is to stimulate cell division by attaching to receptor sites on cell surface membranes and triggering the formation of transcription factors inside the cell.

Like almost all plant growth substances, cytokinins rarely simply stimulate cell division alone, but undergo many interactions with other substances. In the case of cytokinin this is often an interaction with auxin (see Table 11.3).

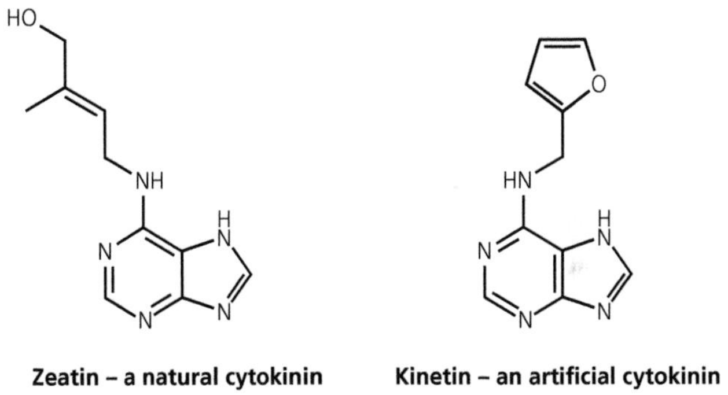

HO
NH
N
H
N
N
N
Zeatin – a natural cytokinin

O
HN
N
H
N
N
N
Kinetin – an artificial cytokinin

Figure 11.7 The structure of the cytokinins zeatin and kinetin

Table 11.3 A summary of some effects of three plant growth substances

Process	Auxin	Gibberellin	Cytokinin
Stem growth	Promotes cell elongation	Promotes cell elongation only with auxin	Promotes cell-division
Root growth	Promotes root formation in cuttings (rooting powder)	Inhibits root formation	No effect
Apical dominance	Promotes apical dominance	Enhances auxin effect	Promotes lateral bud growth (antagonistic to apical dominance)
Bud dormancy	No effect	Breaks dormancy	Breaks dormancy
Leaf fall (abscission)	Inhibits	No effect	No effect

Key term

Micropropagation A form of tissue culture used to grow many copies of a single plant using just a few cells.

One important interaction was first demonstrated by Skoog in the 1950s and is an important feature of plant tissue culture today. Plant breeders spend a long time isolating new varieties that may be more colourful or productive. If successful, the time taken to grow sufficient stock to produce seeds or cuttings on a large scale could be many years. This problem is overcome by using **micropropagation**. Micropropagation is a form of tissue culture that starts by carefully breaking down the meristems of young plants into individual cells and then growing the cells on sterile media. Technically only one cell is needed to form each new plant but usually it is a larger number. To ensure the cells begin to differentiate, the mixture of plant growth substances in the medium must be carefully controlled. Skoog found that the ratio of auxin to kinetin was critical.

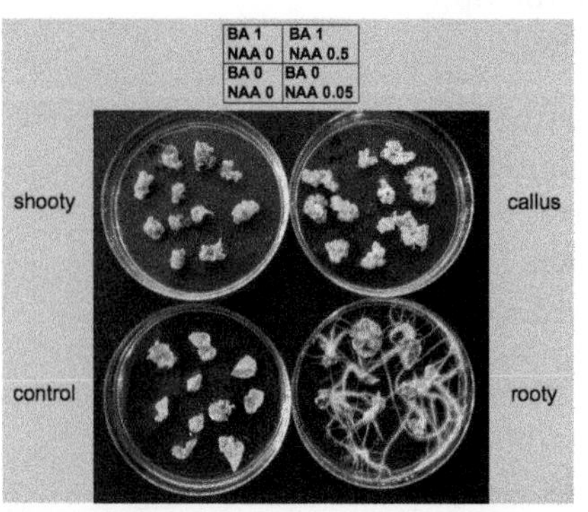

Figure 11.8 The effect of kinetin : auxin ratio on plant tissue differentiation

Figure 11.8 shows that a high kinetin : auxin ratio results in the cells forming shoots. With an intermediate kinetin : auxin ratio only an undifferentiated mass of cells (a **callus**) is formed and a low kinetin : auxin ratio results in the formation of roots. The inference is obviously that to form the correct balance of roots and shoots for a complete plant then the ratio of the two substances needs to be changed in subtle ways throughout development.

Gibberellins

As with auxins and cytokinins, there are a number of different chemical variations of gibberellins. These are usually abbreviated as GA with a number, such as GA_3 in Figure 11.9. Chemically gibberellins are complex terpenoids. Their main functions are control of **internode** length and seed development. You will not be surprised to learn that they are also involved with other growth substances in complex interactions (see Table 11.3 on the previous page).

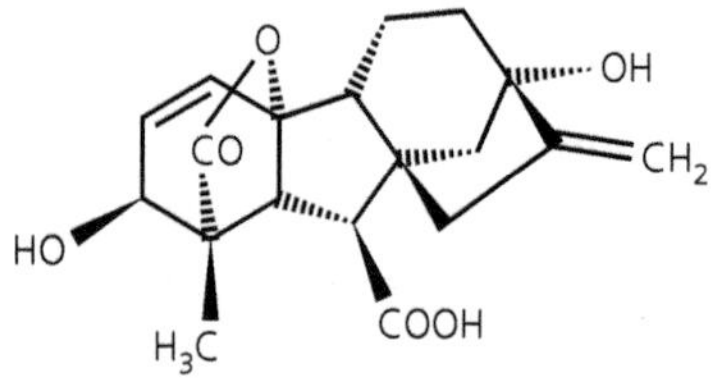

Figure 11.9 The structure of gibberellic acid (GA_3)

Key terms

Callus A mass of undifferentiated plant cells.

Internode The distance along a plant stem between one side branch and the next.

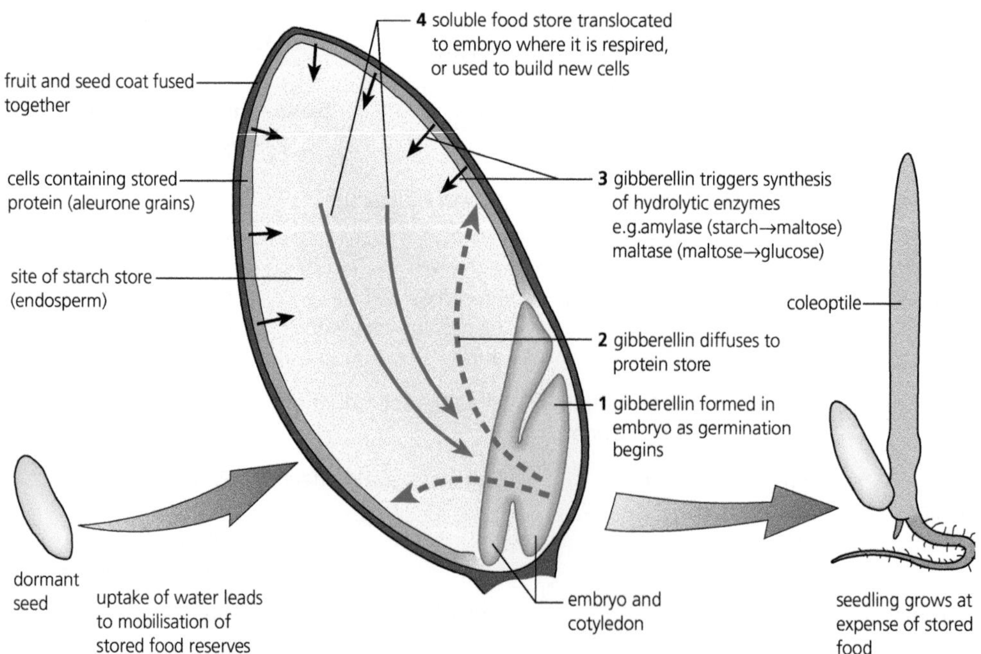

Figure 11.10 The role of gibberellin in the germination of barley fruit

Gibberellins are thought to bring about their effect by controlling transcription factors within the cell. In the cytoplasm they combine with proteins called DELLA proteins. These proteins normally prevent transcription factors from activating several important genes involved in seed germination. When in combination with GA, DELLA proteins no longer prevent these transcription factors from operating and hence the process of germination can begin, as shown in Figure 11.10.

Although there is an important biological difference between a fruit and a seed, and in the case of cereal grains they are a fruit, we will not go into technical details here. However, you might consider the fruits that you eat. The seeds are normally inside the fruit, for example oranges and tomatoes. In the case of cereals the wall of the fruit and the seed are dried out and joined closely together.

Test yourself

10 Name **one** process in plants where cytokinin has:
   a) an antagonistic effect to auxin
   b) a different effect to auxin.

11 Describe the role of a coleoptile in early seedling growth.

12 Suggest the effect on its growth of removing the tip of the coleoptile.

13 Explain how auxin causes an increased elongation of newly formed plant cells.

14 Auxin does not cause elongation in older cells. Explain why.

Core practical 14

Investigate the effect of gibberellin on the production of amylase in germinating cereals using a starch agar assay

Background information

Gibberellins are plant growth substances that have been shown to break dormancy in seeds. Dormancy means that the seeds will not normally begin to germinate until they have received certain treatments, even though all the other conditions may be favourable. This property is particularly important to plants that produce seeds in the autumn. Were the seeds to begin germination immediately, then the small seedlings would not survive the winter. Therefore dormancy delays germination for several months until spring, when the chances of survival are much higher.

The production of gibberellin by the seed embryo triggers the formation of amylases from the aleurone layer. The subsequent hydrolysis of starch to sugars within the endosperm provides the substrate for increased respiration and signals the start of germination (see Figure 11.10).

This assay uses soluble starch suspended in agar gel. When a fruit (seed) containing amylases is inserted into the gel the enzymes will diffuse outwards, hydrolysing the starch. If the plate is then flooded with iodine, any areas containing starch will turn blue-black in colour but any areas of hydrolysed starch will not. The greater the concentration of amylase, the larger the clear area will be.

Carrying out the investigation

Aim: To investigate the effect of increasing concentrations of gibberellin (GA) on the production of amylase in cereal grains.

Risk assessment: The GA solutions are extremely weak and pose no hazard. Many cereal grains will have been treated with anti-fungal seed dressings before sale, so it is essential to handle them with disposable gloves before rinsing them thoroughly before use. Even though bacterial cultures are not used, good sterile technique is needed to avoid contamination. Plain agar, not nutrient agar should be used. Nutrient agar grows bacteria and some fungi, plain agar contains no nutrients and so poses less microbiological risk. The starch agar plates and seeds should be disposed of by wrapping well and placing in the normal refuse immediately after the practical.

1 First of all you will need to make up a suitable range of concentrations of GA solutions. GA is quite expensive but fortunately you need only very dilute solutions. The relative molecular mass of GA is 346 and therefore a 1 M solution would need 346 g dissolved in 1 dm^3. This is far too strong, so a stock solution of 10^{-3}M (or 0.001 M) is advised (0.346 g dissolved in 1 dm^3).
Some research will show that you actually need a good range of concentrations from 10^{-3}M to 10^{-6}M. You can do this by taking 1 cm^3 of your stock solution and adding 9 cm^3 of deionised water. This will dilute the solution ×10 so you will now have a 10^{-4}M solution. Repeating this process of serial dilution will provide you with a range of GA solutions to use. To avoid repeated random errors you must use the most accurate measuring apparatus you have available and mix each dilution well before starting the next.

2 You can use wheat, oat or barley fruits for this investigation. Rinse them several times in distilled water to wash off any dressing and peel the brown outer covering (the husk) off 20 fruits. Cut each one laterally in two as shown in Figure 11.11 and discard the bottom half containing the embryo. It is important that you use only the half without the embryo.

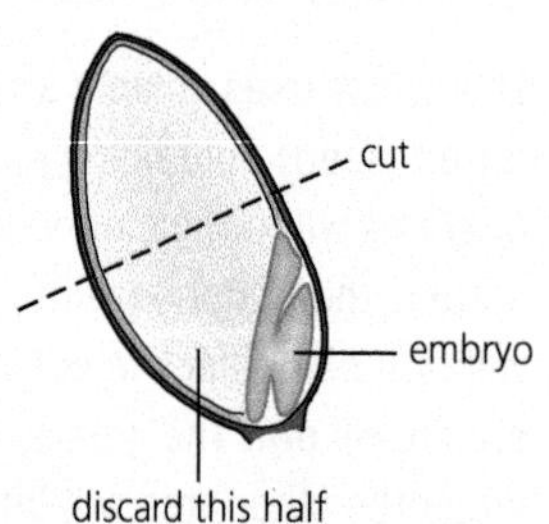

Figure 11.11 Discarding the embryo

3 Soak four halves of fruits in each of your prepared dilutions of GA and four in deionised water, for about 24 hours. During this time, prepare a starch agar (5 g soluble starch, 14 g nutrient agar and 500 cm^3 sterile water) Petri-dish plate for each of the dilutions you have chosen and label them clearly.

4 To prepare each plate, rinse the soaked half fruits in a dilute hypochlorite solution (Milton is fine) to prevent growth of fungi on the plates.

5 Dry the half fruits carefully and push each one gently into the starch agar until it just breaks the surface, spacing them as evenly as possible. Place the lid on the Petri dish and leave it in a warm place for a further 24 hours.

6 Finally, remove the lid and flood the plate with just enough iodine to cover the surface. Leave it until a dark blue colour develops. Rinse off the iodine and measure the diameter of the clear area around each fruit four times, calculating a mean for each one.

7 Record and present your data in a suitable way.

Questions

1 Why are the concentrations written in the form 10^{-3}M rather than 0.001 M?

2 Why do you need to make certain you only use the half fruit that does not contain the embryo?

3 Is it reasonable to assume the diameter of the clear area is proportional to the concentration of amylase in the fruit?

4 Why measure the diameter on one circle four times and take the average?

5 Would it be better to calculate the area of the cleared circles?

6 How can you treat your data to take account of any amylase produced by the distilled water control?

7 If you test your data statistically for a significant correlation (see Chapter 15) would this mean that you have shown that gibberellin does actually stimulate amylase production in cereals?

This is an excellent opportunity to practise some of the mathematical skills that will make up 10 percent of your final papers (see Chapter 15).

Interactions between plant growth substances

Interactions between substances can be of two main types. When the result of the interaction is that the overall effect is greater than the sum of the effects of the substances acting alone we say that the effect is **synergistic**. When the overall effect is less than the effects of the substances alone we say the effect is **antagonistic**.

One example of antagonistic interaction is seen in the control of **apical dominance** in plants. Apical dominance is a very common feature in the growth of many plants. The main shoot of a plant has a growing point (meristem) at its tip. In most plants this point continues to grow upwards in order to gain more light for photosynthesis while the growth of side shoots, which would keep the plant much lower to the ground, is inhibited. In other words the apex is dominating the growth form. So how is this brought about?

> **Key terms**
>
> **Synergism** When the effect of two substances acting together is greater than the sum of the effects of the substances acting alone.
>
> **Antagonism** When the effect of two substances acting together is less than the sum of the effects of the substances acting alone.
>
> **Apical dominance** The inhibition of lateral bud growth in plants by the presence of a terminal bud at the main growing point.
>
> **Axillary bud** A small dormant bud found in the angle (axil) between a main stem and side shoot of a plant.

The classical model

This suggests that the inhibition of side shoots is a result of the action of auxin produced in the growing tip, which is transported downwards. Side shoots will normally grow from **axillary buds** found in the angle between the branches and the main stem. Note that these are axillary not auxiliary buds as the angle between the stems is called the axil. Auxin is transported downwards, possibly via phloem, and prevents the activation of the genes leading to the production of cytokinins, which would stimulate the axillary buds to begin cell division and the growth of new stems, so the bud remains dormant.

As gardeners will know, one simple way of producing more bushy plants (with more side shoots) is to pinch out the main growing point (terminal bud). This removes the main source of the inhibitory auxin and therefore the axillary buds below begin to grow under the influence of cytokinin. This does not continue for very long since the new side shoots have growing points, which also produce auxin. The upper shoots therefore restore apical dominance. In this example auxin and cytokinin act as antagonists. Careful pruning of trees and shrubs seeks to make use of this idea to produce a desired shape and form.

Many plants have a shape that is wider at the base than it is at the top. This is a consequence of incomplete apical dominance. The buds at the very base of the plant are a long way away from the auxin-producing tip. Therefore the concentration and inhibitory effect of the auxin are much less, so there tends to be more lateral growth lower down the plant. This has the advantage of allowing lower branches to grow outwards further to avoid shading of their leaves by higher branches.

Evidence to support the classical model

There is some evidence that cytokinin levels do rise when terminal buds are removed and that the presence of auxin inhibits the biosynthesis of cytokinin. Application of cytokinin to dormant axillary buds can stimulate them into growth. However, this is not always true.

Evidence lacking or contradictory

The levels of auxin found in axillary buds are not always sufficient to account for inhibition. The predicted effects (such as the effect of applying cytokinin) do not always take place.

Other scientists have suggested different models concerned with the transport of auxin, which is partially contradictory to the classical model.

This is a very good example of the way in which scientific advances are made. Models are suggested and can be used to make predictions. It is these predictions that need to be tested experimentally in order to support (or undermine) the model. As more evidence accumulates in support of the model, it becomes more and more accepted.

In the case of apical dominance the classical theory appears to be such a logical explanation that it has become a very common feature in many text books. However, it is a very long way from becoming a widely accepted 'fact' amongst many plant physiologists and it seems that, at the present time, it must be accepted as only a partial explanation at best. As a scientist a key question is always 'How do I know that?' and at A level it is expected that you will begin to question what you see reported in more depth.

Test yourself

15 The effect on stem growth of auxin and cytokinin together can cause a greater effect than the two individually. State the name of this type of effect.

16 Suggest why auxin with cytokinin might cause greater stem elongation than if the two acted alone.

17 The amylase genes are inactive in dormant seeds. Explain why.

Phytochrome

Phytochrome is a blue–green pigment present in green plants in very low concentrations. The amount of phytochrome is not sufficient to mask chlorophyll, and it has been difficult to isolate and purify the substance from plant tissue, although this has been done. Phytochrome is a very large conjugated protein (protein molecule and pigment molecule, combined) and it is a highly reactive molecule. It is not a plant growth substance, but it is a photoreceptor pigment, able to absorb light of a particular wavelength and change its structure as a consequence. It is likely to react with different molecules around it, according to its structure.

We know that phytochrome exists in two inter-convertible forms. One form, referred to as P_R, is a blue pigment that absorbs mainly red light of wavelength 660 nm (this is what 'R' stands for). The other form is P_{FR}, a blue–green pigment that absorbs mainly far-red (FR) light of wavelength 730 nm. When P_R is exposed to light (or red light on its own), it is converted to P_{FR}. In the dark (or if exposed to far-red light alone), it is converted back to P_R:

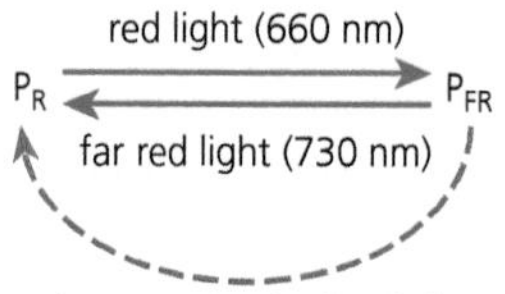

P_R —red light (660 nm)→ P_{FR}

P_{FR} —far red light (730 nm)→ P_R

P_{FR} - - slow reaction in the dark - -→ P_R

Key terms

Phytochrome A conjugated protein in plant cells thought to be responsible for detecting changes in red and far-red light illumination.

Photomorphogenesis The influence of light on plant growth and development.

The influence of light on plant growth and development is known as **photomorphogenesis**. Phytochrome is the pigment system involved in photomorphogenesis. We know this because the red/far-red absorption spectrum of phytochrome corresponds to the action spectrum of some specific effects of light on development.

Phytochrome and the control of flowering

It appears that it is P_{FR} that is the active form of phytochrome in photomorphogenesis, stimulating some effects in plant development and inhibiting others.

One effect of light on plant growth and development is its role in determining the switch from vegetative growth to the production of flowers (reproductive growth). You will be aware that most plants flower at different and particular times of the year. In fact, most species have a precise season when flowers are produced. How is flowering switched on by this environmental condition? The answer is that day length provides important signals and these are mediated by phytochrome (Figure 11.13).

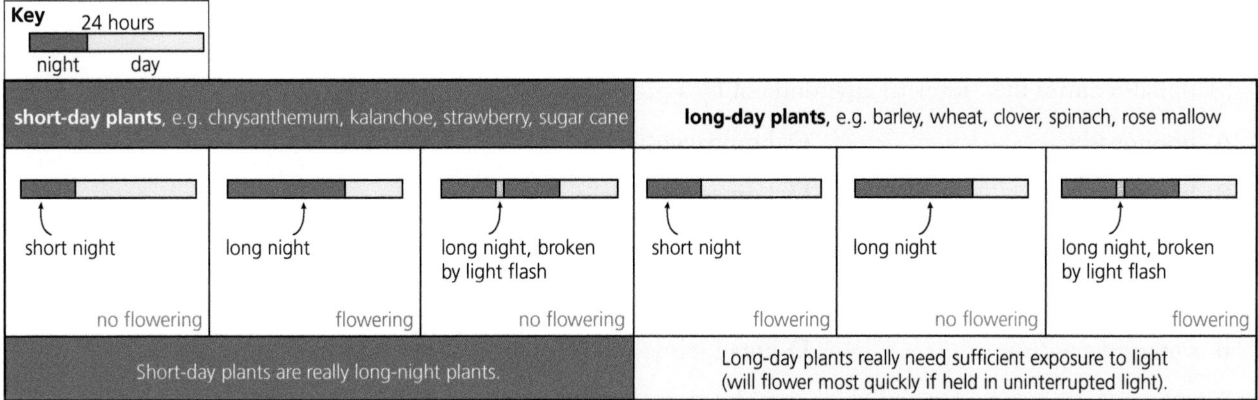

Figure 11.13 Flowering related to day length

Figure 11.14 shows the current model of how phytochrome may detect changes in day-length and so influence flowering. It is known that mRNA molecules and proteins, coded for by specific genes, can also function as growth substances. It is molecules of this sort that might be transported about the plant via the plasmodesmata and the symplast pathway. Currently it is suggested that a gene ('flowering locus' – *FT*) is activated in leaves of photoperiodically-induced plants. As a consequence, it is *FT* mRNA that then travels from induced leaves to stem apex. In the cells there, the *FT* mRNA is translated into *FT* protein. This protein, bonded to a transcription factor, activates several flowering genes and switches off the genes for vegetative growth.

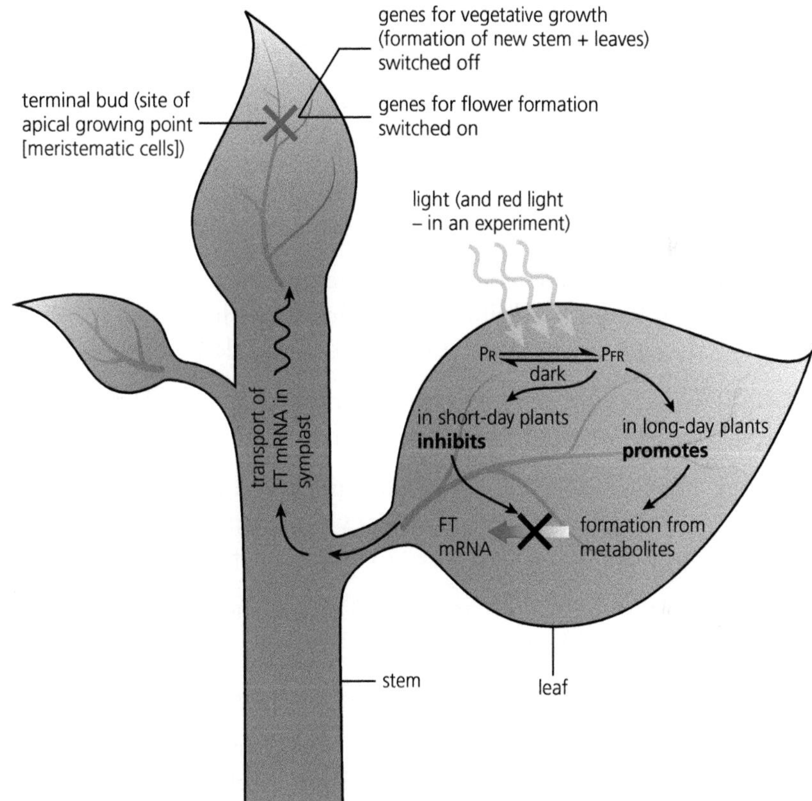

Figure 11.14 Phytochrome and flowering – the suggested model

Other responses to light are also likely to use phytochrome as their receptor:

- The synthesis of chlorophyll – plants kept in the dark have little chlorophyll but when exposed to light they quickly become green.
- The germination of some seeds is heavily influenced by exposure to red and far-red light.
- Plant leaves of the same species often have different size and shape when grown in low or bright light conditions.

Test yourself

18 What is 'far-red' light?

19 State which form of phytochrome will build up when the nights are much longer than the days.

20 Suggest what will cause shade leaves to have more chlorophyll content than brightly lit leaves.

Exam practice questions

1 Mammals control their internal environment by a process called:

A homiopathy **C** homozygosity

B homeostasis **D** homeotrophy *(1)*

2 The actions of insulin and glucagon can be described as:

A agonistic **C** symplastic

B synergistic **D** antagonistic *(1)*

3 Which of the following pairs of plant growth substances act synergistically in their effects on cell enlargement?

A auxin and gibberellin **C** cytokinin and auxin

B cytokinin and phytochrome **D** gibberellin and phytochrome *(1)*

4 **a)** Explain how plant leaves can detect changes in the relative lengths of day and night. *(4)*

b) Explain why it is an advantage for some plants to flower as the nights become shorter and some plants to flower as nights become longer. *(3)*

> **Tip**
>
> Part (a) of Question 4 is simple recall (AO1) but good practice in organising your explanation to match the marks available. Part (b) asks you to apply your knowledge (AO2) with a little more thought.

5 **a)** Steroids are compounds closely related to lipids but without their long fatty-acid chain. Explain why steroid hormones are able to pass through cell surface membranes but peptide hormones are not. *(2)*

b) Explain how peptide hormones can exert their effect without passing through the cell surface membrane. *(3)*

c) Adrenaline (epinephrine) is a peptide. The diagram on the right shows two amino acid molecules. Draw a diagram to show how these two molecules would be joined to form a peptide. *(3)*

d) Hormone receptors are integral membrane proteins that are each specific to one molecule.

i) Explain what is meant by 'integral membrane proteins'. *(1)*

ii) Explain how the properties of proteins make them particularly suitable molecules to form large numbers of highly specific receptors. *(3)*

> **Tip**
>
> Question 5 is a synoptic question that illustrates the need for knowledge from all parts of the specification in a single question, and also encourages you to keep refreshing your knowledge of earlier sections of *Edexcel A level Biology 1* rather than relying on the much more difficult task of attempting to revise everything at once.

6 An investigation was carried out to determine the effect of reduced oxygen concentration on the action of gibberellin (GA_3), leading to the metabolism of starch stored in the endosperm of seeds such as oats. Ten identical oat seeds, with their embryos removed, were placed in each of five GA_3 solutions of different concentrations. These solutions were then incubated at 25 °C for 48 hours with a stream of oxygen bubbled through them. Samples of each solution were then taken and analysed for reducing sugar content. The investigation was then repeated with the solutions in a sealed container and no oxygenation. The results of the investigation are shown in the graph.

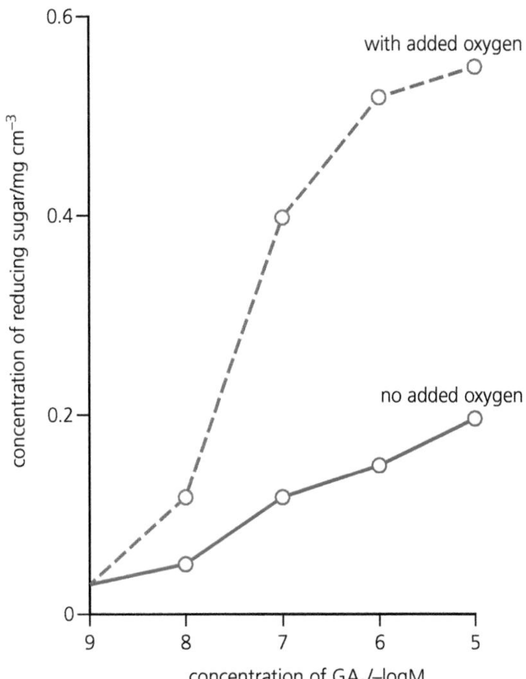

Tip

Question 6 is typical of the type of question you might expect in Paper 3. It requires some knowledge of Core practical 14 from this chapter but also other practical and mathematical skills. It does have a different application and it is less likely that you will get simple recall questions based on core practicals, rather than a test of your understanding of the principles. It is also a synoptic question as knowledge of other parts of the specification are needed for part c) ii). If you are uncertain about logarithms and scale this is also a good opportunity to check the data units in the graph.

a) Explain why it was necessary to remove the embryos from the seeds for this investigation. *(2)*

b) Explain how the action of GA_3 causes an increased concentration of reducing sugar in the solution. *(4)*

c) i) Use the data to calculate the percentage increase in reducing sugar in the solutions of log-5 molar GA_3 caused by the addition of oxygen. *(3)*

ii) Explain why a lack of oxygen will have the effect shown by these data. *(3)*

Stretch and challenge

7 Since the discovery of plant growth substances, many variations with similar effects have been produced artificially.

One of the most well-known examples of this is an artificial auxin (IAA) called 2-4,D (2-4 dichlorophenoxyacetic acid). This has been widely used as a selective weedkiller for grassland and lawns.

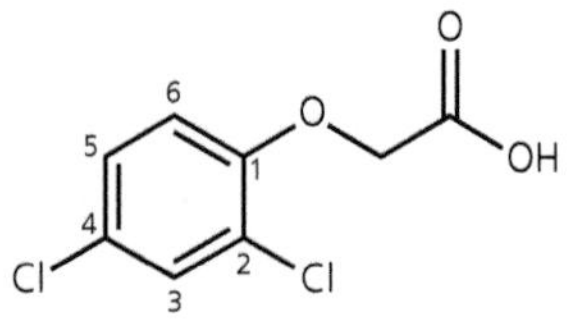

a i) What similarities are there between the two molecules that could result in similar effects in plants?

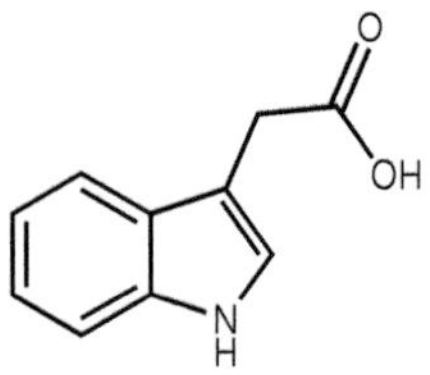

ii) Why produce artificial plant growth substances instead of using the original substance?

iii) Why is it important that there are similarities in the molecules if they are to act in a similar way?

iv) Why does watering a lawn with moderate doses of 2-4,D kill off broad-leaved weeds such as plantain and dandelion but not the narrow-leaved grass?

v) What was 'agent orange' and why has it caused concern?

b) i) Over half of the commercial varieties of grapes grown for worldwide consumption are now seedless. This is good for the preference of consumers but how can this be achieved biologically?

ii) 'Thompson' white grapes are the most popular variety found in supermarkets. Why would growers spray their crops with dilute solutions of gibberellins?

O
6
5
O
1
OH
4
Cl
2
Cl
3

2-4,D

O
OH
N
H

IAA

12 Coordination and control in action

Prior knowledge

In this chapter you will need to recall that:

- the mammalian eye has three layers – the inner retina, choroid and sclerotic coat
- the retina is the light-sensitive layer with connections to the optic nerve
- the retina is made up of rod and cone cells
- rod cells respond to black and white images, cones to colours
- rod cells contain a pigment called rhodopsin
- most images are focused on an area of the retina, opposite the centre of the lens, called the fovea
- the kidneys are responsible for controlling the concentration of body fluids and the excretion of urea
- urea is synthesised in the liver but removed by the kidneys
- kidneys are paired organs with a high-pressure blood supply directly from the main dorsal aorta
- kidneys are connected to the bladder by a tube called the ureter
- mammalian heart rate is controlled by nerves and hormones (*Edexcel A level Biology 1*, Chapter 11)
- countercurrent flow increases total exchange by extending diffusion gradients (*Edexcel A level Biology 1*, Chapter 10).

Test yourself on prior knowledge

1. State the function of the choroid layer of the eye.
2. Name the vitamin important for good vision.
3. State the name given to the point where the optic nerve joins the retina.
4. Explain how the eye is able to regulate the amount of light falling on the retina.
5. State what happens to rhodopsin when it absorbs light.
6. Describe the difference between urine and urea.
7. State the name given to the whole process of adjusting the contents of the blood in the kidney.
8. Suggest why it is important that kidneys are supplied with high-pressure blood.
9. Which hormone has the most important effect of increasing heart rate?
10. Give an example of a countercurrent exchange mechanism in gas exchange.

Introduction

In this chapter you will be looking at how several of the important concepts that you have met in previous chapters are employed to bring about control and coordination of major physiological activities. This relies on your knowledge of the content of Chapters 10 and 11 in this book and it is useful to review these carefully before continuing.

To understand how the kidney selectively reabsorbs substances before they are excreted, discussed at the end of this chapter, you will also need to be familiar with the principles of active transport, osmosis, diffusion and facilitated diffusion from *Edexcel A level Biology 1*, Chapter 9. Once again, this provides an ideal opportunity to reinforce your knowledge of all of the course material in easy stages.

Detection of light by mammals

All control and coordinating mechanisms require information if they are to be effective. Sense organs are needed to detect any need for change and to monitor the results of any actions. This may be a need to respond extremely quickly to a threat from a predator or to coordinate long-term growth and development. You will look at examples of different types of sense organs later in this chapter but first you will consider the way in which mammals detect light.

Our example will be the human eye but it is important to remember that the eyes of other animals have even more remarkable adaptations according to their niche. Birds of prey, such as eagles, are able to focus on small prey from a height of several hundred metres and nocturnal animals have eyes adapted to the detection of very low light levels (but even they cannot 'see' in total darkness!).

Apart from obvious functions such as reading and navigating, vision is also a vital part of other functions such as balance. Just try balancing on one leg with your eyes closed!

Detection of light by the retina

The structure of the retina

Sense organs respond to certain stimuli and need to initiate action potentials if they are to communicate via the nervous system. This is the function of the retinal cells.

The retina of each eye is sensitive to light in the wavelength range 380–760 nm – that is, the visible range of the electromagnetic spectrum (the radiation from the Sun). The retina has two types of light-sensitive cell, the rods and cones, shown in Figure 12.1. These very elongated cells have an outer part called the outer segment. This consists of flattened membranous vesicles housing a light-sensitive pigment. An inner segment contains many mitochondria (the site of ATP formation).

Rods are far more numerous than cones; the human retina contains about 120 million rods compared with 6 million cones. Rods are distributed evenly throughout the retina, while cones are concentrated at and around a region called the **fovea**. This is an area where vision is most accurate – here there is the greatest density of photoreceptors. (Note that light passes through the neurones, synapsing with the rod and cone cells before reaching the outer segments of these cells. Because of this feature, the retina is described as 'inverted' (Figure 12.1).)

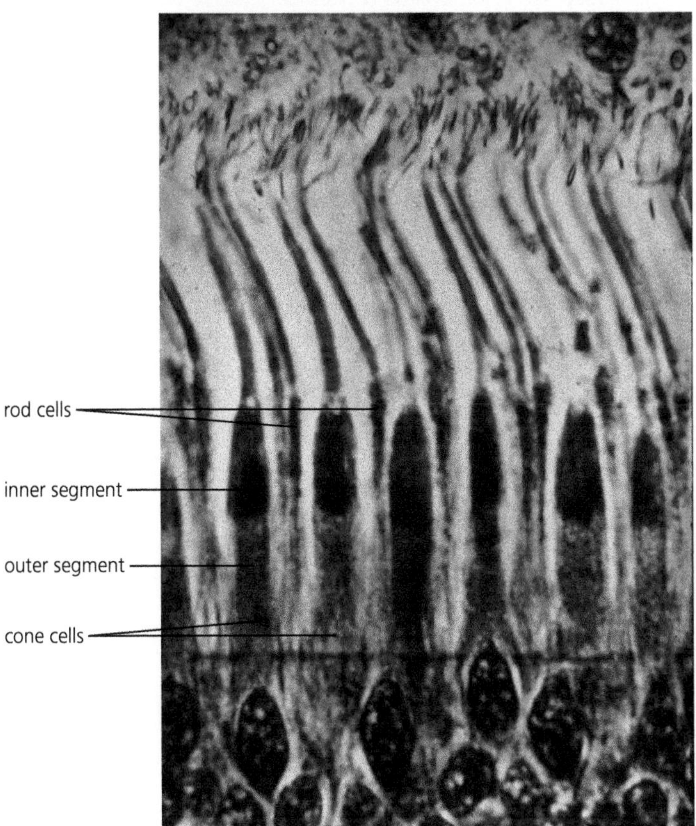

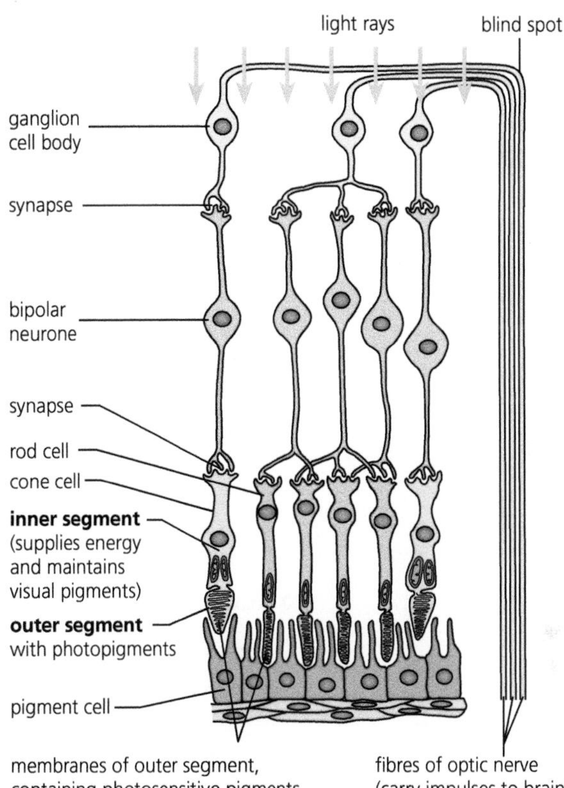

Figure 12.1 The structure of the retina

Rod cells are extremely sensitive to light, much more sensitive than the cones, but rod cells do not discriminate colours. Since they respond to lower light intensities than cones, they are principally used for dim light and night vision. The type of visual pigment molecules housed in the rods is called 'visual purple' or **rhodopsin**. This molecule is a combination of a protein (**opsin**) and a light-absorbing compound derived from vitamin A, called **retinal**. Remember, a diet deficient in vitamin A causes 'night blindness' – the inability to see in low light intensities. Figure 12.1 also shows that each cone cell synapses with only one optic nerve fibre whereas a single optic nerve fibre has synapses connecting it to several rod cells, a phenomenon known as **convergence**. This means that in bright light, cone cells can not only provide a coloured image but also a much more accurate image. However, convergence does have advantages as it means that light can be gathered from a larger area to produce an action potential in dim light.

The accuracy of the image is expressed as **visual acuity**. This is the ability of the eye to distinguish between two points close together. The visual acuity of the cones is much higher than that of rods. As points become increasingly closer and closer, there comes a time when they are so close that despite stimulating two different rods, these rods share the same optic nerve fibre so only one signal is sent to the brain, which interprets this as one point. Stimulation of two cones will always result in two separate impulses until the points are so close that they cover the same cone cell. This, and the tightly packed cones in the fovea, account for the sharp image formed in bright light.

Test yourself

1 Describe the position of the fovea on the retina.
2 Explain the term 'convergence'.
3 Suggest why convergence produces a low visual acuity.
4 State the breakdown products of rhodopsin.
5 If trying to observe a very faint star at night, it is a good idea to look to one side of the star, not directly towards it. Explain why.

Initiating action potentials in rod cells

The ways in which a rod cell responds to stimulation by light, and to its absence, are detailed in Figure 12.2.

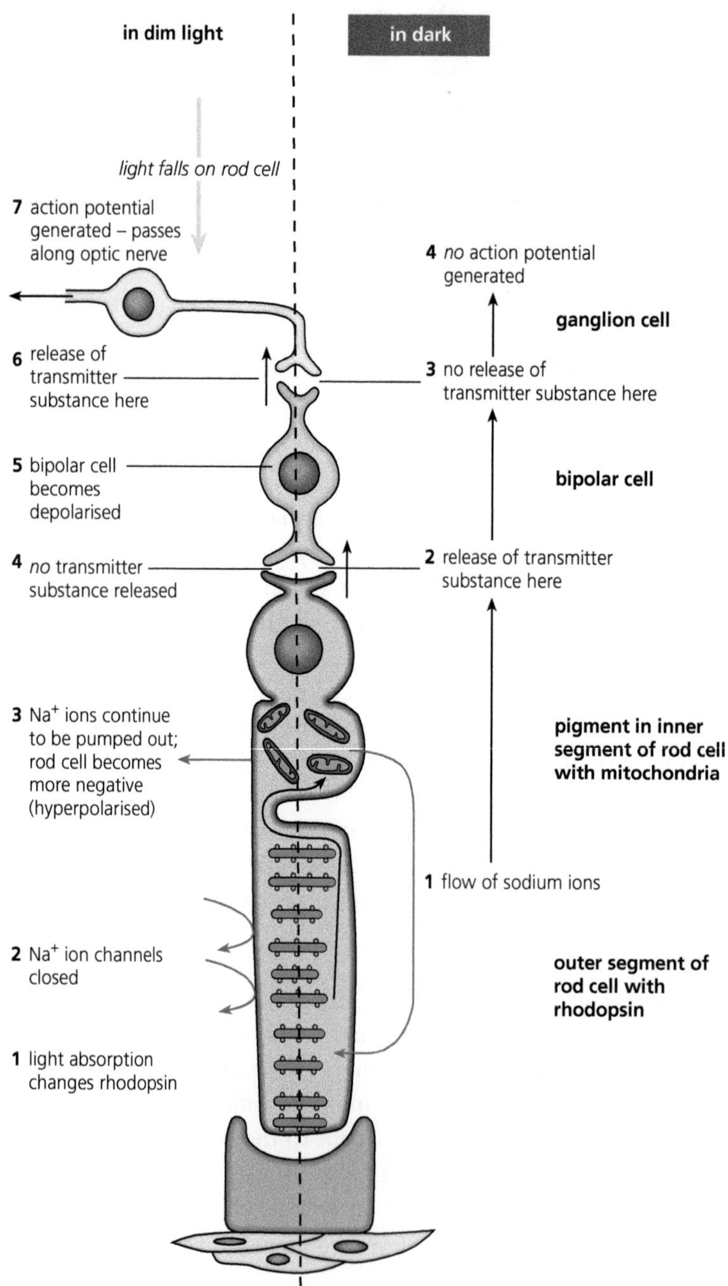

Figure 12.2 The working rod cell

Follow the numbered steps in Figure 12.2 carefully as you read the text below.

Rod cells in the dark

In the dark, there is a steady flow of sodium ions into the outer segment, through open cation channels, located in the cell surface membrane. At the same time, the cell surface membrane of the inner segment reverses this movement of ions – sodium ions are continuously pumped out of the cell at this point. This efflux occurs at the expense of ATP formed in the mitochondria of the inner segment. Consequently a concentration gradient is maintained between outer and inner segments, down which the sodium ions flow. Meanwhile, the influx of sodium ions at the inner segment slightly depolarises the cell, and the potential difference across the cell surface membrane is about −40 mV (compared with a resting potential of −70 mV).

Under these conditions, the rod cell releases a neurotransmitter substance (glutamate) that binds to the bipolar cell and prevents its depolarisation. The consequence is that no action potential is generated in the optic nerve that synapses with that bipolar cell.

Rod cells in the light

When light falls on the retina, it causes reversible structural change in rhodopsin (called 'bleaching'), breaking it down into retinal and opsin. Opsin now functions as an enzyme that activates a series of reactions resulting in the closing of the cation channels of the outer segment, and so the influx of sodium ions is blocked. Meanwhile the inner segment continues to pump out sodium ions. This causes the interior of the rod cell to become more negative – a state described as **hyperpolarisation**. In this condition, no neurotransmitter is released by the rod cell, and the bipolar cell becomes depolarised.

The bipolar cell releases a transmitter substance. An action potential is generated in a neurone of the optic nerve serving the rod cell. This action potential is transmitted to the visual cortex of the brain.

Meanwhile, the structure of rhodopsin is rebuilt, using energy from ATP. In very bright light, all the rhodopsin is bleached. In these conditions you are using cone cells, and the state of the visual pigment in rod cells is not of immediate consequence. In fact, you are not aware your rods cells are temporarily non-functional. But if you move suddenly from bright to very dim light it takes time for sufficient reversing of bleaching to occur, and you are temporarily blinded. We say your eyes are 'adapting to the dark'.

The presence of both rods and cones means that the brain is able to form an image in a wide range of light intensities. Even though the image in low-light conditions is less accurate and not coloured, a retina of cones only would leave you blind in anything other than bright daylight so this is a good compromise.

The role of cone cells

Animals that have cone cells in their retinas are able to distinguish colours. It is not the case for all mammals, but the human eye does contain cones, concentrated in the fovea where light is most sharply focused. Cone cells operate on the same principle as the rod cells, but with a different pigment, called **iodopsin**. This is less readily broken down; it needs more light energy. Cones work only in high light intensities; we cannot see colours in dim light.

According to the **trichromatic theory** of colour vision, there are three types of cone cell present in the retina, each with a different form of iodopsin. These absorb different wavelengths of light – in the blue, green and red regions of the spectrum. White light stimulates all three types equally, but different colours are produced by the relative degree of stimulation of the three types of cone.

Processing action potentials from the retina

Observations of the three-dimensional world around you are reduced to two-dimensional images on the surface of the retina. As a consequence, action potentials generated in the rods and cones are carried by neurones of the optic nerves to the visual cortex of the brain. While each eye views left and right sides of the visual field, the brain receives and interprets action potentials from the right and left visual fields on the opposite side of the visual cortex. This is known as **contralateral processing** (Figure 12.3).

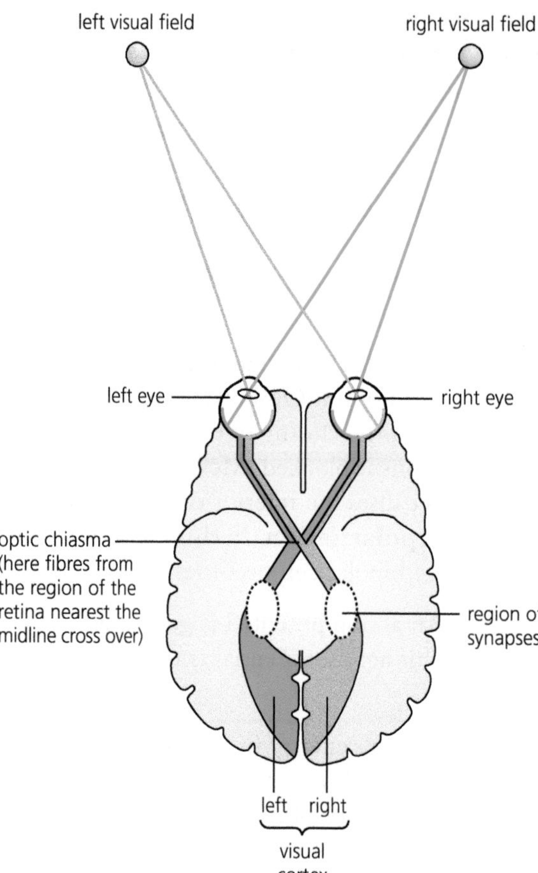

Figure 12.3 The eye and the visual cortex - the pathways of impulses

The messages from interpretation of these action potentials are combined by the brain to produce a single impression – your sight. 'Seeing', therefore, occurs largely in the brain, and the seeing process, known as **perception**, is complex. This is because perception involves the interpretation of sensory data from the retina in terms of existing and past experiences and your expectations. The phenomenon of perception has implications for the nature and reliability of visual sense data, too (and, therefore, for the processes of science).

Test yourself

6 Glutamate binds to receptors to produce small IPSPs (inhibitory post synaptic potentials) (see nervous transmission, Chapter 10). Explain why these IPSPs prevent depolarisation of the bipolar cell.

7 Name the breakdown product of iodopsin that triggers reactions which block the influx of Na^+ ions into the outer segment.

8 In normal daylight almost all visual information received by the brain comes from cones in the fovea. Explain why this is.

9 Suggest why iodopsin is more difficult to break down than rhodopsin.

Control of heart rate in mammals

The control of heart rate is an example of control using the autonomic nervous system. You have looked at the detailed structure of this system in Chapter 10 of this book and this is a good time to refresh your memory.

The heart beats rhythmically throughout life, without rest, apart from the momentary relaxation between beats. Even more remarkably, the origin of each beat is within the heart itself – we say that heart beat is myogenic in origin.

The heart beat originates in a structure in the muscle of the wall of the right atrium, called the **sino-atrial node** (**SAN**), also known as the natural pacemaker. Muscle fibres radiating out from the SAN conduct impulses to the muscles of both atria, triggering atrial systole (contraction).

Then a second node, the **atrio-ventricular node** (**AVN**) situated at the base of the right atrium picks up the excitation and passes it to the ventricles through modified muscle fibres, called the **Purkyne fibres** (Figure 12.4). Ventricular systole is then triggered.

After every contraction, cardiac muscle has a period of insensitivity to stimulation, known as a **refractory period** (in effect, a period of enforced non-contraction, which we may call a 'rest'), when the heart refills with blood. This period is a relatively long one in heart muscle, and doubtless an important feature, enabling the heart to beat throughout life.

The heart's own rhythm, set by the SAN, is about 50 beats per minute, but it is essential that this can be modified according to the ever-changing demands of the body. In humans this can vary from about 70 beats per minute at rest, to about 200 beats per minute during very strenuous exercise.

Tip

Although we shall give a brief summary of the transmission of impulses through cardiac muscle by the SAN and AVN here, more details of this process and the sequence of events in one heartbeat have been discussed in *Edexcel A level Biology 1*, Chapter 11. This is another ideal opportunity to revise work covered earlier in your course and to make links with other parts of the specification in preparation for synoptic questions in the examination papers.

Key term

Myogenic Activity that originates within muscles rather than through the nervous system.

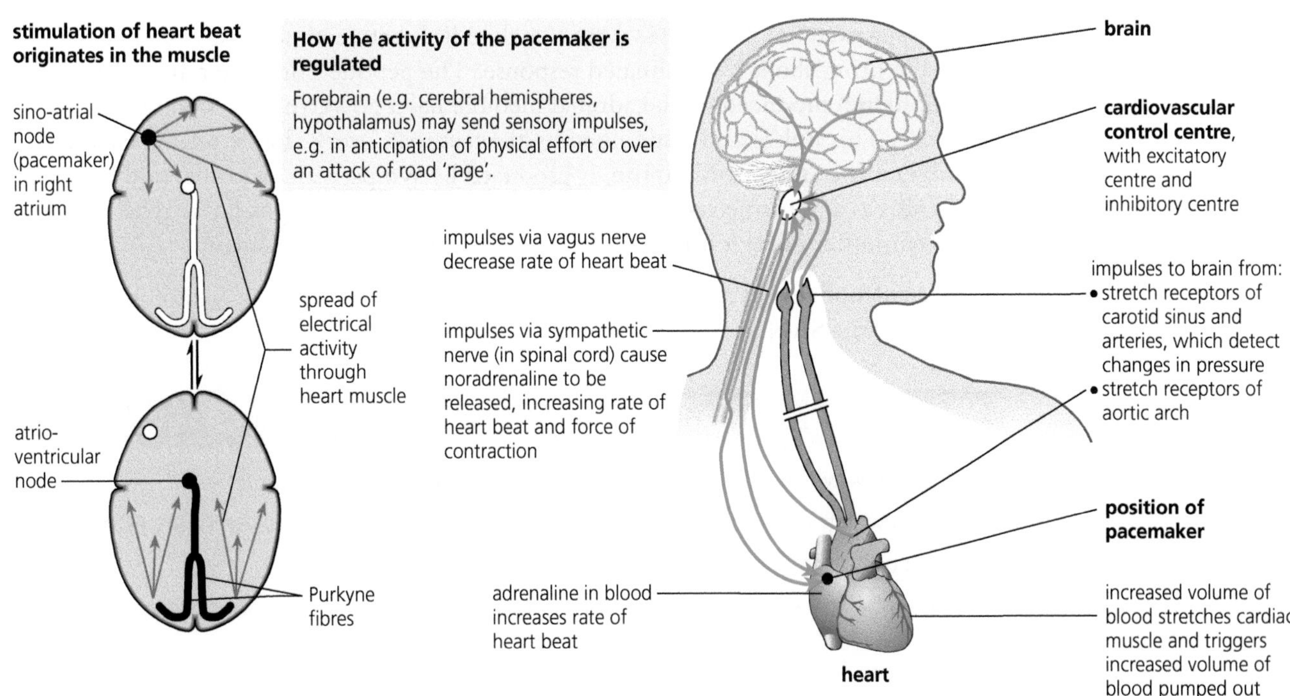

Figure 12.4 Control of heart rate

The main centre for control of heart rate is situated in the **cardiovascular centre** in the **medulla** of the hindbrain (see Chapter 10). This has both a cardiac inhibitory region and a cardiac accelerator region.

The heart receives impulses from the cardiovascular centre via two nerves:

- a sympathetic nerve, part of the sympathetic nervous system, from the accelerator centre
- a branch of the **vagus nerve**, part of the parasympathetic nervous system, from the inhibitory centre.

Since the sympathetic nerve and the vagus nerve have opposite effects in this matter of regulation of heart beat, we say they are **antagonistic** – a typical feature of the sympathetic and parasympathetic systems.

Key terms

Stretch receptors
Sensory receptors that initiate action potentials in response to changes in tension. Also called baroreceptors.

Chemoreceptors
Sensory receptors that initiate action potentials in response to changes in their chemical environment.

In order to respond correctly the cardiovascular centre needs information from sense organs. For example, it receives impulses from **stretch receptors** (baroreceptors) located in the walls of the aorta, in the carotid arteries and in the wall of the right atrium, when changes in blood pressure occur at these positions.

When blood pressure is high in the arteries, the rate of heart beat is lowered by impulses from the cardiovascular centre, via the vagus nerve. When blood pressure is low, the rate of heart beat is increased.

The oxygenation of the blood is also a strong factor in controlling heart rate. **Chemoreceptors** in the main arteries also provide information to the cardiac centre through sensory nerves. These are sensitive to pH levels. Rising carbon dioxide levels in the blood lower the pH during exercise and trigger impulses, which stimulate the cardiac accelerator centre and increase impulses in the sympathetic nerve.

The rate of heart beat is also influenced by impulses from the higher centres of the brain. For example, emotion, stress and anticipation of events can all cause impulses from the sympathetic nerve to speed up heart rate.

As was suggested in Chapter 11, there is a close link between nervous and endocrine control to bring about a coordinated response. The peptide hormone **adrenaline** (**epinephrine**) produced by the adrenal medulla has a powerful effect on raising heart rate. This link is also underlined by the fact that a very closely related chemical **noradrenaline** (**norepinephrine**) is produced by sympathetic neurones and has the same effect as adrenaline on the heart. The parasympathetic vagus nerve releases the neurotransmitter **acetylcholine** in bringing about its inhibitory effect.

Both acetylcholine and noradrenaline are released by their respective autonomic neurones at the SAN.

Test yourself

10 Explain why the stimulation of the heart is said to be 'myogenic'.

11 State which division of the autonomic nervous system increases heart rate.

12 Which part of the brain contains the cardiovascular centre?

13 Describe the sensory information received by the cardiovascular centre.

Temperature regulation

The regulation of body temperature, known as **thermoregulation**, involves controlling the amount of heat lost and heat gained through the skin surface. Heat may be transferred between an animal and the environment by **convection, radiation** and **conduction**. These processes are summarised in Figure 12.5.

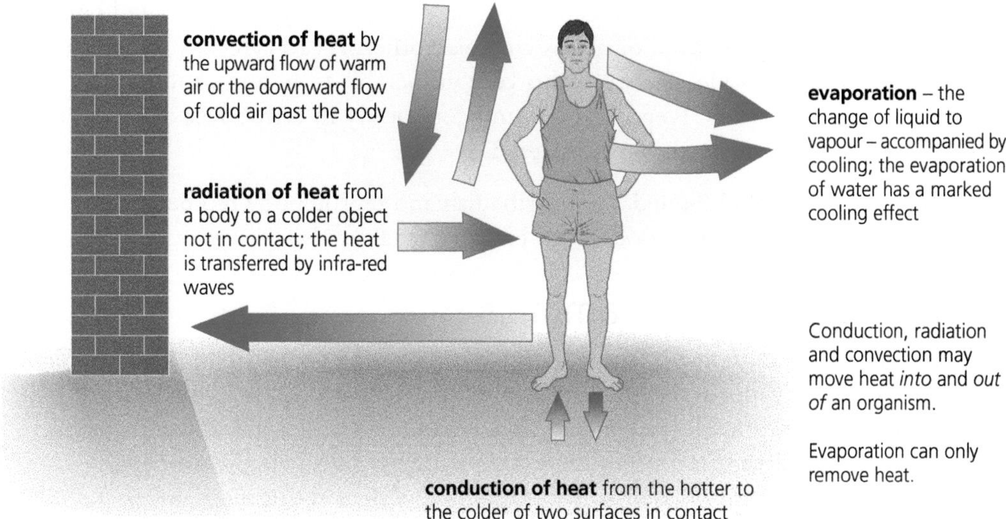

Figure 12.5 How heat is transferred between an organism and its surroundings

Body temperature of fish and reptiles

Fish are unable to regulate their body temperature. The huge gill surface area over which water moves continuously for gaseous exchange (see *Edexcel A level Biology 1*, Chapter 10) is also an efficient heat exchanger! In fish, the temperature inside the body is approximately the same as the temperature of their surroundings. This is because any body heat is quickly lost to the surroundings. In relation to thermoregulation, fish are good examples of 'non-regulators'.

On the other hand, the air-breathing land animals, reptiles, do have a crude form of body temperature control, at least when they are active and alert. Lizards and snakes are good examples (Figure 12.6).

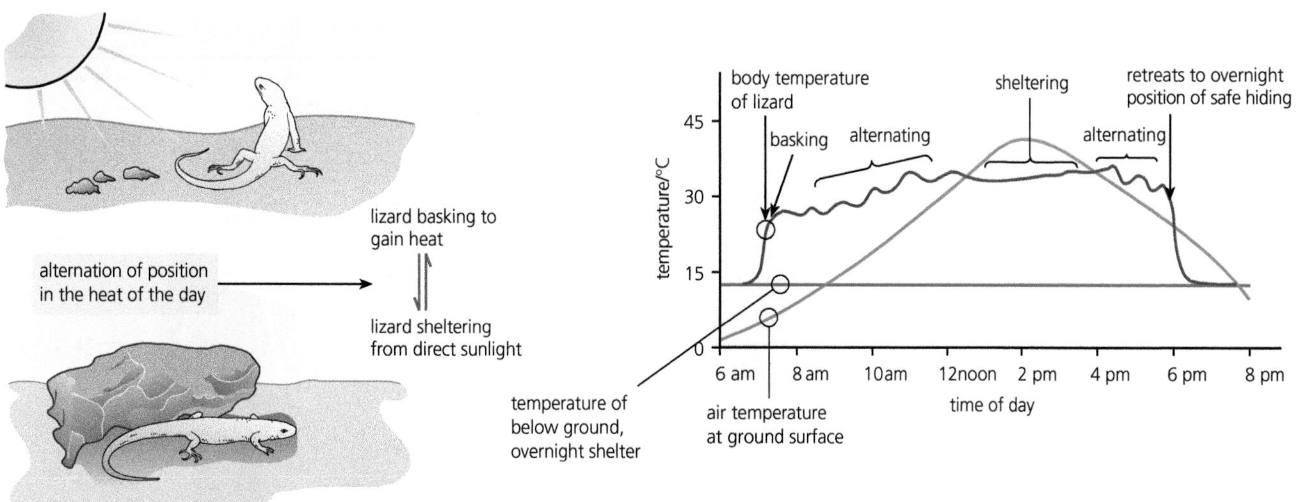

Figure 12.6 Thermoregulation in an ectotherm

Here, control is exercised by behavioural changes to take in heat as needed, exploiting any heat available in the external environment. In the morning, the animals 'bask' in sunlight to warm up. Then, for the remainder of the day, they move into and out of sunlight to absorb more or less heat according to their need. At night and over winter, their body temperature drops with that of the environment, and they become sluggish or even totally inactive.

Key term

Ectotherm An organism that relies on external heat sources for thermoreguation.

An animal with this form of thermoregulation is called an **ectotherm**, meaning 'outside heat'. During warm days an ectotherm may achieve a closely regulated body temperature. However, the kind of places where ectotherms can typically live is more restricted, as is their lifestyle. They are also very vulnerable to mammal and bird predators during cold times. Reptiles are found commonly in only a few of the wide range of habitats that mammals have mastered.

(Note: the old term 'cold blooded' to describe fish and reptile body temperature is now avoided, as is the term 'warm blooded' for birds and mammals.)

Thermoregulation in mammals

Key term

Endotherm An animal that relies on heat energy generated by internal metabolism for thermoregulation.

Mammals maintain a high and relatively constant body temperature. They achieve this by using heat energy generated by metabolism within their bodies, or by generating additional heat in their muscles when cold, and carefully controlling heat loss through the skin. An animal with this form of thermoregulation is called an **endotherm**, meaning 'inside heat'. Birds as well as mammals have perfected this mechanism. For example, humans hold their inner body temperature ('core temperature') just below 37 °C. In fact, in a human who is in good health the body's inner temperature varies only between about 35.5 and 37 °C within a 24 hour period (Figure 12.7). When the external temperature is low, however, only the temperature of the trunk is held constant. The body temperature falls progressively from the trunk towards the end of the limbs.

Body temperature over a 48 hour period
The body temperatures shown were taken with the thermometer under the tongue. Although this is a region close to the body 'core', temperatures here may be altered by eating/drinking, and by the breathing in through the mouth of cold air, for example. More accurate values are obtained by taking rectal temperature.

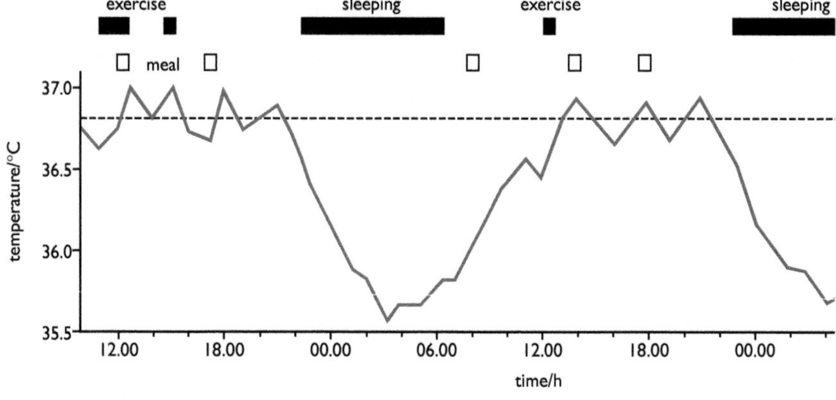

at 20 °C

37 °C
36 °C
32 °C
28 °C
34 °C
31 °C

Temperature distribution in environments at 20 °C and at 35 °C
The lines, **isotherms**, connect sites of equal temperature. The shaded area is the core, and around this the temperatures varies according to the temperature of the surrounding air (**ambient temperature**).

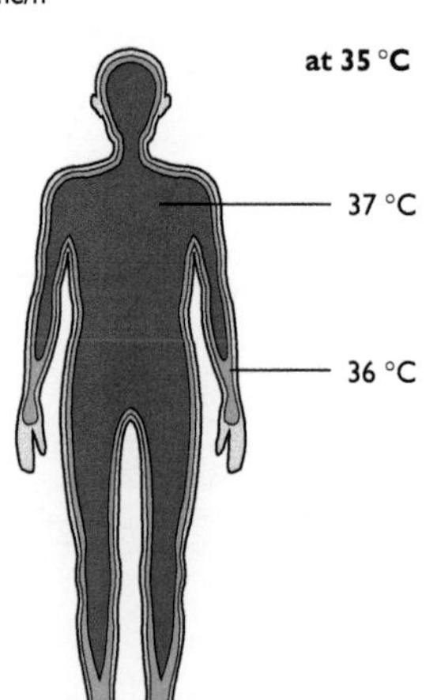

Figure 12.7 Body temperature of a human

Heat production in the human body

The major sources of heat in endotherms are the biochemical reactions of metabolism, which generate heat as a waste product. From the site of production, this heat reaches the rest of the body through the blood vessels. The organs of the body vary greatly in the amount of heat they produce. For example, the liver is extremely active metabolically, but most of its metabolic reactions require an input of energy (that is, they are endothermic reactions), so little energy is lost as heat. In consequence, the liver is more or less thermally neutral.

The bulk of your body heat (over 70 per cent) comes from other abdominal organs - mainly from the heart and kidneys, but also from the brain and lungs. In contrast, when the body is at rest, the skeleton, muscles and skin, which make up over 90 per cent of the body mass, produce less than 30 per cent of the body heat (Figure 12.8).

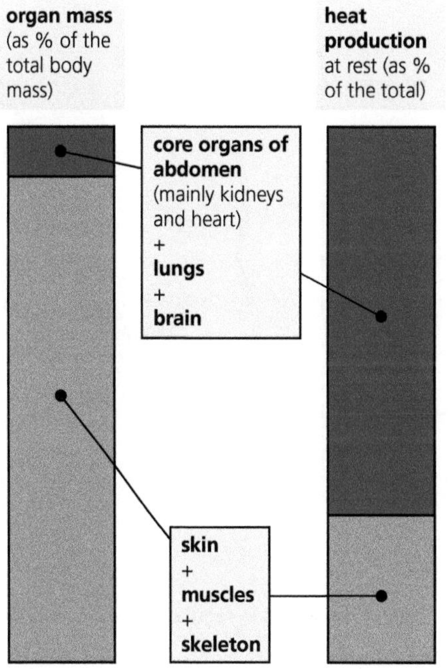

Figure 12.8 Heat production in the body at rest

The role of the skin in thermoregulation

Heat exchanges occur at the skin. The outer layer of the skin, the epidermis, consists of stratified epithelium. The cells in its basal layer (called the **Malpighian layer**) constantly divide, pushing the cells above them towards the skin surface. These upper cells are progressively flattened and the cell contents turn into keratin. The outermost layer of cells is continuously being rubbed off, but replaced from beneath so it does not wear away.

Below this layer is the dermis. This is a much thicker layer that consists of elastic connective tissue. In the dermis are blood capillaries, the hair follicles with hair erector muscles and the sweat glands. The sense receptors and sensory nerve endings are also found in this layer, and these are especially numerous in certain parts of the skin, which are consequently very sensitive.

At the base of the dermis is **adipose tissue**. In a mammal this is one of the major sites of fat storage. This tissue has a limited blood supply and is a poor conductor of heat, so it insulates internal organs against heat loss. Aquatic mammals that inhabit cold waters, such as whales and seals, have an extremely thick layer of fat stored below the skin (known as blubber), which they maintain throughout life. Terrestrial mammals that remain active through the unfavourable season of the year also tend to build up their fat here, mostly as a food store but also to provide insulation.

The amount of heat loss through the skin can be varied to control body temperature, for example:

- **at capillary networks:** the arterioles supplying them are widened (**vasodilation**) when the body needs to lose heat, but constricted (**vasoconstriction**) when it needs to retain heat
- **by the hair erector muscles:** these contract when heat must be retained, raising the hairs to trap a thicker layer of insulating air, but relax when more heat needs to be lost
- **by the sweat glands:** these produce sweat only when heat needs to be lost. The evaporation of the liquid sweat into vapour requires significant amounts of heat energy (specific latent heat of vaporisation), which is taken from the skin.

These mechanisms are shown in Figure 12.9 on the next page.

Key terms

Malpighian layer The basal layer of the epidermis of the skin, in which cells are constantly dividing by mitosis.

Adipose tissue Groups of cells containing large fat stores.

Vasodilation The widening of arteries and arterioles in the circulatory system.

Vasoconstriction The narrowing of arteries and arterioles in the circulatory system.

structure of the skn

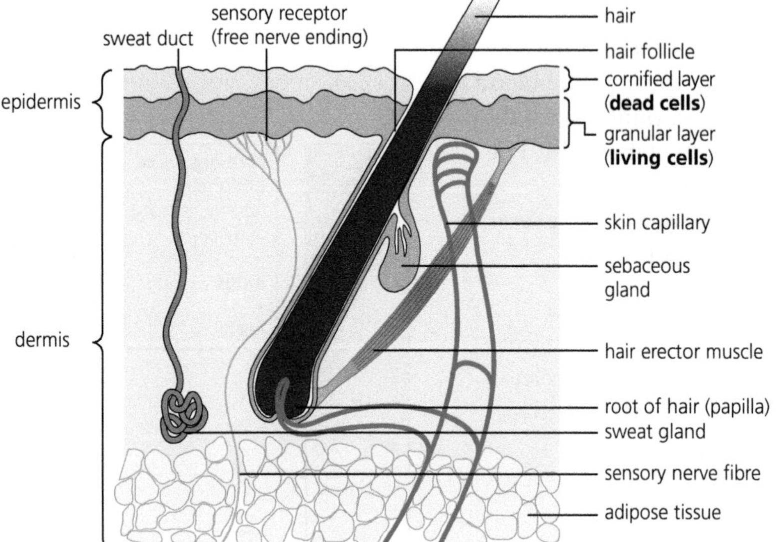

role of the sweat glands in regulating heat loss through the skin

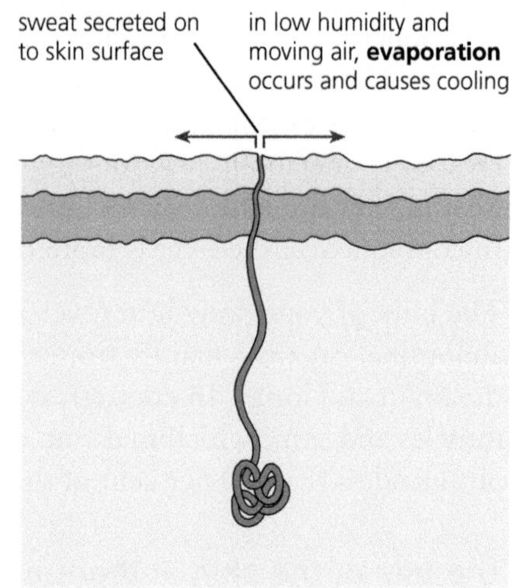

role of capillaries in regulating heat loss through the skin

In skin that is especially exposed (e.g. outer ear, nose, extremities of the limbs) the capillary network is extensive, and the arterioles supplying it can be dilated or constricted.

warm conditions

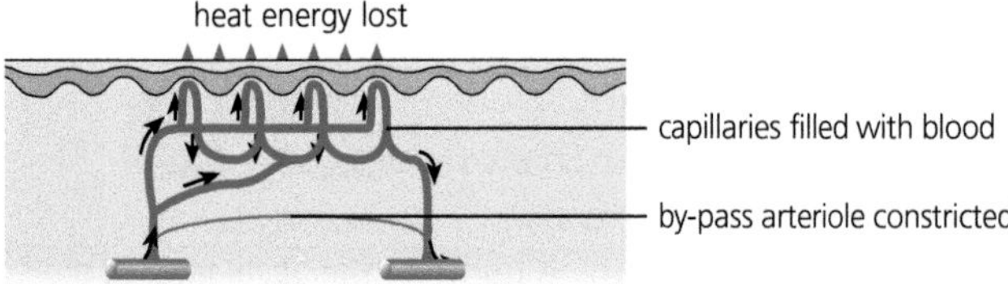

cold conditions

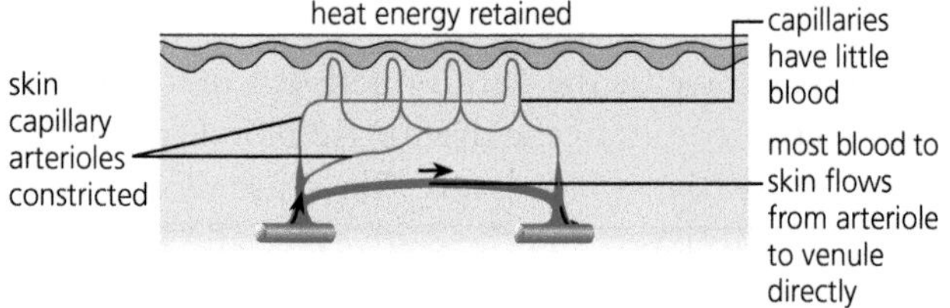

role of the hair in regulating heat loss through the skin

warm conditions

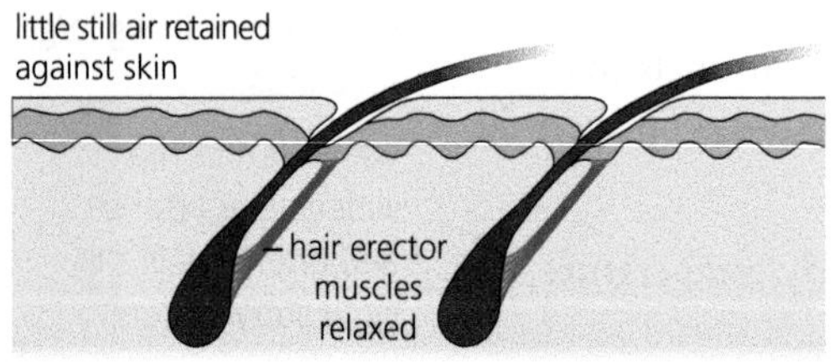

The hair erector muscles may be contracted or relaxed.

Still air is a poor conductor of heat.

cold conditions

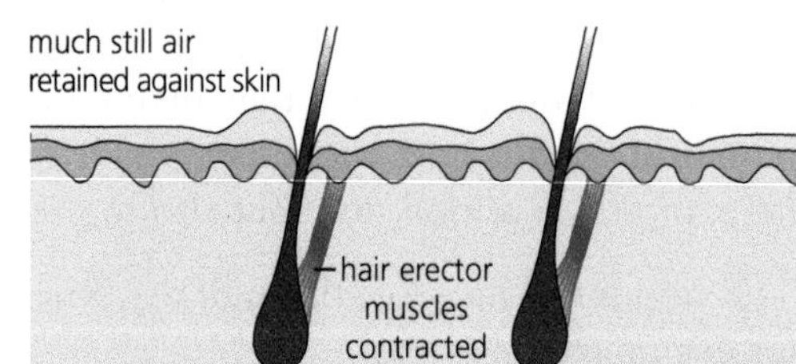

Figure 12.9 The skin and temperature regulation

Test yourself

14 Suggest why 'cold blooded' is a poor scientific term to describe ectotherms.

15 State the source of heat for endotherms.

Other mechanisms of thermoregulation

Changing metabolic rate

If a mammal's body experiences persistently cold conditions then its heat production is increased. The rate of heat release by an organism at rest is dependent on its basal metabolic rate (BMR). This is under the control of two hormones. In the short term it is regulated as another of the roles of the hormone adrenaline. In the longer term it is controlled by the actions of the hormone thyroxine, which is secreted by the thyroid glands (see Figure 11.3, Chapter 11).

A specialised site of metabolic heat production is tissue known as brown fat, which is found in patches in the thorax of many mammals. The role of brown fat is to generate heat. When tissue is stimulated by sympathetic nerves, respiration of glucose formed from surrounding fat reserves is speeded up. The ATP formed in the brown fat cells is immediately hydrolysed to ADP and P_i, and all the free energy of this reaction is released as heat and circulated by the blood.

If conditions are persistently hot the metabolic rate is reduced.

Behaviour changes

Under chilly conditions, heat output from body muscles increases. Live muscle has a firm solid feel, which is known as muscle tone. This is caused by many tiny contractions within the muscle and is normally used to maintain posture. These contractions are distinct from those used to move the skeleton and can be speeded up to increase the production of heat all over the body. Further uncontrolled contractions, known as shivering, are also triggered in cold conditions. This shivering can raise muscle heat production to about five times its resting value.

In contrast, panting is a very efficient method of losing heat. It is used by mammals such as dogs, which have a pronounced snout to make it effective and have fur, which prevents effective sweating. In this method air is drawn in through the nose and mouth and expelled over the moist tongue. As with sweating, the evaporation of moisture requires heat, which is taken from the blood vessels in the tongue, to cool the body.

Other behavioural changes used to regulate heat include simply moving to a hotter or colder place, huddling together with other individuals, or becoming vigorously active to generate even more muscle heat. In addition, humans carefully adjust the type of clothing they wear to match the prevailing environmental conditions.

Key terms

Basal metabolic rate (BMR) The rate of heat release by metabolic reactions when an organism is at rest.

Thyroxine A hormone produced by the thyroid gland that regulates basal metabolic rate and therefore has a long-term effect on growth and development.

Brown fat cells Groups of cells around the thorax of mammals capable of releasing heat using very high rates of respiration.

The hypothalamus as a control centre

In a direct parallel to the role of the medulla in control of heart rate, a region of the forebrain called the hypothalamus contains the thermoregulatory centre. The hypothalamus has a 'heat loss centre' and a 'heat gain centre'. The thermoregulatory centre receives information from temperature-sensitive nerve endings monitoring blood temperature in the hypothalamus itself, as well as others found in the skin and many major organs.

The hypothalamus communicates with the rest of the body using the autonomic nervous system with its antagonistic divisions of sympathetic and parasympathetic neurones. These effects are shown in Figure 12.10 on the next page.

Key term

Hypothalamus A region of the forebrain controlling many important autonomic functions. Closely connected to the pituitary gland.

If the body temperature is lower than normal, the heat-gain centre inhibits the activity of the heat loss centre. Impulses are sent down sympathetic nerves to skin, hair erector muscles, sweat glands and elsewhere to initiate actions to decrease heat loss (for example vasoconstriction, shivering and increased brown fat respiration). When the body temperature is higher than normal the heat-loss centre inhibits the heat-gain centre and impulses are sent to the same organs but through parasympathetic nerves, which initiate actions to increase heat loss (for example vasodilation, sweat production and inhibiting brown fat respiration).

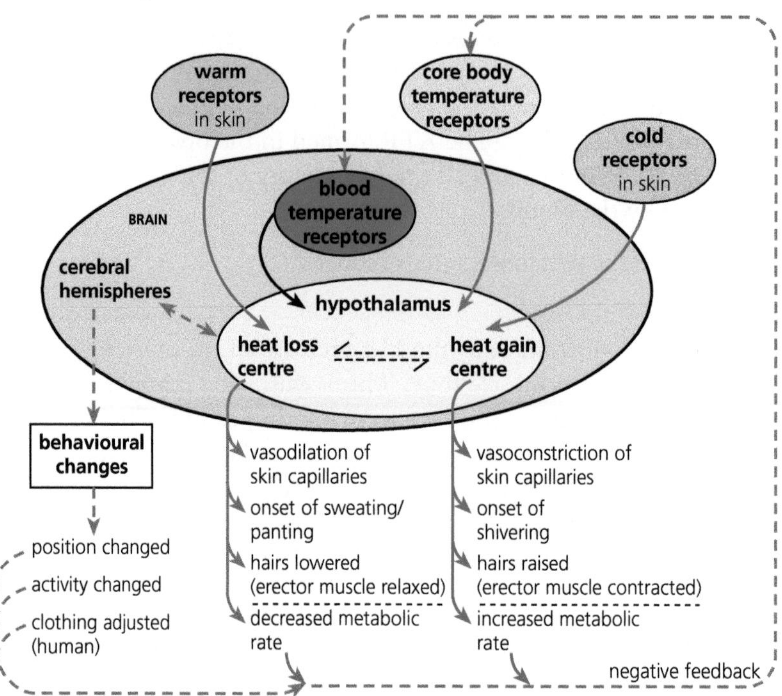

Figure 12.10 Temperature regulation by the hypothalamus

Test yourself

16 Respiration produces ATP. What is the fate of ATP in brown fat cells.

17 Explain why sweating causes cooling.

18 State which division of the autonomic system initiates heat gain.

Activity

How do we know that there are hot and cold receptors in the skin?

Carrying out the investigation

Aim: To demonstrate the effect of hot and cold skin receptors.

This is a very simple method but an excellent demonstration of both detection and perception of stimuli.

Take three small beakers. Fill each about three quarters full of water at different temperatures. Fill the first one with iced water containing an ice cube, the second with water as hot as a hot bath and the third with tepid water that feels just very slightly warm.

Place your left index finger in the cold water and your right index finger in the hot water for 1 minute.

Then quickly transfer both fingers into the tepid water and record the sensation from each one.

Question

1 Why do both fingers give different sensations when in water at the same temperature?

Osmoregulation and excretion in the kidneys

Urea formation in the liver

In animals, excess proteins and amino acids cannot be stored as they would be too disruptive. Instead they are broken down by a process called deamination (because the first step is the removal of the amino groups). This process must happen without the release of free ammonia within the tissues. This is because ammonia is both very toxic and very soluble. To achieve the breakdown safely, cells in the liver convert amino groups into urea, which is a relatively harmless product that can be safely transported in the blood to the kidneys, where it is excreted. The deamination process is summarised in Figure 12.11.

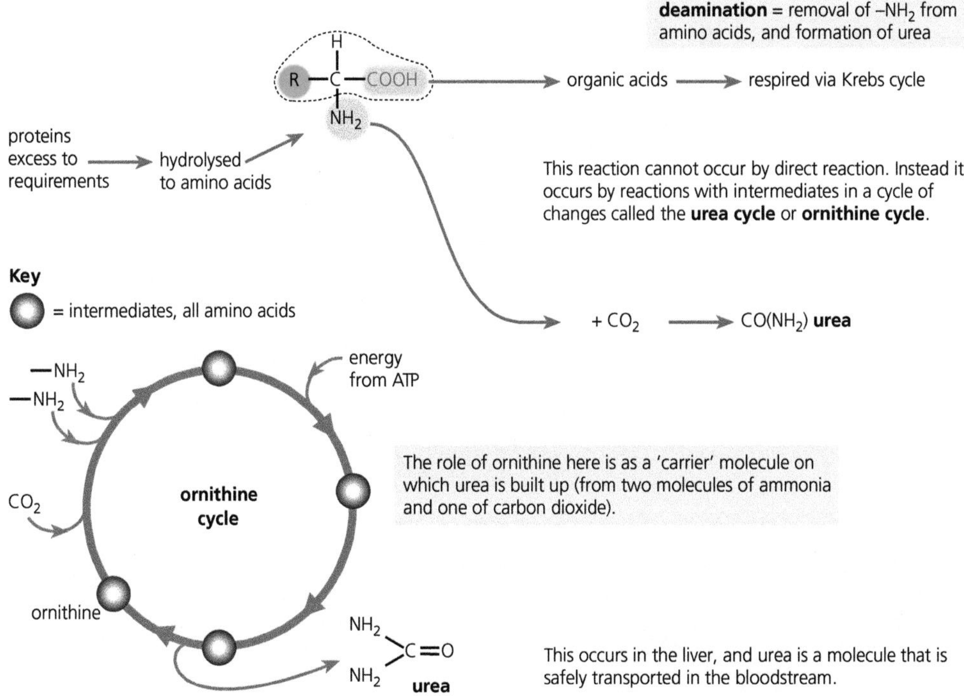

Figure 12.11 Deamination of excess amino acids

Key terms

Deamination The biochemical process of breaking down excess amino acids by removal of amino groups , ultimately producing urea.

Urea An excretory nitrogen-containing compound of low-toxicity produced by deamination of amino acids in the liver of mammals and some other animals (not to be confused with urine).

Urine A solution of urea, mineral ions and some other metabolic wastes excreted by the kidneys (not to be confused with urea).

Ureter The tube carrying urine from the kidney to the bladder (not to be confused with the urethra).

Urethra The tube carrying urine from the bladder to the exterior (not to be confused with the ureter).

Structure and function of the kidneys

The kidneys regulate the internal environment by constantly adjusting the composition of the blood. The waste products of metabolism are transported from the metabolising cells by the blood circulation, removed from the blood in the kidneys, and excreted in a solution called urine. The concentration of inorganic ions such as Na^+ and Cl^- are also regulated by the kidneys.

The position of the kidneys in humans is shown in Figure 12.12. Each kidney is served by a renal artery and drained by a renal vein. Urine from the kidney is carried to the bladder by the ureter, and then, at intervals, to the exterior by the urethra, when the bladder sphincter muscle is relaxed.

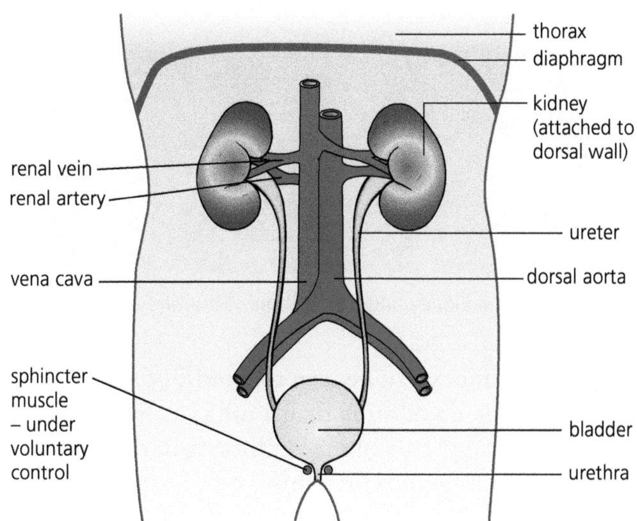

Figure 12.12 The human urinary system

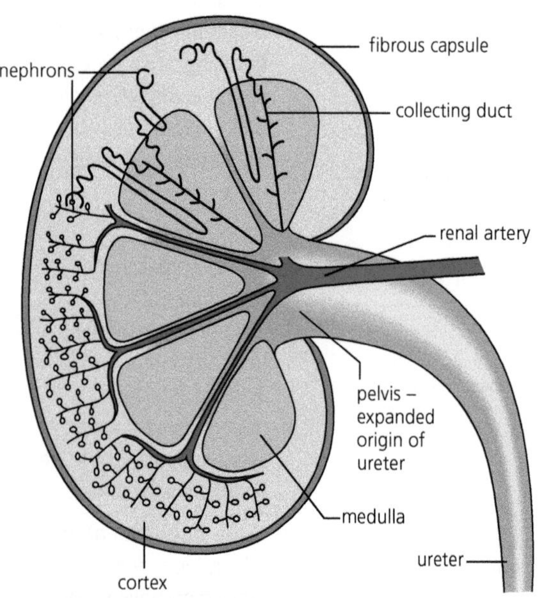

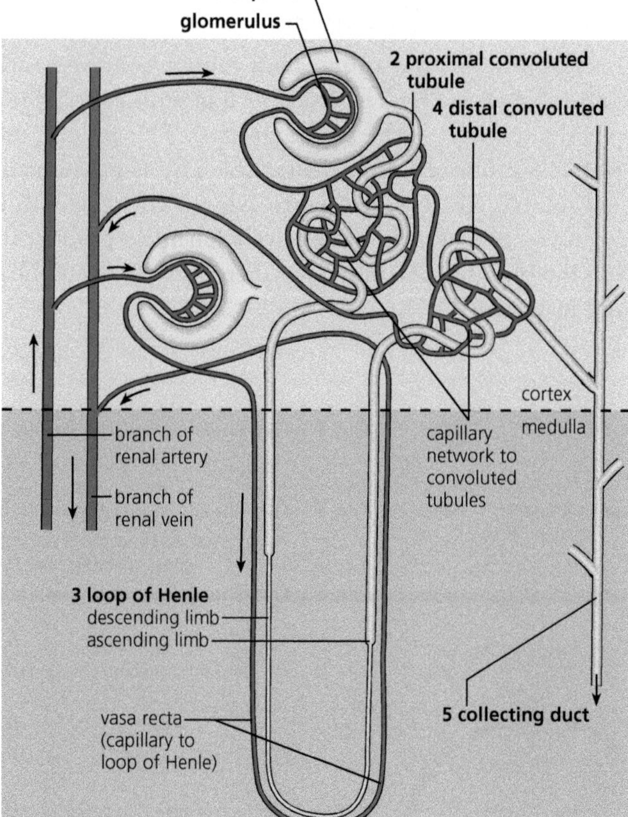

Roles of the parts of the nephron:

1 Bowman's capsule + glomerulus = ultrafiltration
2 proximal convoluted tubule = selective reabsorption from filtrate
3 loop of Henle = water conservation
4 distal convoluted tubule = pH adjustment and ion reabsorption
5 collecting duct = water reabsorption

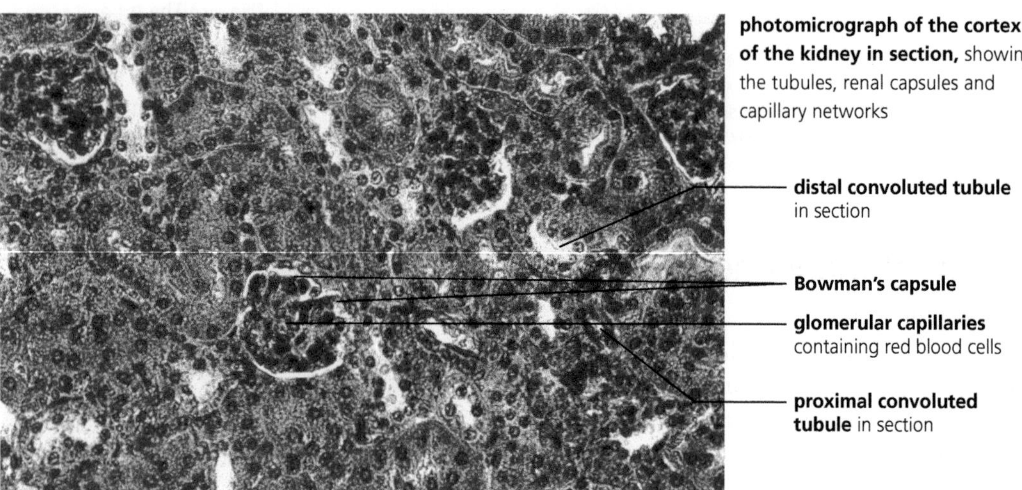

Figure 12.13 The kidney and its nephron – structure and roles

In section, a kidney can be seen to consist of an outer **cortex** and inner **medulla**. These are made up of more than a million tiny tubules called **nephrons**, together with their blood supply. Part of the nephron is in the cortex and part is in the medulla (see Figure 12.13). A nephron is a thin-walled tube about 3 cm long. Capillary networks associated with the nephron are crucial to its function.

Key terms

Cortex (kidney) The outer part of the kidney.

Medulla (kidney) The inner part of the kidney.

Nephrons Fine tubules that make up the functional units of the kidney.

The formation of urine

In humans, about 1.0–1.5 litres of urine is formed each day, typically containing about 40–50 g of solutes, of which **urea** (about 30 g) and **sodium chloride** (up to 15 g) make up the bulk. The nephron produces urine in a continuous process, which can be conveniently divided into five steps to show how the blood composition is so precisely regulated.

Step 1: Ultrafiltration in the renal capsule

In the glomerulus, water and relatively small molecules of the blood plasma, including useful ions, glucose and amino acids, are forced out of the capillaries, along with urea, into the lumen of the capsule. This is described as ultrafiltration because it is powered by the pressure of the blood, which drives substances through an extremely fine sieve-like structure.

> **Key term**
>
> **Ultrafiltration** The process by which small molecules are forced out of the capillaries in the capsule of the kidney by high blood pressure.

The **blood pressure** here is high enough for ultrafiltration because the input capillary (afferent arteriole) is wider than the output capillary (efferent arteriole). The 'sieve' is made of two layers of cells (the endothelium of the capillaries of the glomerulus and the epithelium of the capsule), between which is a basement membrane. You can see this arrangement in Figure 12.14.

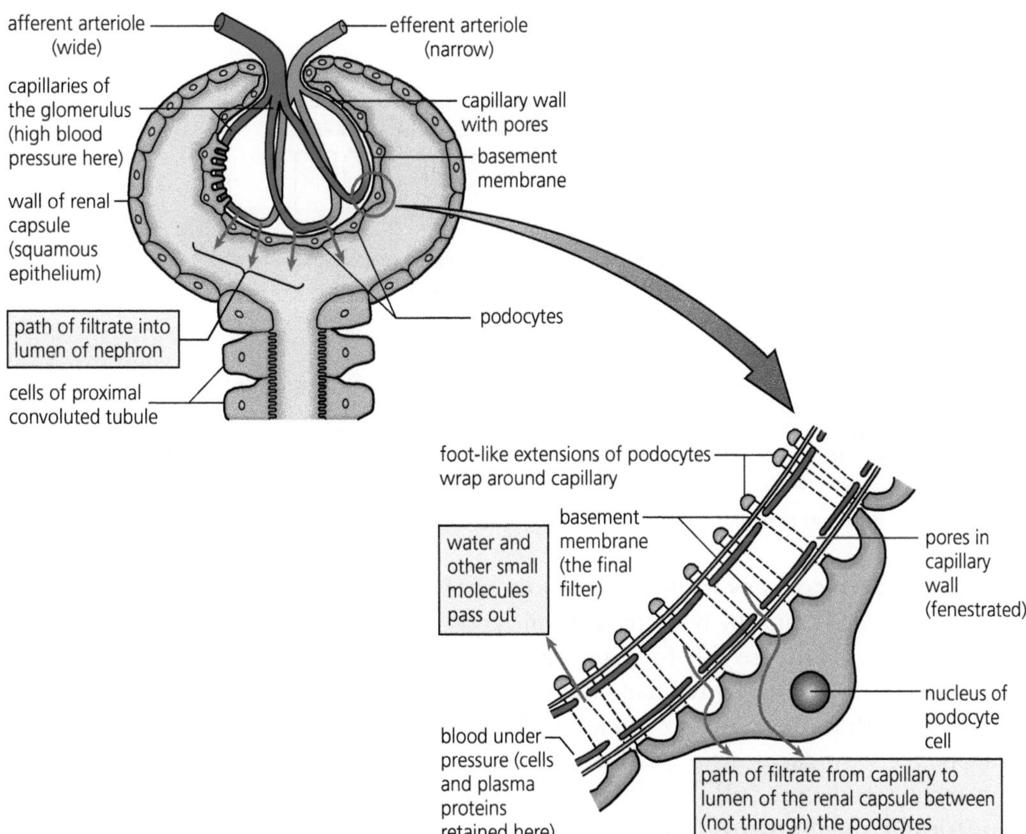

Figure 12.14 The site of ultrafiltration

Notice that the cells of the capsule wall are called podocytes because they have foot-like extensions that form a network with tiny slits between them (a situation we call **fenestrated**). Similarly, the endothelium of the capillaries has pores, too. This detail has only become apparent from studies using the electron microscope – these filtration gaps are very small indeed.

> **Key term**
>
> **Podocytes** Specialised cells found in the wall of the kidney capsule.

The entire contents of blood are not forced out. Not only are blood cells retained, but the majority of blood proteins and polypeptides dissolved in the plasma are also retained in the circulating blood. This is because of the presence of the **basement membrane**.

Step 2: Selective reabsorption in the proximal convoluted tubule

The proximal convoluted tubule is the longest section of the nephron. The walls are one cell thick and are packed with mitochondria (ATP is required for the active transport). The cell membrane in contact with the filtrate has a brush border of microvilli which enormously increase the surface area for reabsorption. A large part of the filtrate is reabsorbed into the capillary network here (Figure 12.15).

The individual mechanisms of transport are:

- movement of water by **osmosis**
- **active transport** of glucose and amino acids across membranes
- movement of mineral ions by a combination of **active transport**, **facilitated diffusion**, and some **exchange of ions**
- **diffusion** of urea
- movement of proteins by **pinocytosis**.

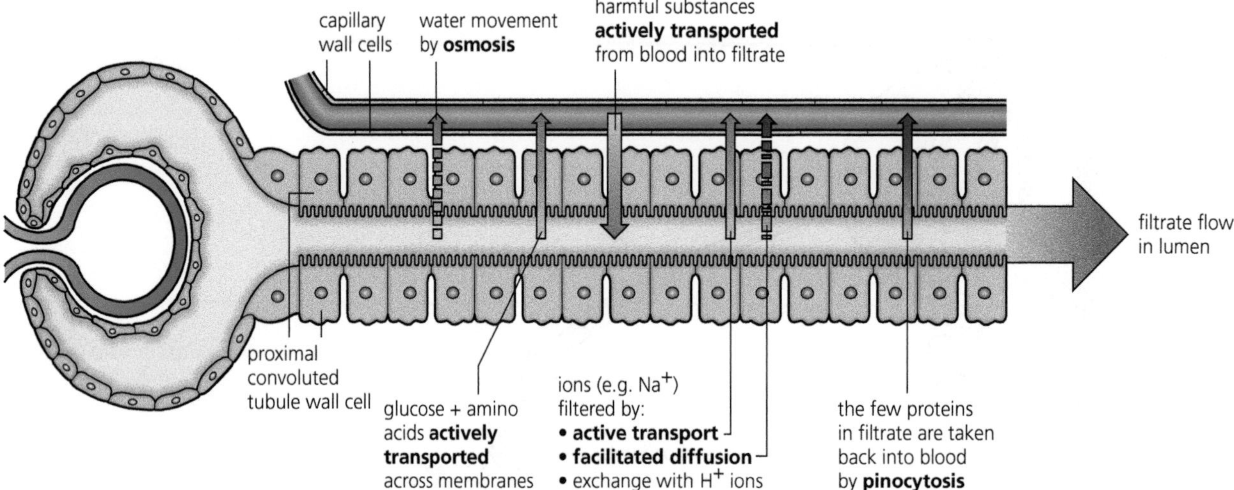

Figure 12.15 Selective reabsorption in the proximal convoluted tubule

Step 3: Water conservation in the loop of Henle

Urea is expelled from the body in solution, so water loss in excretion is inevitable. However, mammals are able to form urine that is more concentrated than the blood (when necessary), thereby reducing the water loss to a minimum. The role of the loop of Henle with its **descending** and **ascending limbs**, together with a parallel blood supply, the vasa recta, is to create and maintain a high concentration of salts in the tissue fluid in the medulla of the kidney. This is brought about by a **countercurrent multiplier mechanism**. It is the building up of a high concentration of salts in the tissue of the medulla that causes water to be reabsorbed from the filtrate in the collecting ducts. The collecting ducts run through the medulla.

Key terms

Vasa recta Blood vessels found adjacent to the loop of Henle in the kidney.

Kidney medulla The inner part of the kidney.

The roles of the vasa recta are to:

- absorb water that has been absorbed into the medulla at the collecting ducts
- remove carbon dioxide and deliver oxygen to the metabolically active cells of the loop of Henle without removing the accumulated salts from the medulla.

Figure 12.16 explains how the countercurrent mechanism works. Notice that the descending and ascending limbs lie close together.

Look first at the second half of the loop, the ascending limb.

Here, sodium and chloride ions are pumped out into the medulla but water is retained inside the ascending limb. Opposite, the descending limb is permeable here, so sodium and chloride ions diffuse in. Water passes out into the medulla tissue, due to the salt concentration in the medulla. As the filtrate flows down the descending limb, this water loss increases the salt concentration in the loop, making the filtrate more concentrated.

Consequently, sodium ions and chloride ions diffuse out down their concentration gradient, around the 'hairpin' zone at the base of the descending limb, adding to the concentration of ions in the medulla. How this concentration helps in the formation of concentrated urine is explained in step 5 on the next page.

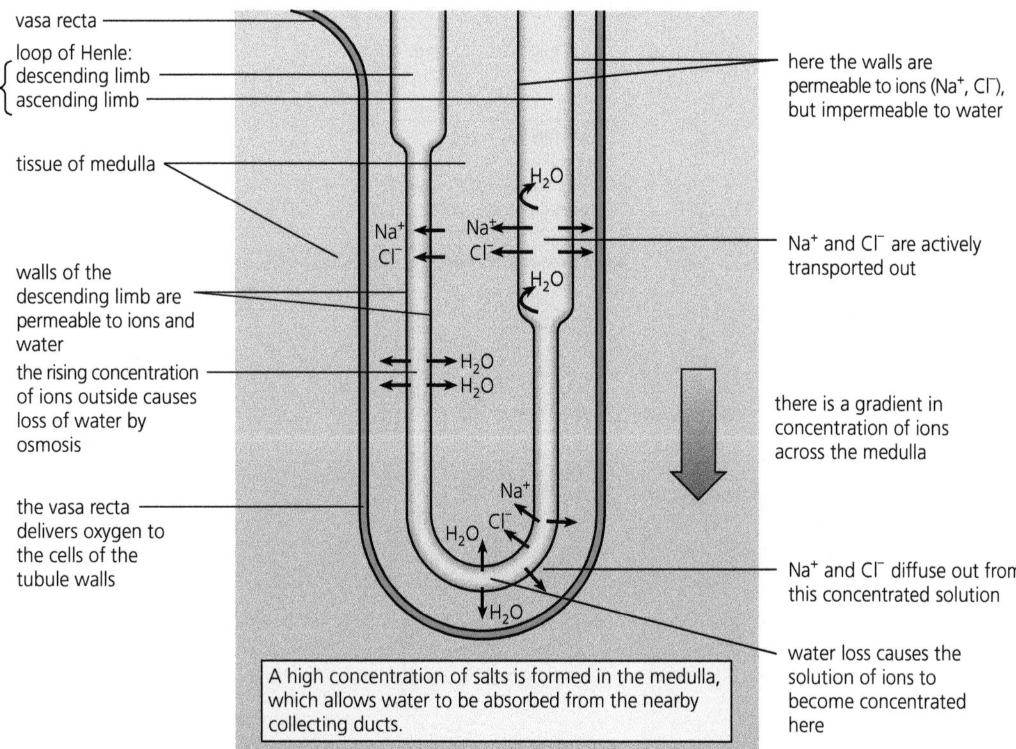

Figure 12.16 Countercurrent mechanism

Step 4: Blood pH and ion concentration regulation in the distal convoluted tubule

Here the cells are of the same structure as those of the proximal convoluted tubule, but their role is to adjust the composition of the blood, and in particular the **pH**. An initial tendency for the pH of the blood to change is buffered by the blood proteins, but if the blood does begin to deviate from pH 7.4, then the concentration of hydrogen ions and hydroxyl ions in the blood is adjusted, along with the concentration of hydrogencarbonate ions. Consequently, blood pH does not vary outside the range pH 7.35–7.45, but the pH of urine varies from pH 4.5 to pH 8.2.

Also in the distal convoluted tubule, the selective reabsorption of ions useful in metabolism occurs from the filtrate.

Key term

Antidiuretic hormone (ADH) A hormone synthesised in the hypothalamus and released by the posterior pituitary gland, which increases the permeability of the walls of the collecting ducts in the kidney.

Step 5: Water reabsorption in the collecting ducts

The collecting ducts are where the **water content of the blood** (and therefore of the whole body) is regulated (Figures 12.17 and 12.18). When the water content of the blood is low, **antidiuretic hormone (ADH)** is secreted from the posterior pituitary gland. When the water content of the blood is high, little or no ADH is secreted.

The permeability of the walls of the collecting ducts to water is variable (a case of facilitated diffusion) – the presence of ADH causes the walls of the collecting ducts to be fully permeable. This allows water to be withdrawn from the filtrate of the tubule into the medulla, due to the high concentration of sodium and chloride ions there (see step 3 on page 228). This water is taken up and redistributed in the body by the blood circulation, and only small amounts of concentrated urine are formed. Meanwhile, the ADH circulating in the blood is slowly removed at the kidneys.

When no ADH is secreted, the walls of the collecting ducts become less permeable. The result is that large quantities of very dilute urine are formed.

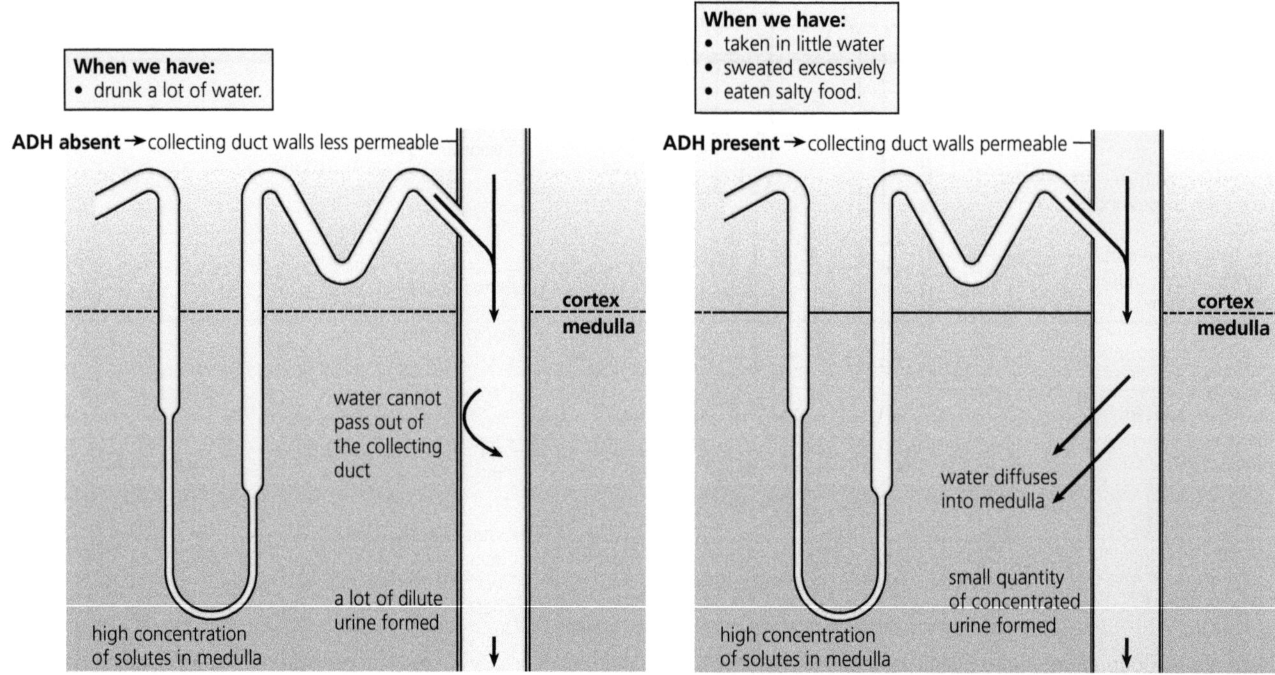

Figure 12.17 Water reabsorption in the collecting ducts

Test yourself

19 Cells of the walls of the proximal convoluted tubule have a brush border. Describe what this means, and explain how it helps in tubule function.

20 Predict in what circumstances in the body ADH is released.

Figure 12.18 Homeostasis by osmoregulation and regulation of ion content in the kidneys – a summary

Hypothalamus, pituitary and neurosecretion

The control of osmoregulation provides us with another excellent example of the close coordination between the nervous and endocrine system. This is shown very clearly by the links between the **pituitary gland** and the hypothalamus. The hypothalamus is part of the floor of the forebrain. It is exceptionally well supplied with blood vessels and is the site of specialised neurones. The hypothalamus has a key role in monitoring and control of many aspects of homeostasis and information from the composition of the blood as it flows through its capillary networks provides important sensory information. This information, and that which comes from sensory neurones via the spinal cord enables the hypothalamus to regulate many body activities concerned with homeostasis, such as thermoregulation discussed in the earlier part of this chapter.

The pituitary gland is situated below the hypothalamus but is connected to it. This gland consists of two parts, the anterior and posterior lobes. Just like the hypothalamus, the pituitary has a key role in homeostasis but its actions are mediated through release of hormones. It is sometimes referred to as the 'master gland' because, not only does it release hormones that have a direct effect, it also releases hormones that stimulate other endocrine glands. However, systems do not work well with two different managers and it is the hypothalamus, with its superior sensory information, that controls the activity of the pituitary by releasing a number of controlling hormones from its special **neurosecretory cells** into the portal vein that supplies the anterior lobe of the pituitary as shown in Figure 12.19 on the next page.

Key terms

Pituitary gland An endocrine gland attached to the hypothalamus at the base of the forebrain. Often known as the 'master gland' because it produces and releases many hormones controlling growth, development and homeostasis.

Neurosecretory cells Modified neurones producing and transporting hormones.

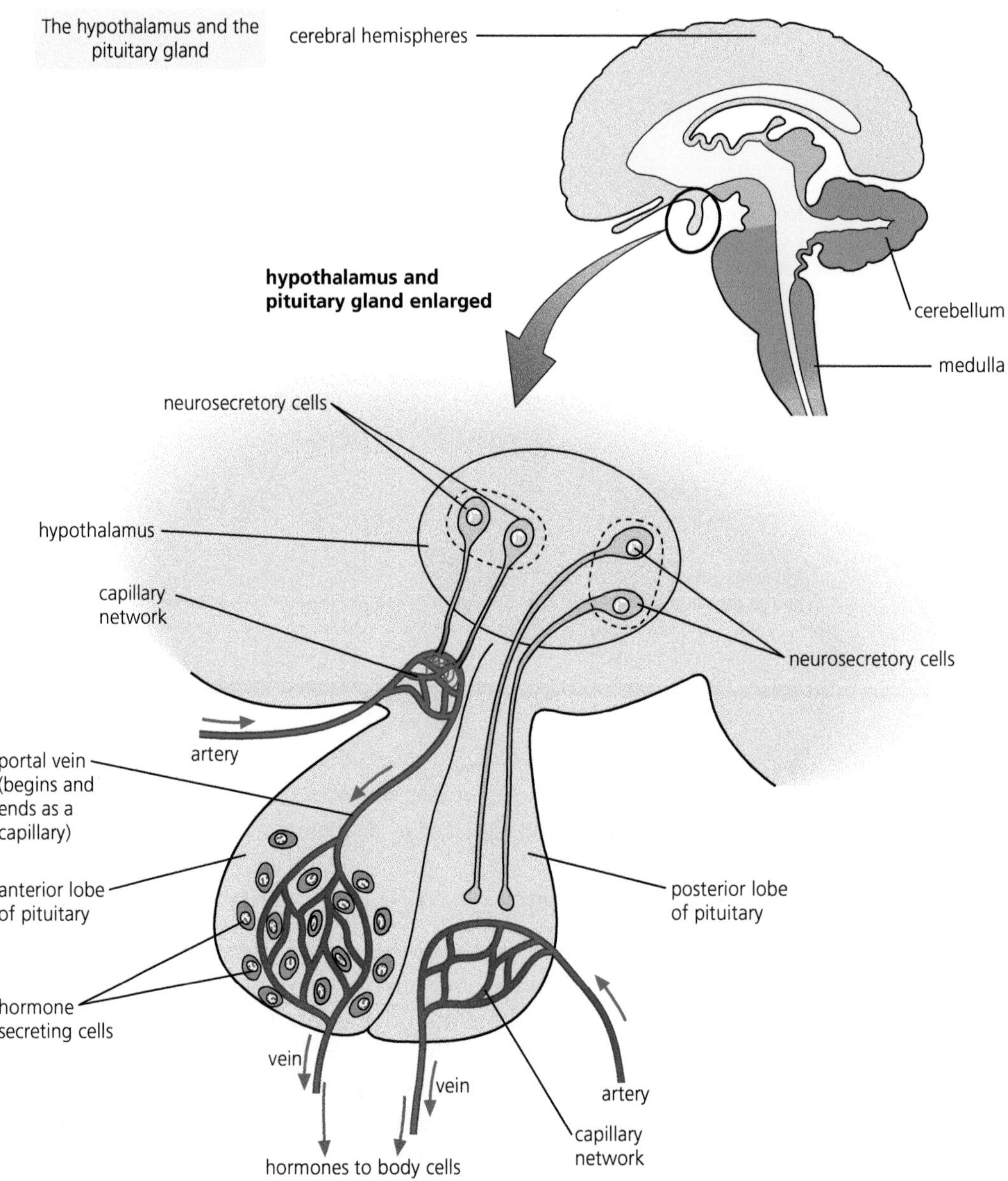

Figure 12.19 The hypothalamus and the pituitary gland

The posterior lobe of the pituitary does not synthesise hormones but stores and releases hormones synthesised by the hypothalamus. These hormones are transported to the posterior pituitary along modified neurones and stored in the vesicles found at the end of these specialised neurosecretory fibres within the gland (Figure 12.20). Impulses from the hypothalamus cause the release of specific hormones into the blood in the capillary network in a process very similar to the release of neurotransmitters from the presynaptic membrane of synapses.

The secretion of ADH and its mode of action

You have seen that osmoregulation by negative feedback is brought about by the secretion of ADH and its action on water reabsorption in the collecting ducts (Figure 12.18). This is yet another example of the nervous and endocrine systems working in harmony, and you can illustrate this using your knowledge of the links between the hypothalamus and pituitary shown in Figure 12.19.

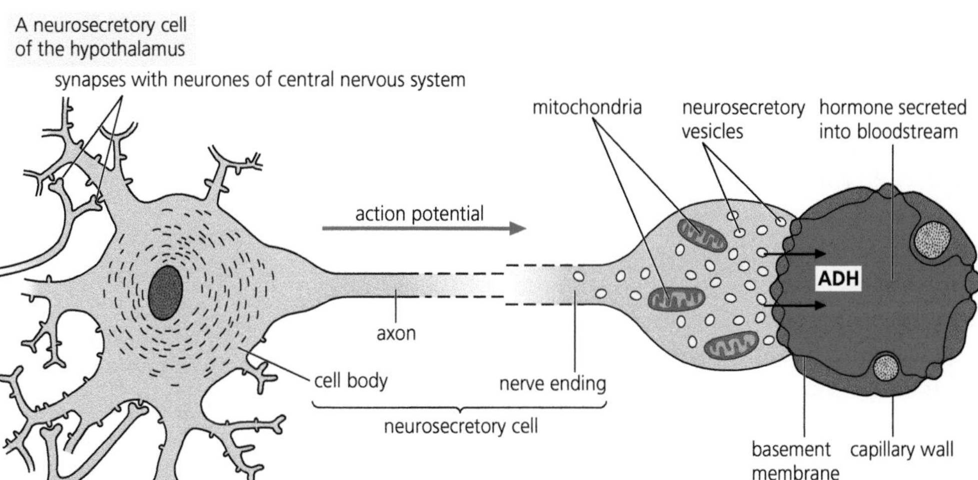

Figure 12.20 The working neurosecretory cell of the hypothalamus

ADH is one of the hormones synthesised in the hypothalamus and transported by neurosecretory cells to the posterior pituitary as shown in Figure 12.20.

ADH is a peptide hormone and therefore acts by binding to receptor sites on the cells of the collecting ducts (as you saw in Chapter 11) and by the use of a second messenger triggers the insertion of integral membrane protein channels called **aquaporins**. Aquaporins are selectively permeable to water molecules and therefore increase the rate of water diffusion (osmosis) through the membrane. This is because the water molecules no longer have to pass through the lipoprotein membrane structure that restricts their flow. In this way the presence of ADH ensures that a maximum volume of water can be reabsorbed into the medulla of the kidney and back into the bloodstream.

Key term

Aquaporins Integral membrane proteins that form channels to allow the free passage of water molecules through the phospholipid bilayer.

Hormones such as ADH are removed from the bloodstream and broken down into inactive compounds by the liver. This means that continual secretion is needed to retain its effect on the kidney and allow the hypothalamus to retain accurate control. Figure 12.18 also shows you that the action of ADH is another example of a negative feedback control system, which was described at the beginning of Chapter 11.

Adaptations of the kidney to dry environments

Mammals are a successful group of animals that have been able to colonise many different habitats. Extremely dry environments, such as deserts, pose particular challenges, and there are only a limited number of mammals that can survive the high temperatures and arid conditions.

Excretion of urea inevitably involves water loss that a desert mammal can ill-afford, and to survive, adaptations are needed to keep this to a minimum.

One mammal that is able to thrive in arid conditions is the kangaroo rat, *Dipodomys* sp., shown in Figure 12.21.

Figure 12.21 A kangaroo rat – *Dipodomys* sp.

Dipodomys is able to survive because of a number of water-conservation features, both behavioural and physiological. It does not have sweat glands, nor does it evaporate moisture off the tongue in panting to cool itself. It lives in burrows which, in addition to providing a shelter, allow it to modify its behaviour to avoid the hottest times of day.

It forages for seeds, mainly at night, to further limit exposure to the hot sun, and it has large eyes with good nocturnal vision to find food and to escape predators, using its speed and agility provided by long hind legs and a balancing tail.

The kidneys of *Dipodomys* are able to reduce water loss by producing a very concentrated urine, approximately 20 times more concentrated than humans. To achieve this by active transport would be far too energy-demanding, so *Dipodomys* utilises the counter-current multiplier effect (see Gas exchange in fish, *Edexcel A level Biology 1*, Chapter 10) of the loop of Henle to raise the concentration of the medulla to very high levels. As you can see from Figure 12.16 (page 229), Na^+ and Cl^- ions are lost from the fluid inside the descending limb of the loop of Henle as it passes into the medulla. The further into the medulla the descending limb passes, then the more ions are lost from the fluid and the more concentrated the medulla becomes. The proximity of the ascending limb to form a **counter-current system** means that there are always ions available to do this. A very highly concentrated medulla is just what is needed to draw water from the collecting ducts by passive osmosis. Examination of the kidneys of *Dipodomys* shows that they have, comparatively, very long loops of Henle that pass deep into the medulla to enhance this effect.

Key term

Counter-current system An arrangement of two fluid movements in opposite directions, which maintains a diffusion gradient along its whole length.

Test yourself

21 State in which part of the nephron, glucose is reabsorbed.

22 Explain why proteins are not found in the capsule of the glomerulus.

23 Name **three** substances that are transported by active transport in the reabsorption process.

24 State the **two** ions that are responsible for the high concentration in the medulla.

25 In which part of the kidney is most water reabsorbed back into blood capillaries?

26 State which part of the kidney is extended in length in a kangaroo rat.

Exam practice questions

1 Antidiuretic hormone is produced in the:

A anterior pituitary gland

B medulla

C posterior pituitary gland

D hypothalamus *(1)*

2 When there is no light shining on the retina:

A rod cells release a transmitter substance

B bipolar cells release a transmitter substance

C rhodopsin forms retinal and opsin

D Na^+ ion channels in the outer rod segment close *(1)*

Tip

Despite being multiple choice, Question 2 is a good test of your detailed knowledge of the events taking place on the retina.

Tip

Question 3 is a simple question that is largely AO1. However, take care to read it carefully. This is about showing you understand negative feedback, not about detailed descriptions of the methods mammals use to increase or decrease heat loss, so make sure you keep to the point. You don't get marks knocked off for irrelevant material but neither do you gain any credit. You do, however, penalise yourself by wasting a lot of precious time.

3 a) Explain what is meant by the term 'negative feedback control'. *(3)*

b) Explain how negative feedback control plays an important part in thermoregulation by the autonomic nervous system in mammals. *(4)*

4 a) Explain why the structure of the retina is described as 'inverted'. *(1)*

b) During periods of low light intensity, the iris of the eye causes the pupil to dilate. This allows the image on the retina to expand over the surface of the retina. Explain how this helps to provide better low light vision. *(3)*

c) Explain how light falling on a rod cell in the retina can produce an action potential in an optic nerve fibre. *(6)*

Tip

Question 4 is a mixed question where you need to apply your knowledge of the retina and show that you know all the details of the events leading up to the formation of an action potential in the optic nerve. A good exercise in testing your revision and emphasising the level of detail you will need in the examination papers.

5 The table below shows the relative concentration of the blood plasma and urine in a number of different mammals.

Mammal	Habitat	Urine concentration mmols dm^{-3}	Urine concentration : blood plasma concentration ratio
Common brown rat	Temperate	2900	9
Kangaroo rat	Desert	5500	16
Beaver	Freshwater/land	520	1.7
Human	Temperate	1400	4.5
Camel	Desert	2800	8

a) Estimate the approximate blood plasma concentration in these mammals and explain why this might be an expected trend in mammals. *(3)*

b) Explain why the beaver might be expected to have the lowest urine concentration:blood plasma concentration ratio compared to the other mammals. *(2)*

c) Drugs that cause an increase in urine production are called diuretics. A trial was conducted to test the effects of two different diuretics, acetazolamide and frusemide on urine production in patients. Two groups of 50 patients were each given identical concentrations of the drugs and the total volume of urine produced was measured for 24 hours. A third group of 50 patients were given identical doses of isotonic saline solution as a control. Isotonic saline is a dilute salt solution of exactly the same concentration as the blood plasma. All groups had exactly the same water and food intake throughout the investigation.

The results of this investigation are shown in the graph below.

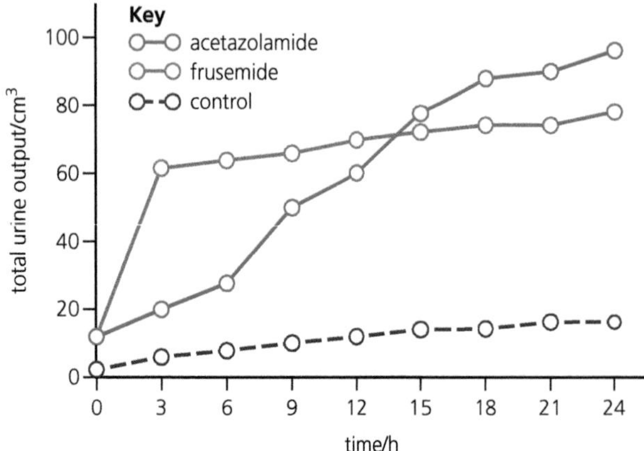

i) Explain why it was important that the control group were given an isotonic solution. *(2)*

ii) Diuretics are sometimes used to help control blood pressure. Analyse the data to explain why frusemide might reduce high blood pressure and be a more effective treatment. *(5)*

d) Sailors stranded at sea without drinking water are sometimes tempted to drink seawater. Explain why drinking sea water will neither reduce thirst nor aid survival. *(3)*

Tip

Question 5 is a very common type of question that you are likely to face. Not only are you expected to understand the working of the kidney but also apply your knowledge to a novel situation by interpreting data and carrying out some simple calculations. Note: 'estimate' means you do not have to carry out all the calculation, just mentally check the approximate values.

Stretch and challenge

6 Seals are extremely well-adapted to life in cold water and they are capable of diving to great depths for periods of up to an hour. As mammals, they breathe through lungs and maintain a high body temperature. One adaptation is known as the 'dive reflex' and this can also be demonstrated in humans. The strongest dive reflexes occur when the face is fully immersed in cold water.

What is the 'dive reflex', how is it controlled and how can it benefit the seal on a long dive?

7 Humans and other animals use urea as a nitrogenous excretory product but this is by no means the only compound used in this way.

The table below shows some different excretory products and their properties.

Animal	Habitat	Excretory compound	Chemical properties
Trout	Freshwater	Ammonia	Highly soluble and highly toxic
Locust	Desert	Uric acid	Almost insoluble in water
Cod	Salt water	Trimethylamine oxide	Very soluble; less toxic than ammonia
Human	Terrestrial	Urea	Moderately soluble

Explain how each nitrogen-containing compound can be excreted and how this is an adaptation to the environment of the animal.

8 The temperature of the Arctic and Antarctic oceans is below the freezing point of fish blood. Many species, including the Arctic cod, have 'antifreeze' in their blood.

a) If fish blood is likely to freeze, why does the sea water not freeze?

b) The relative molecular mass (RMM) of these antifreeze agents is between 5000 and 30 000. The glomerular membrane is normally impermeable to compounds with a RMM of greater than about 50 000+. What problems does this pose to these fish and how might they be overcome?

13 Ecosystems

Prior knowledge

In this chapter you will need to recall that:

→ food chains and food webs can be used to represent relationships in ecosystems
→ most food chains and food webs start with energy from sunlight trapped by photosynthesis
→ food chains represent the transfer of energy from organism to organism
→ the number and total mass of organisms gets smaller as you move down the food chain
→ pyramids of number show the decrease in organisms as you move down a food chain
→ quadrats and transects can be used in ecological investigations
→ ecosystems are groups of plants and animals that live together in fixed area, interacting to form a balanced system
→ organisms can be divided into feeding types such as producers, consumers, herbivores, carnivores, detrivores and decomposers
→ autotrophs are able to synthesise food molecules from simpler substances whilst heterotrophs cannot
→ carbon and nitrogen cycles recycle nutrients within the ecosystem.

Test yourself on prior knowledge

1 Name an example of an ecosystem.
2 Food chains contain less and less organisms as you move down them. Explain why.
3 Explain the term 'quadrat'.
4 List **three** ways in which organisms interact with each other in an ecosystem.
5 State the reason why nitrogen is so important to living organisms.
6 Explain how food chains and webs show that an ecosystem is 'balanced'.
7 State the most common method of autotrophic nutrition.
8 Describe the difference between an autotroph and a heterotrophy.

What is an ecosystem?

The whole area of the Earth that can be inhabited by living organisms is quite limited. If we simply consider its height above and below sea level then it extends for approximately 30 km, forming a sphere known as the **biosphere**. Living within the biosphere are lots of individual organisms. Members of one species are usually found living together in groups called **populations** and each species has a particular set of adaptations and behavioural patterns. The whole biosphere has countless different smaller areas known as **habitats** and within each habitat there will be a number of populations forming a **community**.

An **ecosystem** is more than just a collection of communities. It also concerns how these communities are linked to each other and to their surroundings to form a balanced system.

Key terms

Biosphere The total area of the Earth and its atmosphere that is inhabited by living organisms.

Population A group of members of the same species that interact with each other.

Habitat A division of the biosphere with its own unique characteristics and conditions.

Community A group of populations of different species living together in one habitat.

Ecosysytem A collection of communities and their non-living surroundings, which form a stable, self-perpetuating system in which there is an energy flow and nutrients are recycled.

It is much more difficult than you might think to give undisputed examples of single ecosystems if you are to meet all the conditions of the definition. First of all, most ecosystems are influenced by organisms that have links to many ecosystems. For example, a small isolated lake forms a separate ecosystem, but predation of fish by birds such as herons is an important factor, yet the herons almost certainly have this effect on several different lakes and streams. Swallows and other migratory birds play an important role in different ecosystems many thousands of miles apart.

However, one thing is certain, and that is, that ecosystems can vary in size enormously. Large areas of desert or semi-arid scrub can form ecosystems of thousands of square kilometres, with animals needing to roam widely to find scarce food. In contrast they can be small, isolated pools of less than 1 square metre. The important part is that most of the organisms are present for a significant period of time and that nutrients and energy can flow freely throughout.

Unusual ecosystems

Key terms

Hydrothermal vents Fissures on the deep ocean floor where the formation of tectonic plates causes hot water, minerals and gases to escape.

Chemosynthesis The use of chemical energy from the oxidation of inorganic compounds to synthesise organic compounds.

For many years ecologists considered that solar energy was the start point of energy flow in all ecosystems. The discovery of deep-sea **hydrothermal vents** and their associated ecosystems necessitated a revision of this model.

Hydrothermal vents are formed in the deep ocean at mid-ocean ridges where the tectonic plates diverge and new plates forming the Earth's crust are made from the magma beneath. At these points, hot water and minerals leak out onto the ocean floor. The minerals solidify as they come into contact with the cold water, forming large chimney-like structures. Many of the minerals are sulphides, which are dark in colour and their escape from vents gives rise to their common name 'black smokers' (Figure 13.1). Others have mixtures of gases such as carbon dioxide and are known as 'white smokers'.

These vents are found at depths of more than 3000 m and the gases emerging from them can have temperatures as high as 300 °C. Remarkably, they are often surrounded by ecosystems of previously unknown species, with extreme adaptations that allow them to survive in an environment with no light, enormous pressure and temperatures varying from 200–300 °C. Even more surprising was the discovery that the energy within these ecosystems is not linked to sunlight but to a special type of bacteria. These bacteria are able to use chemicals such as the sulfide minerals as a substrate for **chemosynthesis** to gain the energy they need to grow and begin new food chains.

Food chains

Figure 13.1 A 'black smoker' hydrothermal vent with associated ecosystem

A feeding relationship in which a carnivore eats a herbivore, which itself has eaten plant matter, is called a **food chain**. Of course, light is the initial energy source in most food chains (except our example of hydrothermal vents). Note that in a food chain, the arrows point to the consumers, and so indicate the direction of energy transfer.

oak leaf → caterpillar → beetle → shrew → owl

A food chain tells us about the feeding relationships of organisms in an ecosystem, but they are shown as entirely qualitative relationships (we know which organisms are present as prey and as predators) rather than providing quantitative data (we do not know the numbers of organisms at each level). In a whole ecosystem food chains are interconnected in many different ways to form a complex **food web**. There are also significant drawbacks to simple food webs. In most cases they tell us only a very limited part of the story. In Figure 13.2 on the next page for example, there are literally hundreds of different insects or their larvae that feed on oak leaves. Similarly, the fox and the shrew

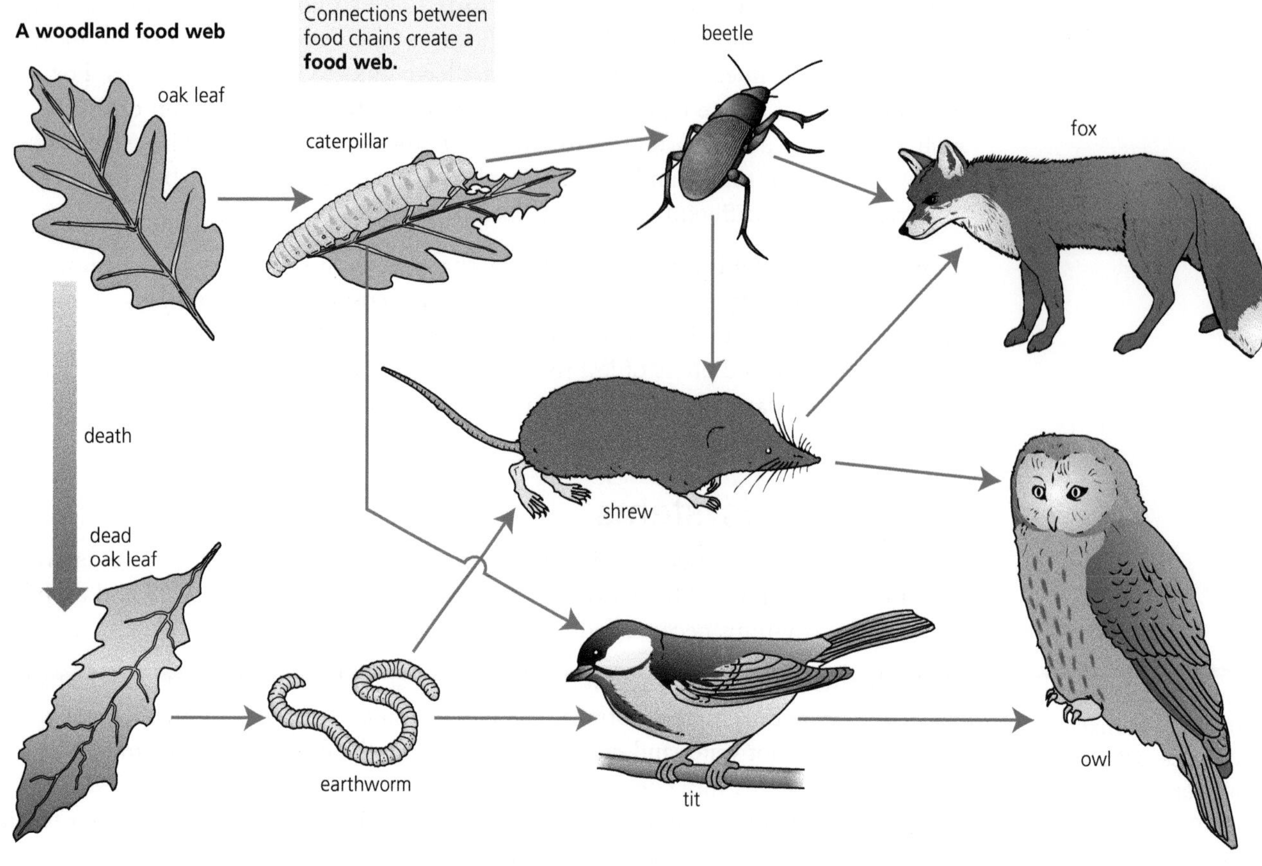

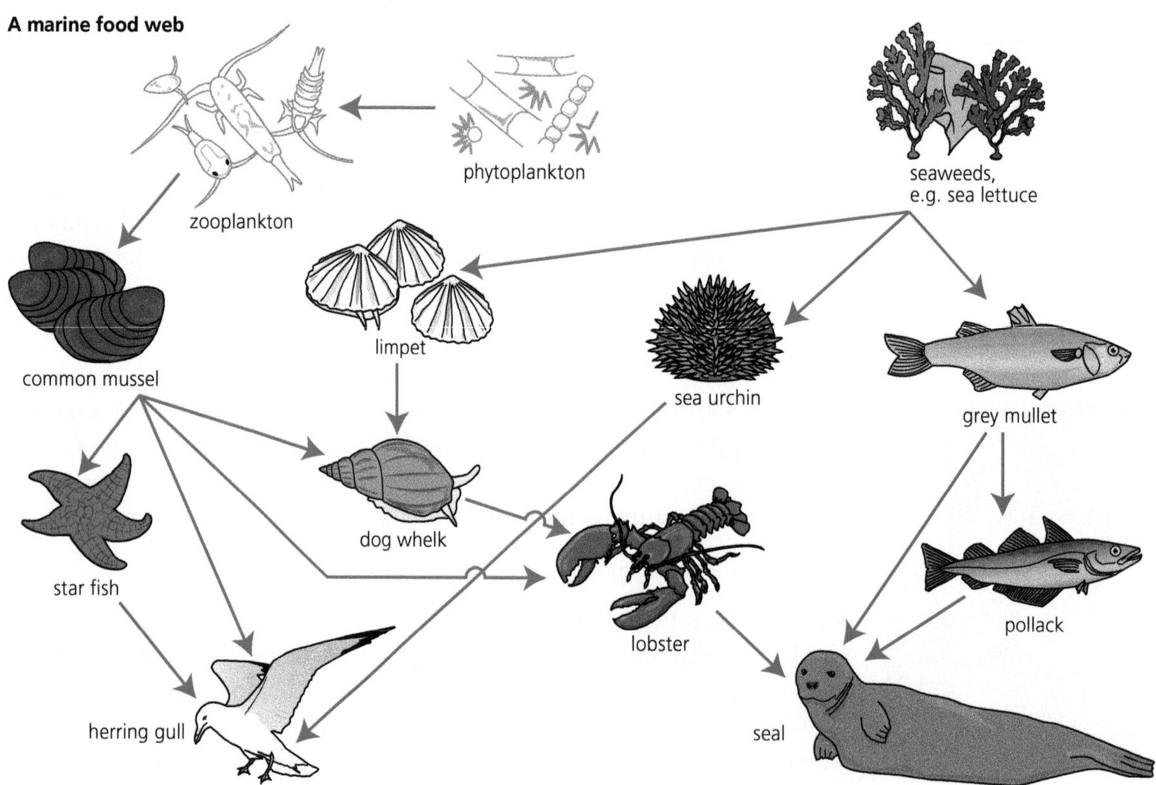

Figure 13.2 Examples of food webs

have very varied diets, often determined by what food they can find. Figure 13.2 shows two different food webs but these are only a tiny portion of the ecological relationships in these ecosystems.

The level at which an organism feeds in a food chain is called its **trophic level**. In this way of classifying feeding relationships, the producers are designated as trophic level 1 because their energy has been transferred once, from Sun to plant. All herbivores are in level 2, because here energy has been transferred twice, and so on.

The trophic levels of some woodland organisms are classified in Table 13.1. Note that there is no fixed number of trophic levels to food chains, but rather they are typically of three, four or five levels only. There is an important reason why stable food chains remain quite short, which we shall consider later.

Key term

Trophic level A level in a food chain defined by the number of energy transfers an organism is from the primary energy source.

Table 13.1 An analysis of trophic levels

Trophic level	Woodland	Rainforest	Savannah
Producer	Oak	Vines and creepers on rainforest trees	Grass
Primary or first consumer	Caterpillar	Silver-striped hawk moth	Wildebeest
Secondary consumer	Beetle	Praying mantis	Lion
Tertiary consumer	Shrew	Chameleon lizard	
Quaternary consumer	Fox	Hook-billed vanga-shrike	

Test yourself

1 Explain why a lion feeding on an antelope is described as the third trophic level.

2 Suggest what eventually happens to the energy in any animal at the top of a food chain.

3 Use Figure 13.2 to suggest three possible effects on the marine ecosystem if the zooplankton were to be severely reduced by a pollution incident.

4 State the main difference between a population and a community.

Energy flow through ecosystems

The productivity of photosynthesis

Only about half the energy emitted by the Sun and reaching the Earth's outer atmosphere gets through to soil level and to plant life – the remainder is reflected into space as light or heat energy. Of the radiation reaching green plants, about 45 per cent is in the visible wavelength range (400–700 nm) and can be used by the plant in photosynthesis. Much of the light energy reaching the plant is reflected from the leaves or transmitted through them, however. Furthermore, of the energy absorbed by the stems and leaves of plants, much is lost in the evaporation of water (Figure 13.3).

A small quantity of the energy reaching a green leaf is absorbed by the photosynthetic pigments and used in photosynthesis; only one quarter of this light energy ends up as chemical energy in molecules like glucose. The remainder is lost as heat energy in the various reactions of the light-dependent and light-

Key terms

Gross primary productivity (GPP)
The total amount of light energy fixed by photosynthesis in a given area and fixed time, usually expressed as $kJ\,m^{-2}\,y^{-1}$.

Net primary productivity (NPP)
The total amount of light energy fixed by photosynthesis in a given area and fixed time, which remains available to the next trophic level after losses and uses by a producer, usually expressed as $kJ\,m^{-2}\,y^{-1}$.

independent reactions. In green plants, the total amount of light energy fixed through photosynthesis in a given period of time is known as the **gross primary productivity** (**GPP**). GPP is usually expressed as units of energy per unit area per year, typically either $kJ\,m^{-2}\,y^{-1}$, or $MJ\,ha^{-1}\,y^{-1}$.

Finally, much of the energy in glucose is lost as heat energy in cellular respiration and other reactions of metabolism. The remainder is retained in new materials, either in the form of new structures (cells and tissues) or as stored food, and represents the **net primary productivity** (**NPP**) of the plant. The value given in Figure 13.3 for net primary productivity of about 5.5 per cent applies to a fully grown crop plant. It is achieved for only a short period in the growth cycle of the crop plant – for a significant part of the year agricultural land is uncultivated.

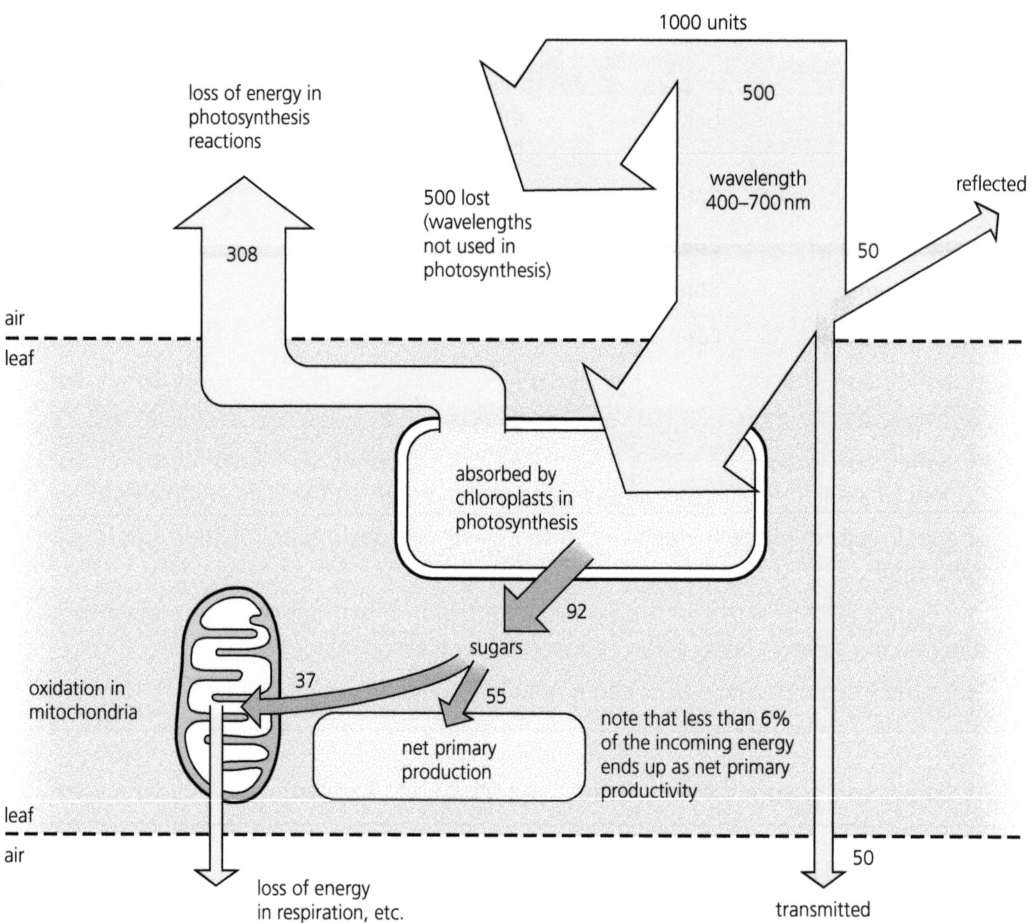

Figure 13.3 The fate of light energy that reaches the green leaf of a crop plant (taken as 1000 units per unit time)

Of course, in many natural habitats you might study, a part of the organic matter that makes up 'net primary productivity' is directly available to sustain browsing herbivores, and indirectly available to other organisms in the environment around the plant. We will examine energy transfer between organisms in a food chain shortly. Meanwhile, the relationship between net primary productivity (NPP), gross primary productivity (GPP) and respiration (R) is summarised in the equation:

$$NPP = GPP - R$$

Applying this equation to data from a temperate ecosystem where GPP was found to be 43 510 $kJ\,m^{-2}\,y^{-1}$, and R to be 23 930 $kJ\,m^{-2}\,y^{-1}$, then NPP here was 9580 $kJ\,m^{-2}\,y^{-1}$.

The fate of energy within and between trophic levels

At the base of the food chain, green plants transfer light energy to the chemical energy of sugars, in photosynthesis. Of this, while some is transferred in the reactions of respiration that drive metabolism (and is then lost as heat energy), much is transferred to essential metabolites used in the growth and development of the plant. In these reactions, energy is locked up in the organic molecules of the plant body. Then, when parts of the plant are consumed by herbivores (or parasites), energy is transferred to other organisms. Finally, on the death of the plant, the remaining energy passes to detritivores and saprotrophs when dead plant matter is broken down and decayed. We will describe the activities of these microorganisms later in this chapter (page 247).

Similarly, energy is transferred in the consumer when it eats, digests and then absorbs nutrients. The consumer transfers energy in muscular movements by which it hunts and feeds, and as it seeks to escape from predators (and is then lost as heat energy). Some of the food eaten remains undigested and is lost in the faeces. Also, heat energy – a waste product of the reactions of respiration and of the animal's metabolism – is continuously lost as the consumer grows and develops, and forms body tissues. If the consumer itself is caught and consumed by another, larger consumer, energy is again transferred. Finally, on the death of the consumer, the remaining energy passes to detritivores and saprotrophs when dead matter is broken down and decayed.

Energy transfers within and between trophic levels are summarised in Figure 13.4 on the next page.

So, only a limited amount of the energy transferred between trophic levels is available to be transferred to the next organism in the food chain. In fact, only about 10 per cent of what is eaten by a consumer is built into the organism's body, and so is potentially available to be transferred on in predation. There are two consequences of this:

- The energy loss at transfer between trophic levels is the reason why food chains are short. Few transfers can be sustained when so little of what is eaten by one consumer is potentially available to the next step in the food chain. Consequently, it is very uncommon for food chains to have more than four or five links between producer (green plant) and top carnivore.
- Feeding relationships in a food chain may be structured like a pyramid. At the start of the chain is a very large amount of living matter (**biomass**) of green plants. This supports a smaller biomass of primary consumers, which in turn supports an even smaller biomass of secondary consumers.

Key term

Biomass The total mass of a living organism in a given area.

Test yourself

5 Name **three** ways in which energy in any one trophic level is lost so that it is not available to the next level.

6 The GPP for an area of grassland in a fixed time is $100\,kJ\,m^{-2}$ and losses of energy are $71\,kJ\,m^{-2}$. Calculate the NPP of the grassland.

7 Suggest why it is more efficient to produce protein from soya beans than it is to produce beef from cattle.

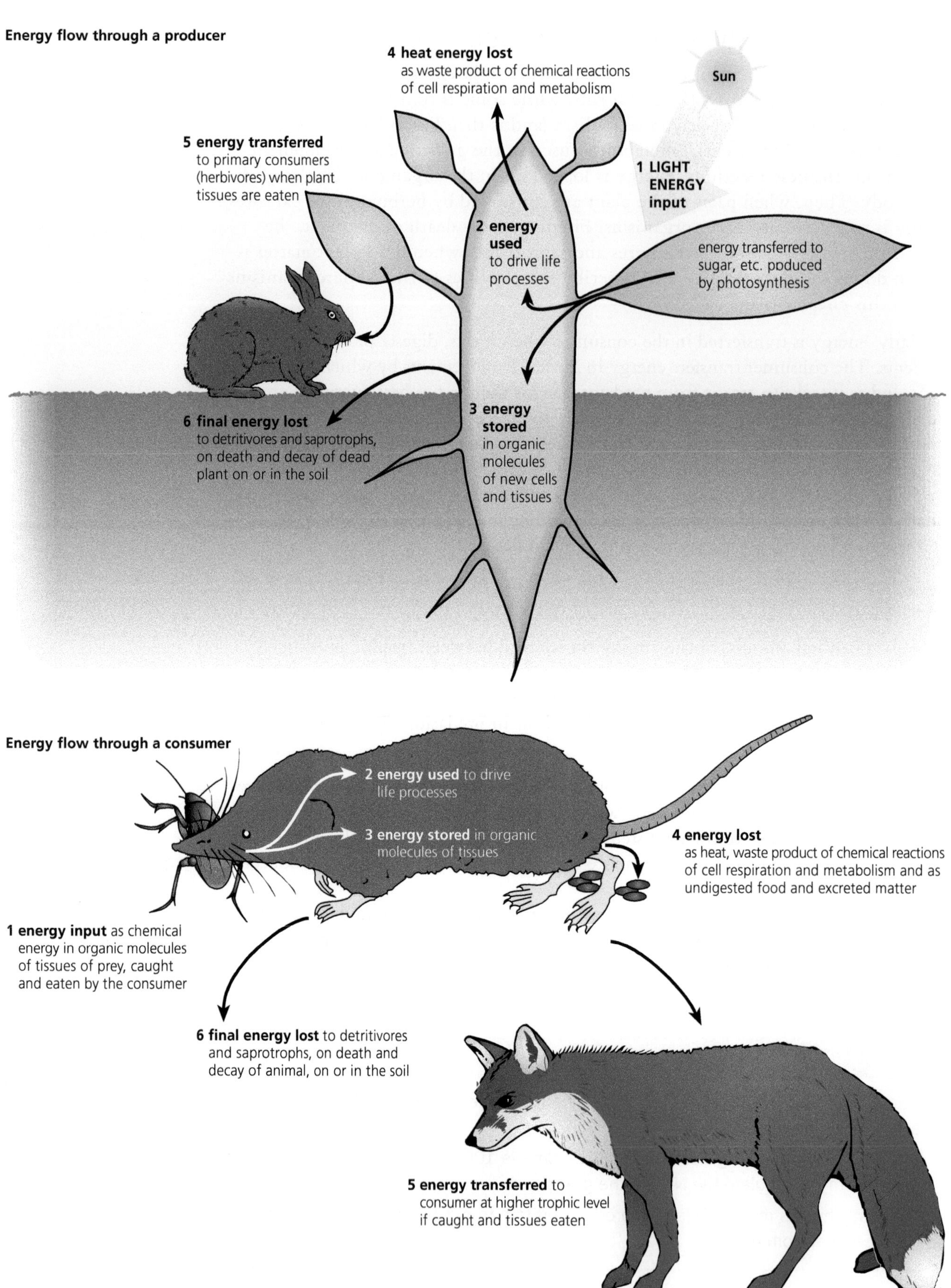

Figure 13.4 Energy flow through producers and consumers

Illustrating feeding relationships quantitatively

Pyramids of numbers

The simplest way to illustrate the feeding relationships in a whole ecosystem is to count the number of organisms at each trophic level and represent them by horizontal bars on top of each other (Figure 13.5). This is known as a pyramid of numbers.

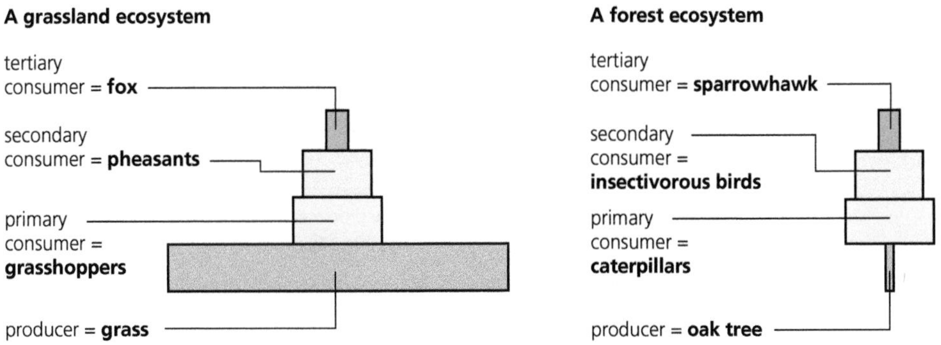

A grassland ecosystem

tertiary consumer = **fox**

secondary consumer = **pheasants**

primary consumer = **grasshoppers**

producer = **grass**

A forest ecosystem

tertiary consumer = **sparrowhawk**

secondary consumer = **insectivorous birds**

primary consumer = **caterpillars**

producer = **oak tree**

Figure 13.5 Examples of ecological pyramids of numbers

Unfortunately, if you were to attempt to draw a real pyramid of numbers for any of the food chains or the food webs in Figure 13.2 you would quickly encounter a confusing picture as shown in Figure 13.5. In addition, accurately estimating the actual numbers for a whole ecosystem is very difficult indeed. Try counting the numbers of insects on an oak tree!

The main drawback is that the bars simply compare counts of organisms, but these organisms are of very different sizes and therefore they distort the actual picture. In this case 'one oak tree = one caterpillar' which is obviously not representative of the real ecosystem. To add to this confusion, trees such as oaks live for many years and build on previous growth year after year, whilst almost all the insects will have life cycles lasting less than one year. In addition, the scale would be impossible to select on one axis as you would have to cope with one oak tree and many thousands of primary consumers.

Pyramids of biomass

To solve the problems of simply counting numbers you can measure the total biomass (Figure 13.6). Biomass is simply the mass of all the organisms at any one trophic level. But this, too, has its major drawbacks. The biomass of almost all living organisms is made up of over 80 per cent water and this is not helpful in illustrating relationships. Water content can also vary enormously over relatively short periods of time. To overcome this problem it is important to find the dry biomass in each trophic level. Even then you still have a further problem in that different components of biomass have very different energy contents per kilogram. For ecologists, studies finding dry biomass are highly undesirable as all the organisms need to be killed. Fortunately there are tables to allow conversion from fresh to dry biomass in different organisms but you still need to find and weigh all the organisms concerned.

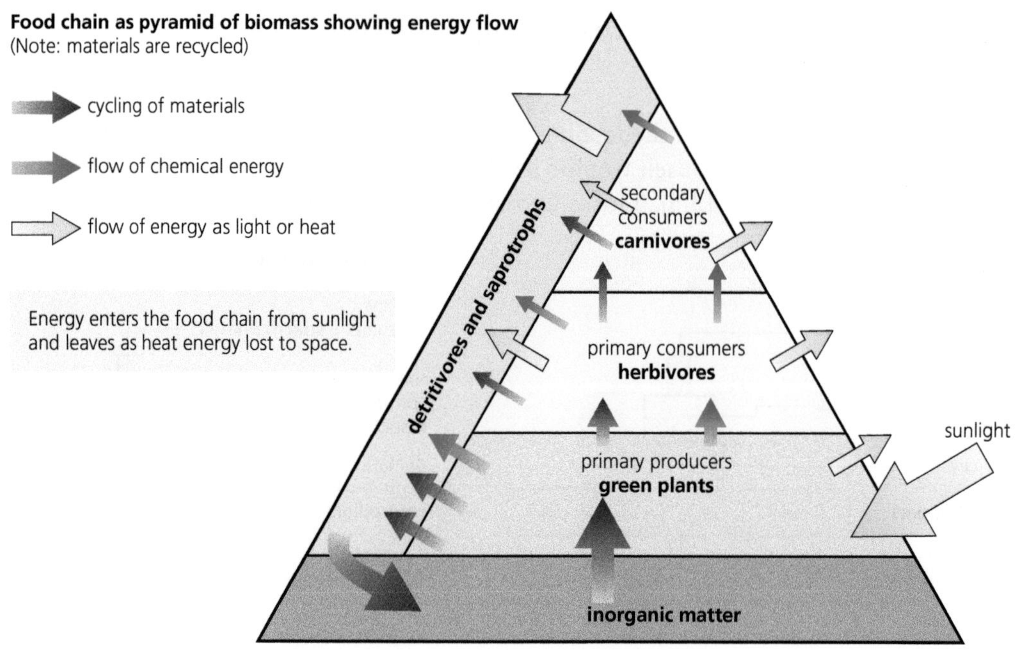

Figure 13.6 Energy flow through a food chain – the pyramid of biomass

Pyramids of energy

Given the problems of pyramids of biomass then the next logical step is to estimate the amount of energy available in each trophic level and to draw a pyramid of energy (Figure 13.7). Once again there are tables that will convert simple dry biomass measurements into energy content for each organism. The units for energy are expressed as $kJ\,m^{-2}\,yr^{-1}$ and you will notice that the final term includes a time element, in this case yr^{-1}. This is a vital part of the total energy calculation and overcomes the difficulties of different life-spans, even though it makes accurate measurement even more difficult.

Only energy taken in at one trophic level and then built in as chemical energy in the molecules making up the cells and tissues is available to the next trophic level. This is about 10% of the energy. The reasons are as follows.

- Much energy is used for cell respiration to provide energy for growth, movement, feeding, and all other essential life processes.
- Not all food eaten can be digested. Some passes out with the faeces. Indigestible matter includes bones, hair, feathers, and lignified fibres in plants.
- Not all organisms at each trophic level are eaten. Some escape predation.

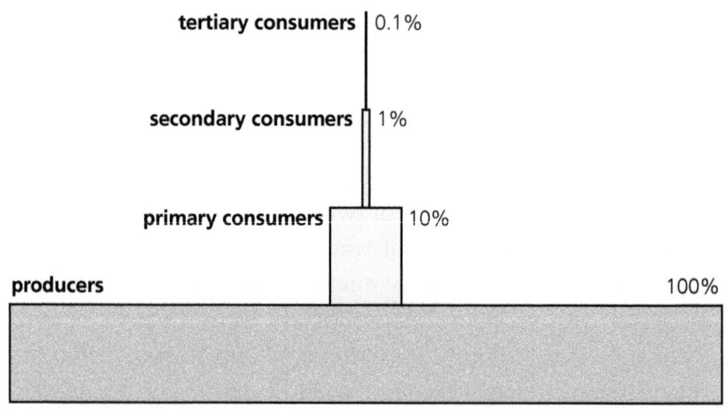

Figure 13.7 A generalised pyramid of energy

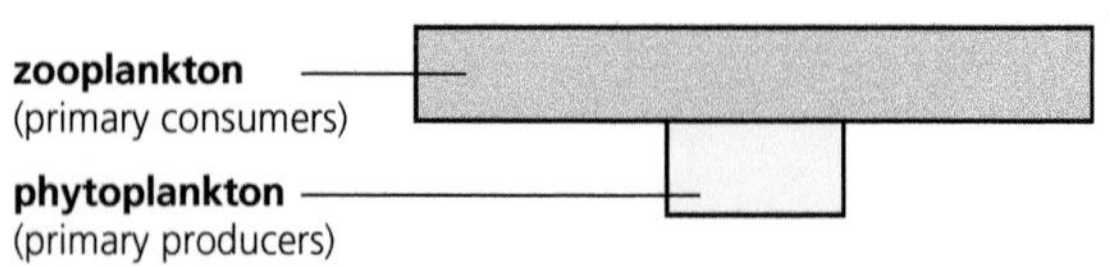

Figure 13.8 Phytoplankton and zooplankton – an inverted pyramid

Why this is so important can easily be demonstrated by considering another difficulty with pyramids. A single set of measurements only gives us a snapshot at one instant in time, but we know that ecosystems are subject to many fluctuations. We can show the problem by looking at yet another pyramid. Figure 13.8 shows the pattern of a very simple pyramid (dry biomass) obtained from data measured in the sea just off the coast of the British Isles.

Phytoplankton are the microscopic algae found in the oceans that photosynthesise and are therefore primary producers. Although often overlooked, they play a vital part in maintaining oxygen balance in the atmosphere as two thirds of the Earth's surface is ocean. Zooplankton are a diverse group of tiny marine invertebrates that feed off the phytoplankton, so are classic primary consumers. Zooplankton are a crucial step in a large number of marine food chains. The data have repeatedly been shown to be accurate, but why are things upside down?

The explanation is simple but a cautionary reminder that you must think carefully about exactly how data have been collected before you jump to conclusions. It is quite true that the biomass of the consumers is higher than that of the producers in this sample. However, this is just a single sample at one time. If we study things more carefully we find that the reproduction rate of the plankton is very high, especially in summer conditions when days are longer and sea temperatures are higher. Therefore at any one time, although the population of plankton appears to be too small, they are rapidly reproducing, fast enough to pass on sufficient energy to maintain a very high population of consumers, even though, logically, it appears that the ecosystem is about to collapse. However, if you express the energy content available in a fixed time you return to a predictable pyramid shape.

Test yourself

8 State the main drawback of using pyramids of numbers to illustrate feeding relationships.

9 Explain why it is essential to consider dry mass of organisms.

10 Suggest why simple dry mass or energy content of an oak tree is not directly comparable with that of the insects living upon it.

11 Explain how ecologists overcome the problem of having to kill organisms in order to determine their dry mass.

The role of microorganisms in an ecosystem

Microorganisms and the recycling of energy

All organisms in all trophic levels produce waste and eventually die. Although, as you have seen, a good deal of the energy contained within each trophic level is lost during transfer to the next, significant amounts of energy remain in the organic molecules, such as cellulose, lignin, lipids, carbohydrates and proteins of dead organisms. Much of this energy is recycled by microorganisms, especially bacteria and fungi. Microorganisms can produce extracellular enzymes that break down dead tissues in decay processes. The products of this enzymic digestion are then reabsorbed into the cells of the microorganisms to be used as respiratory substrates or raw material for synthesis. This method of nutrition is known as saprotrophic (or saprobiontic) nutrition.

You may not be aware of the activities of **saprotrophs**, but without them our world would be covered in a layer many hundreds of metres thick, consisting of leaf litter and woody tissue mixed with dead animal remains and their waste products. A walk in woodland during the autumn will provide you with some clues about the activities of many fungi. At this time of year it is possible to see many different mushroom-like growths. This is the 'tip of the iceberg' as these are only the spore-producing bodies of the fungus. What lies beneath is a large network of fine threads, known as **hyphae**, forming a network called a **mycelium**, and it is these hyphae that secrete the enzymes and reabsorb the products. The activities of countless bacteria are even less visible to the naked eye.

Key term

Saprotrophs Mainly fungi and bacteria; feed by secreting extracellular enzymes to break down dead organic matter and reabsorb the products; also known as saprobionts.

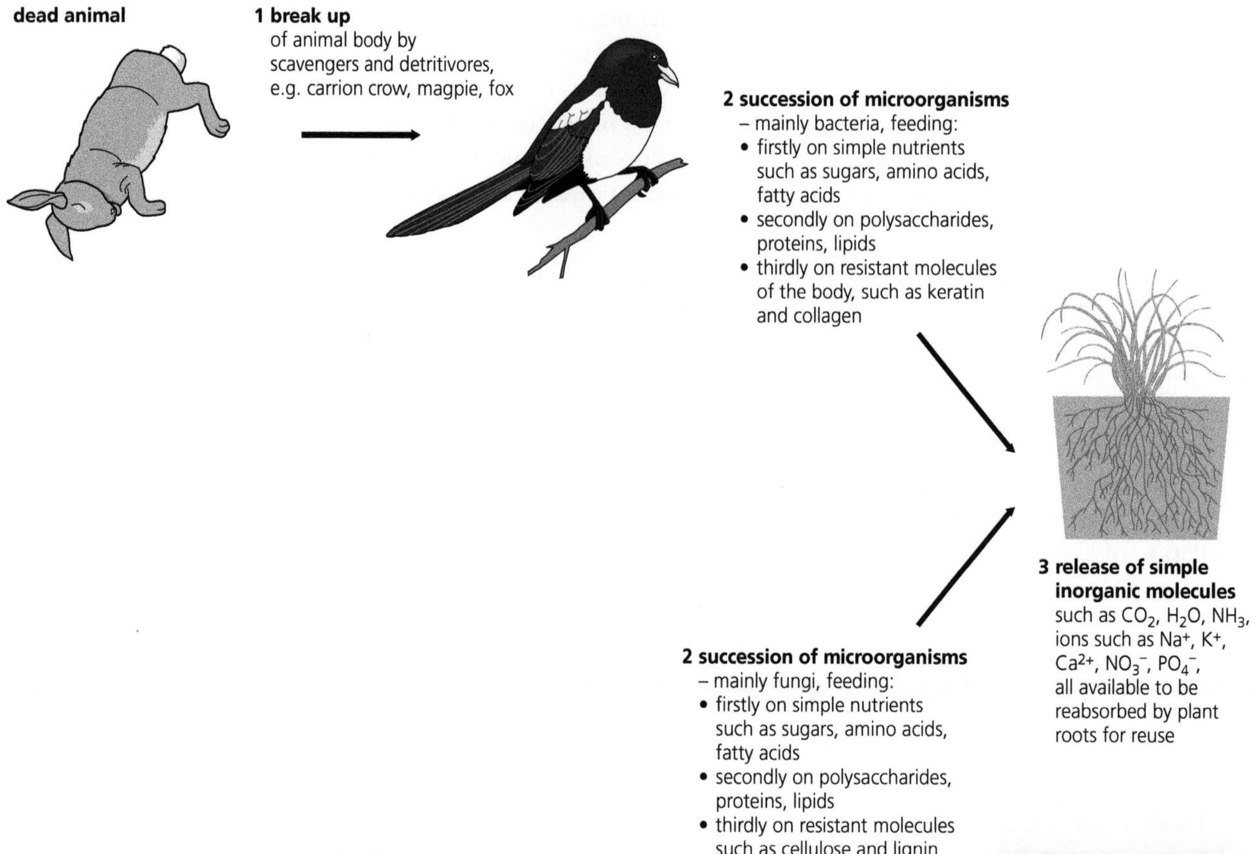

dead animal

1 break up
of animal body by scavengers and detritivores, e.g. carrion crow, magpie, fox

2 succession of microorganisms
– mainly bacteria, feeding:
- firstly on simple nutrients such as sugars, amino acids, fatty acids
- secondly on polysaccharides, proteins, lipids
- thirdly on resistant molecules of the body, such as keratin and collagen

3 release of simple inorganic molecules
such as CO_2, H_2O, NH_3, ions such as Na^+, K^+, Ca^{2+}, NO_3^-, PO_4^-, all available to be reabsorbed by plant roots for reuse

2 succession of microorganisms
– mainly fungi, feeding:
- firstly on simple nutrients such as sugars, amino acids, fatty acids
- secondly on polysaccharides, proteins, lipids
- thirdly on resistant molecules such as cellulose and lignin

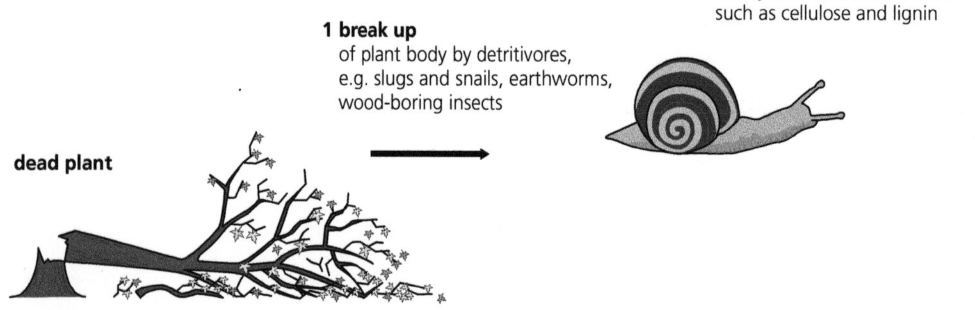

1 break up
of plant body by detritivores, e.g. slugs and snails, earthworms, wood-boring insects

dead plant

Figure 13.9 The sequence of organisms involved in decay

Figure 13.10 A detrivore – *Gammarus pulex*

Key terms

Scavengers Opportunist feeders on the remains of dead animals.

Detritivores Animals that feed off dead and decaying organic materials. They are important recyclers in all ecosystems.

Larger parts of dead organisms form the diet of a group of animals known as **scavengers** (Figure 13.9).

Dead organisms, especially plants, become broken down or decay and form a valuable food source, which is also recycled by a whole range of animals known as **detrivores**. For example, the enormous numbers of leaves that fall into streams and rivers are often broken down by the water flow and rocks under the surface. The particles formed are the main food supply for very large numbers of the freshwater shrimp, *Gammarus pulex* (Figure 13.10). This small shrimp is a vital part of freshwater ecosystems as it is the major food source for many freshwater fish. Earthworms, and slugs and snails are also described as detrivores and they play an important part in the physical breakdown of dead materials as well as recycling them. All ecosystems have their unique pattern of saprotrophs, scavengers and detrivores.

Microorganisms and the recycling of nutrients

Recycling is essential for the survival of living things because the available resources of many elements are strictly limited. The activities of bacteria and microorganisms in decay not only make some of the energy in dead organisms available to other food chains, but also release many of these elements locked up within them. These become part of the soil solution, and some may react with chemicals of soil or rock particles before becoming part of living things again when they are reabsorbed by plants.

Nitrogen is a particularly important element. It is a component of proteins and nucleic acids. There is an abundance of nitrogen in the biosphere since it makes up almost 80 per cent of the atmosphere. However, nitrogen molecules in the air are not directly available to plants and animals. This is because the bonds in molecules of nitrogen (N_2 or dinitrogen) are very strong and a great deal of energy is required to break them. Plants and animals cannot do this. Instead, plants take up nitrogen from the soil in the form of compounds such as nitrate and ammonium ions. From these nitrogen compounds plants are able to manufacture the amino acids they require.

Animals take in 'ready-made' proteins in their diet, digest them to amino acids and absorb them. The amino acids are then used to rebuild the animal's own specific proteins. This is a fundamental difference between plant and animal nutrition.

The cycling processes by which essential elements are released and reused are called biogeochemical cycles, and the nitrogen cycle (Figure 13.11) is one of the most important.

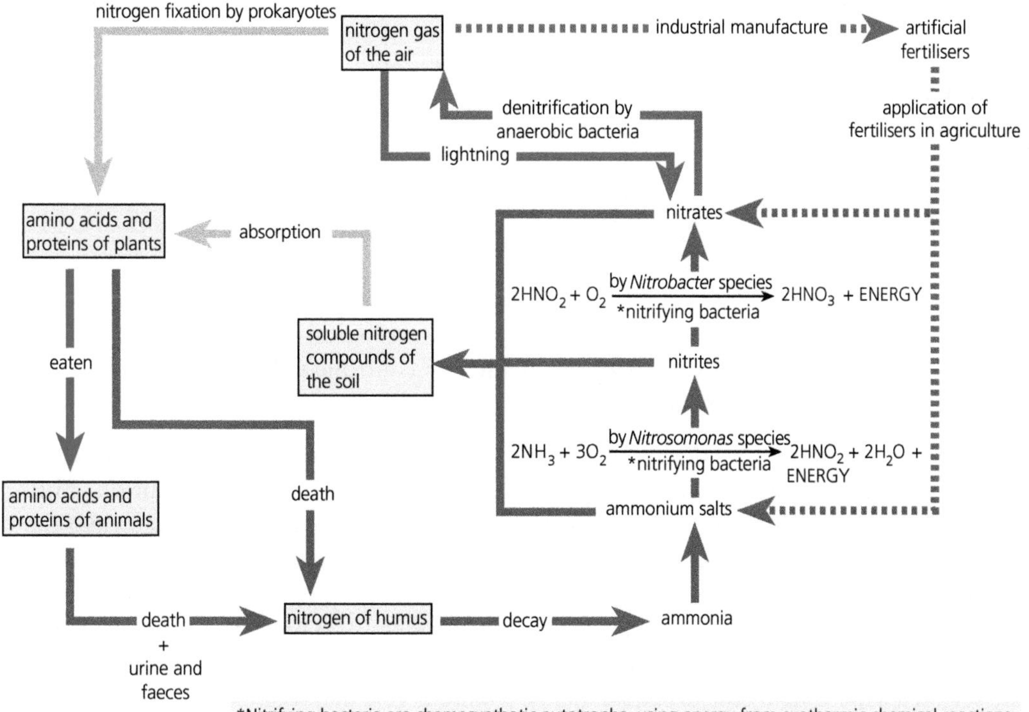

Figure 13.11 The nitrogen cycle

The process of converting ammonium compounds released by the decay of plant and animal materials into nitrates (nitrification), which can be absorbed by plant roots, is an excellent example of the role of bacteria in these cycles. Notice that these are oxidation reactions, which take place inside the bacterial cells (intracellular) where the enzymes are located, and these reactions are the way in which the bacteria gain their energy. A look back at the description of the bacteria in the hydrothermal vent ecosystems will remind you that this is **chemosynthesis** and that it is different from the saprotrophic nutrition involved in the decay process.

A careful look at the nitrogen cycle will also reveal that, because ammonium and nitrate compounds are very soluble, they can often be washed out of the soil and lost to the ecosystem by **leaching**. Terrestrial ecosystems cannot thrive without some means of replacing this loss. **Nitrogen-fixing bacteria**, often of the species **Rhizobium**, are found in root nodules of leguminous plants. They are able to convert nitrogen gas from the atmosphere into ammonium compounds and therefore play a vital role in maintaining levels of nitrogen available to the ecosystem.

The natural nitrogen cycle cannot keep pace with the demands of modern agriculture, where intensive farming of high yields of a single crop (monoculture) removes large quantities of nitrogen-containing ions from the soil. To compensate, the addition of industrially produced nitrogen-containing compounds is essential.

Key terms

Leaching The loss of soluble materials in soil caused by water flow.

Nitrogen-fixing bacteria Bacteria capable of converting atmospheric nitrogen in to ammonium ions using ATP.

***Rhizobium* sp.** An important group of nitrogen-fixing bacteria found in the root nodules of leguminous plants.

Test yourself

12 Describe the difference between chemosynthesis and saprotrophic nutrition.

13 Explain why nitrogen is such an important element to living organisms.

14 What is meant by 'nitrogen fixation'?

15 Nitrogen-fixing bacteria are essential to maintain levels of nitrogen circulating in an ecosystem. Explain why.

Estimating population size and distribution in a habitat

Unlike a highly controlled laboratory investigation, ecological investigations often involve many variables that cannot all be adequately controlled. This means they need even more careful planning and cautious interpretation.

We will begin here by considering how you might, first of all, obtain accurate data about what organisms are present in a habitat and what is the pattern of their distribution. This will give you many clues about the complex interrelationships within this habitat and the whole ecosystem. The very first problem you encounter is exactly what method would be suitable to count your chosen organisms. Your choice will need to be very different if you wish to include mobile animals such as mice and voles as opposed to the plants in some grassland. Ecologists use several methods of actually recording abundance, which are selected according to the organism being counted. The methods chosen must be both reliable and practical. For example, it is easy to count the number of individual crabs found on a rocky shore within a few square metres but an almost impossible, and extremely time-consuming task to assess the huge number of barnacles in the same way.

Methods of assessing abundance

- **Individual counts** – the simplest method to use provided numbers are reasonably low and each individual is easily distinguished.
- **Percentage cover** – used where individual organisms form continuous cover on a surface, for example lichens covering a rock surface or dense grass cover.
- **ACFOR scales** (Table. 13.2) – these are used for more approximate assessments where abundance is measured on a five-point scale. There are tables defining how these scales might be applied to different organisms to make data comparable across different investigators. ACFOR data are useful in displaying comparative overall patterns but cannot be used statistically. A very similar DAFOR scale is often used for plant species (D = Dominant here).

Table 13.2 An example of an ACFOR scale for small barnacles

ACFOR rating	Numbers per unit area
Abundant	100+0.1 m^{-2}
Common	10–99 0.1 m^{-2}
Frequent	1–9 0.1 m^{-2}
Occasional	1–99 m^{-2}
Rare	<1 m^{-2}

Sampling

It is normally impossible to measure everything in a whole habitat or ecosystem, therefore samples of smaller areas are taken from which inferences can be drawn. This is useful because it makes the investigation possible, but might not tell the whole story. Sampling can also introduce bias. For example, if you are measuring the heights of some plants you are more likely to select taller plants than to search for shorter examples. If you are sampling coloured molluscs you are much more likely to select brightly coloured examples that catch your eye.

Quadrats

A **quadrat** is simply a frame that outlines an area of known size for sampling purposes. The size of the quadrat used is determined by the actual habitat and organisms to be counted. The most common sizes are square quadrats of 1 m × 1 m or 0.5 m × 0.5 m (remember, this is a 0.25 m^2 not a 0.5 m^2 area, so be careful how you record it!). However, much smaller quadrats may be needed, as shown in the barnacle ACFOR scale in Table 13.2.

A **gridded quadrat** is a frame quadrat modified by adding strings to form extra squares within the frame as shown in Figure 13.13 on the next page. This helps in ensuring organisms are not counted twice and is particularly useful when making estimates of percentage cover.

A **point frame** (Figure 13.12) is a adaptation of the same idea which is used where there is particularly dense vegetation that makes estimation of numbers in a whole frame very difficult. It consists of a frame with 10 pointed metal pins. The frame is placed at the chosen sampling points and only the plant touching the ten points of the metal pins is recorded.

Sand deposited by the sea and blown by the wind builds into small heaps around pioneer xerophytic plants, such as marram and couch grass, at coasts where the prevailing wind is on-shore. The tufts of leaves growing through the sand accelerate deposition, and gradually drifting sands gather a dense cover of vegetation. Fixed sand dunes are formed.

point frame quadrat in use

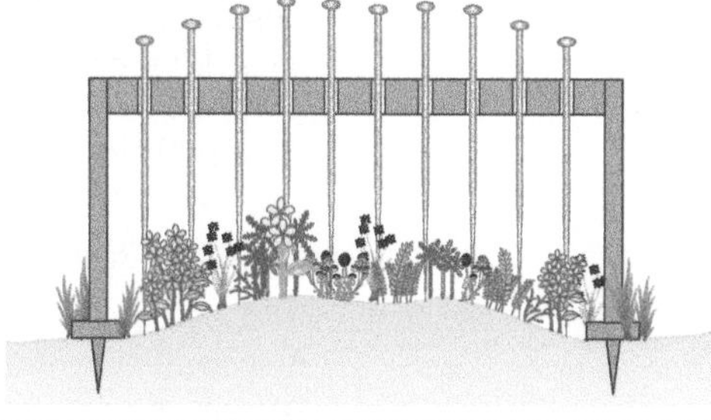

The frame is randomly placed a large number of times, the 10 pins lowered in turn onto vegetation and the species (or bare ground) recorded.

Figure 13.12 A point frame

Random sampling

Having decided upon your method of actual counting you must now think about exactly where to count your samples. This is where you will need to use your planning skills and some thought. Habitats are so varied that you must select the most appropriate type of sampling according to the area you are studying and the hypothesis you are investigating.

One of the most common approaches is to compare two fairly large areas to discover if there are any significant differences between them. In this case we often use a large marked area as a grid and select points to sample within it using a table of random numbers or a random number function on a calculator as shown in Figure 13.13.

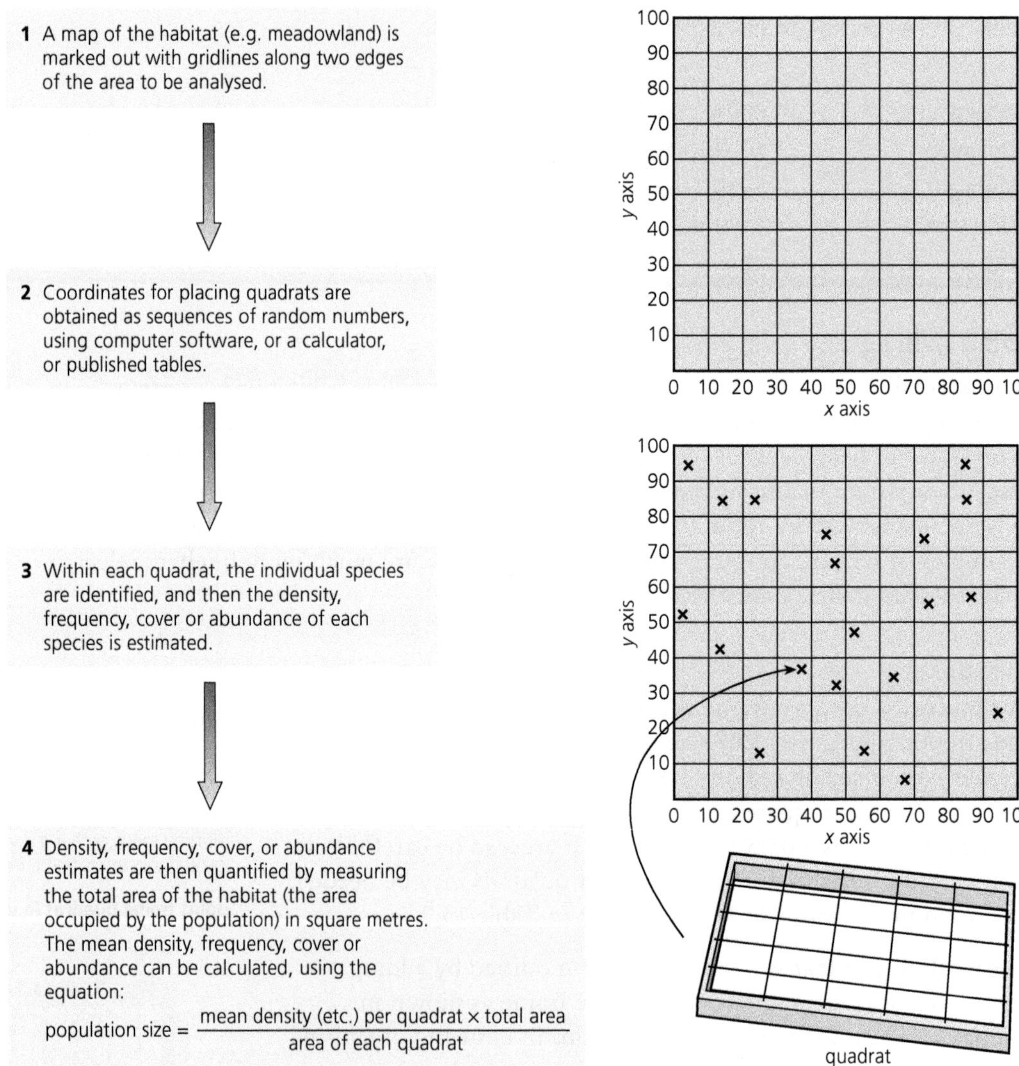

Figure 13.13 Random locating of quadrats

A second very common method of random sampling is to use a **transect**. Transects are simply a graduated line across the area you wish to study. Quadrats or point frames are placed at regular intervals along the line to make the required counts. Transects are most useful when you are investigating the effect of a gradient of factors across the habitat, for example changes in a sand dune system as you move inland, or as you move away from wetter areas on the banks of a river towards drier areas up a slope.

If the transect is short then you might use intervals where the quadrats touch each other; this would form a **continuous belt transect**. For much longer transects you would select intervals that give you a sensible number of quadrats to assess but do not miss any obvious ecological changes along the line. This is an **interrupted transect**.

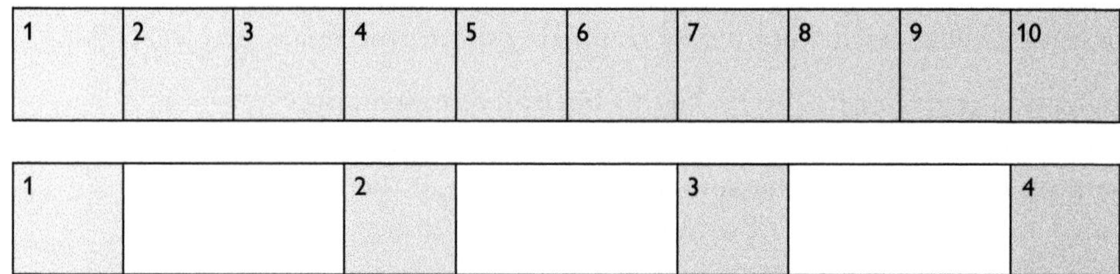

Figure 13.14 Continuous and interrupted transects

Test yourself

16 Spell out the acronyms ACFOR and DAFOR in full.

17 Suggest in what way a scale such as ACFOR is inaccurate.

18 Describe the circumstances in which it is advisable to use a transect method of sampling.

Assessing abundance of mobile populations

The methods described so far depend heavily on the organisms being sedentary. Unfortunately most wild animals are not so cooperative and do not sit around whilst a quadrat is placed over them or a transect line passes by. Therefore it is usually necessary to trap animals in some way in order to count them. Nocturnal insects can be trapped in large numbers by attracting them with a bright light, whilst others such as butterflies can be captured using large soft nets.

However, if you wish to estimate the populations of ground animals that move quickly and over a wide area you need to employ other techniques. There are many designs of traps but most have a chamber where suitable bait is placed and a trapdoor that closes behind the animal as it enters, without injuring it. The most common method of doing this is to set the traps randomly over a wide area, mark any animals caught and release them. This is repeated over a long period and careful records kept of the numbers of marked and unmarked animals recaptured. The principle of the method is that after a suitable period of time the marked animals will mix with the unmarked animals. The results will be that further trapping will give you data where the ratio of marked to unmarked that you recapture will be the same as that in the whole population. Since you know precisely how many you have marked, then the only unknown will be the number in the whole population, so you can calculate that as shown in the equation below. For obvious reasons this is known as the **mark-release-recapture** (MRR method).

$$\frac{N}{n} = \frac{K}{k}$$

Where N = total number of animals in the population

n = total number of animals recaptured

K = total number of animals marked

k = total number of recaptured animals that were marked

Since we know how many we marked it is possible to calculate the number in the whole population as follows:

$$N = \frac{nK}{k}$$

There are a lot of mathematical assumptions in making this calculation that can lead to significant errors, but if carried out over a reasonably short time, many are acceptable.

- The population must be large to make this fraction representative.
- All of the animals must mix randomly and stay together in the test area.
- The marking must not make the animals unacceptable to their population.
- There is no death, migration or births to change the numbers you are using.

Safety and ethics in ecological investigations

Risks

As with all investigations the first consideration in planning must always be, is this safe to perform? There needs to be a serious review in order to answer this question, not just a casual, subjective decision. This review takes the form of a **risk assessment**.

As the name suggests this must be an assessment that reaches conclusions, not just a simple listing of possible dangers. There are many forms of risk assessment but they tend to follow a common pattern.

- What are the principle risks involved?
- How can I plan my investigation to eliminate all risks?
- If I cannot eliminate all risks, what measures must I take to reduce these risks to an acceptable level?

When making final decisions, the level of risk will depend on two things: (a) the likelihood of it happening and (b) the consequences if it were to happen.

If both these are high, then the investigation should not be undertaken unless the likelihood can be reduced to a very low level.

Field investigations mean you will encounter very different and often more serious risks than in the laboratory. First and foremost, personal safety should be a priority, as many habitats such as rocky shores and exposed locations such as moorland mean that there are significant dangers. Obvious precautions, such as not working alone, and consulting weather forecasts and tide tables, are vital. Similarly, having a means of communication is important should help be required. Mobile phones are very useful but make sure their batteries are fully charged!

Field study centres often use numbers to quantify the dangers and likelihood of risks, where the totals for any investigation can indicate the need for more precautions or refusal of permission to go ahead.

Ethical issues

Ethical issues are not simply about deciding between right and wrong; they very often include complex decisions to which there is more than one obvious logical answer. Above all, there are many different and often very logically argued, ethical viewpoints, which are justifiably held by different people.

In any ecological study you would at least expect biologists to adhere to two main ethical ideas:

1 Any investigation should do minimal harm to the environment.

2 All living things should be treated with respect.

Therefore, planning an ecological investigation should always involve both these principles, ensuring that such damage as trampling or sampling is kept to the absolute minimum and that simple measures such as removing all litter are strictly adhered to.

Just to illustrate how complicated things can become, what do we mean by treating all living things with 'respect'? The criminal law lays down some strict guidelines about the use of animals in scientific investigations. Special licences are needed to carry out investigations using mammals for example, but this does not apply to many other animals, generally ectotherms. So you can be prosecuted for using inhumane traps for wild animals that may cause them suffering, but Britain's most popular outdoor sport, angling, involves hauling fish from the water by a hook through the sensitive tissues in their mouth. So why should a fish be treated differently from a vole? You may be an angler and argue that the fish suffer no significant harm, or you may dislike the practice and hold different views.

This type of dilemma and differing ethical stances are very common subjects for debates.

Test yourself

19 In assessing a population using a method involving marking organisms, it is important that the marking does not stimulate a response from other members of the population. Suggest why.

20 Describe an instance when you might consider using an interrupted transect rather than a continuous transect.

21 Describe a situation when you would use a point frame in preference to an open quadrat.

22 State what fraction of 1 m^2 is covered by the area of a 0.5 m × 0.5 m quadrat.

Core Practical 15

Investigate the effect of differing sampling methods on estimates of the size of a population, taking into account the safe and ethical use of organisms

Background information

Whilst this practical (and Core practical 16) are designed to give you practice in using the ecological techniques described, this is not their main purpose. This practical aims to give you the opportunity to plan a whole investigation and to analyse your data statistically. You will find a simple summary of some of the important points about statistics at the end of this chapter, but the mathematical skills section (Chapter 15) gives much more detail.

The suggested method on the next page is given as an illustration, as the investigation you will perform will depend on the habitat that is available and the apparatus you can use. The principles, however, will be exactly the same no matter what populations you assess.

Remember that you are trying to compare two methods to see if they give the same result or, if they differ, is this statistically significant or simply expected variation?

In this method you are using any area of mown grassland and comparing assessment by a transect with assessment by random sampling, but you can also try comparing the effect of using a large quadrat with using a small quadrat (say 1 m × 1 m vs 0.5 m × 0.5 m) or compare using a quadrat with using a point frame. Comparing the methods of counting individuals and assessing percentage cover can also be interesting. The plants you are using are common to most grassland – plantains or dandelions.

If you are using plantains you need to be able to identify the two main species (Figure 13.15) that you are likely to find and make sure you count only one species. You can find many sources to help you identify these on the web, but remember to check the sites you use carefully. Many of the websites originate in the USA or other countries and whilst they are often very good, they do not always describe the same species you might find in this country. The Science and Plants in Schools website (SAPS) is a good place to start. The illustrations below will help you.

Figure 13.15 The two common species of plantain. Left: greater plantain (*Plantago major*); right: ribwort plantain (*Plantago lanceolata*)

Dandelions are much easier to identify but do make sure that you can distinguish them from the plantains, especially the ones without flowers. (Check the edges of the leaves and the main veins in the leaf if you are not sure.)

Carrying out the investigation

Aim: To compare the assessment of population numbers using random sampling and a transect.

Risk assessment: If you use the school field then the risks will probably be very limited, but there are possible dangers that need very simple checks. Are there any groups practising athletic field events? Has the field been treated recently with fertilisers (or weedkillers). If you are using any other habitat you must check with your teacher. Always wash your hands thoroughly after working outside, and before any hand to mouth contact.

1 First of all, make sure that you and any partners are confident in identifying your chosen plants. Then start with some basic observations of the whole area. Make sure that you select an area where there are plenty of plantains and that the area you select does not have any other obvious factors that could affect their distribution, for example large trees giving big patches of light and shade, a large slope that might be far wetter at one end, a part of a football field that gets heavy wear or has large bare patches. In other words, try to make sure you select your large sample area to mitigate the effect of other variables.
2 The next step is to decide on a sensible number of samples for each method. You might make a quick trial to see how long it takes for each quadrat reading and match your sampling to the time you have available.

3 To test random sampling, use two tapes, about 100 m long, at right angles and set up a grid system as shown in Figure 13.13. If you do not have tapes this long then lay a metre ruler on the ground at a suitable start point to indicate your first axis, and generate a random number. (You can generate random numbers on most scientific calculators or use printed copies of random number tables.) Walk the number of paces indicated by the first two random digits along this axis then turn 90 degrees and walk the number of paces shown by the second two random digits. At this point lay the quadrat in front of you and count the plantains inside. Repeat this at least 10 times.

4 Using the same start point, lay out a single tape in one direction and count the plantains in another 10 quadrats at points along the tape indicated by random numbers.
Remember, you can choose to investigate other interesting variables, as indicated in the background information, in which case you would simply use the grid system for both.

Data processing

- Make sure that you have at least 10–15 measurements for each method of assessment. If time is short, consider pooling your data with other groups.
- Calculate the mean and standard deviation for each of your two data sets.
- Use a suitable graph to display your data.
- Try to make a subjective judgement to decide if you think the two data sets show there is a difference or are about the same.
- Use your data to test for a significant difference at the 5% significance level (see Chapter 15). You will need to use either a *t*-test or a Mann-Whitney U test.

Questions

1 How do I count a plantain that is partly inside and partly outside the quadrat?
2 What is the easiest way to tell the different between *Plantago lanceolata* and *Plantago major*?
3 Is there any scientific reason why a transect should give a different result from a random sample?
4 Could I just throw the quadrat in a random manner in the area?
5 What would be a null hypothesis for this investigation?
6 What does a '5% confidence level' mean?

Example

A worked example of a *t*-test

Most biologists use computers to carry out the detailed calculations needed to arrive at the test statistic necessary to find the overall probability. You are likely to do the same, but you will be required to carry out some part of a calculation and interpret the results in examination papers. A full detailed calculation such as this will take some time and it is unlikely this amount of time will be available, but you will certainly need to be familiar with substituting values in parts of an equation and be able to manipulate a formula.

Don't panic!

The *t*-test is, by far, the most complicated calculation of any of the main statistical tests; others are much more straightforward (see the correlation test in Chapter 14).

Make sure you learn the principles of statistical testing outlined in Chapter 15, Mathematics for biology. An understanding of these will assist you in gaining a good proportion of the marks available in this type of question. Don't simply switch off because it contains a little more mathematics. Use your core practical work and other examples to become familiar with the calculations.

Try the specimen examination question at the end of this chapter and, as you will see, such questions are often well-structured to guide you through.

Table 13.3 A worked example of a t-test based on some typical results of Core practical 15

	Numbers of plantains in transect quadrats (x_1)	Numbers of plantains in random quadrats(x_2)	
	10	8	
	3	9	
	9	7	
	7	11	
	5	3	
	8	5	
	5	4	
	6	7	
	4	9	
	7	9	
Σx	64	72	Total (= sum of the 10 quadrat values)
n	10	10	sample number
$\bar{x}$	6.4	7.2	Mean (= $\frac{\text{total}}{n}$)
Σx^2	454	576	Sum of the squares of each quadrat value
$(\Sigma x)^2$	4096	5184	Square of the total (Σx). It is not the same as Σx^2
$\frac{(\Sigma x)^2}{n}$	409.6	518.4	
Σd^2	44.4	57.6	$\Sigma d^2 = \Sigma x^2 - \frac{(\Sigma x)^2}{n}$
σ^2	4.9	6.4	$\sigma^2 = \frac{\Sigma x^2}{(n-1)}$
$\sigma d^2 = \frac{\sigma_1^2}{n_1} = \frac{\sigma_2^2}{n_2}$	$4.9 + 6.4 = 11.3$	σd^2 is the variance of the difference between the means	
σd	3.36	σd^2 (the standard deviation of the difference between the means)	
$t = \frac{\bar{x}_1^2 - \bar{x}_1^2}{\sigma d}$	$\frac{6.4 - 7.2}{3.36}$ $= 0.238$	ignore minus sign	

So you now have a *t* value which you can use to check how likely it is that your results come from the same population.

To do this you look up the value in *t*-test tables as shown in Table 13.4.

Table 13.4 Table of critical values of *t*

Degrees of freedom	Significance level					
	20% (0.20)	10% (0.10)	5% (0.05)	2% (0.02)	1% (0.01)	0.1% (0.001)
1	3.078	6.314	12.706	31.821	63.657	636.619
2	1.886	2.920	4.303	6.965	9.925	31.598
3	1.638	2.353	3.182	4.541	5.841	12.941
4	1.533	2.132	2.776	3.747	4.604	8.610
5	1.476	2.015	2.571	3.365	4.032	6.859
6	1.440	1.943	2.447	3.143	3.707	5.959
7	1.415	1.895	2.365	2.998	3.499	5.405
8	1.397	1.860	2.306	2.896	3.355	5.041
9	1.383	1.833	2.262	2.821	3.250	4.781
10	1.372	1.812	2.228	2.764	3.169	4.587
11	1.363	1.796	2.201	2.718	3.106	4.437
12	1.356	1.782	2.179	2.681	3.055	4.318
13	1.350	1.771	2.160	2.650	3.012	4.221
14	1.345	1.761	2.145	2.624	2.977	4.140
15	1.341	1.753	2.131	2.602	2.947	4.073
16	1.337	1.746	2.120	2.583	2.921	4.015
17	1.333	1.740	2.110	2.567	2.989	3.965
18	1.330	1.734	2.101	2.552	2.878	3.922
19	1.328	1.729	2.093	2.539	2.861	3.883
20	1.325	1.725	2.086	2.528	2.845	3.850
21	1.323	1.721	2.080	2.518	2.831	3.819
22	1.321	1.717	2.074	2.508	2.819	3.792
23	1.319	1.714	2.069	2.500	2.807	3.767
24	1.318	1.711	2.064	2.492	2.797	3.745
25	1.316	1.708	2.060	2.485	2.787	3.725
26	1.315	1.706	2.056	2.479	2.779	3.707
27	1.314	1.703	2.052	2.473	2.771	3.690
28	13.13	1.701	2.048	2.467	2.763	3.674
29	1.311	1.699	2.043	2.462	2.756	3.659
30	1.310	1.697	2.042	2.457	2.750	3.646
40	1.303	1.684	2.021	2.423	2.704	3.551
60	1.296	1.671	2.000	2.390	2.660	3.460
120	1.289	1.658	1.980	2.158	2.617	3.373
∝	1.282	1.645	1.960	2.326	2.576	3.291

When you do so you will find another term that you need to understand. The left-hand column is entitled **'degrees of freedom'**, so what does this mean? When checking what your *t*-value means in terms of the probability of the two sets of measurements coming from the same population, it is very important to know just how many samples you took. The probability will change if you had taken 100 samples and not 10. So you need to look at the row that corresponds to the total number samples, but for reasons not important here, you subtract one from each set of samples.

So, in your example, degrees of freedom = (10 – 1) + (10 – 1) = 18.

The general expression for this is $(n_1 - 1) + (n_2 - 1)$.

If you have consulted the principles of statistical testing in Chapter 15, you will know that biologists use a 5% significance level as the benchmark for deciding whether to accept or reject the null hypothesis. If there are less than 5 chances in one hundred that your data could come from the same population then you will reject the null hypothesis (and be able to state that there is a significant difference).

At last, we reach the final stage. You know that you are looking for the 5% column and that you want the row for 18 degrees of freedom.

The critical value here is 2.101, but your t value is only 0.238. To reject your null hypothesis your value must be *greater* than this critical value. Hence, you must accept your null hypothesis and your conclusion from the t-test is 'There is **no significant difference** between the density of plantains measured using random quadrats and the density of plantains using quadrats on a line transect.'

Notice we have been quite precise to include exactly what the hypothesis was all about.

We have deliberately chosen an example here where there is no significant difference. This is partly because there is a tendency to dismiss such conclusions as a failure or, 'it didn't work!' Scientifically, nothing could be further from the truth. Showing there is no difference can be just as valuable, when amassing evidence to support or challenge a suggested model, as any other conclusion.

Two other technical points

Most biologists now carry out calculations like these on computers and there are many programmes to allow you to do this. However, they will require you to choose exactly which variation of the t-test you want to use and therefore it is important you understand what to select.

- **Paired or unpaired test?** – paired data means that the samples you take are always linked together because by taking one of the samples you always determine the position of the other. For example, if you wish to compare the lichen cover on a tree trunk on the north and south facing sides, you can place a small quadrat on the north side and count the lichens. If you then simply take a second reading at exactly the same position but move around to the south side you will have paired data where the position of one determines the position of the other. This is quite rare so you would normally select 'unpaired'.
- **One- or two-tailed test?** – almost all of the tests you are likely to perform will be two-tailed tests. Therefore choose this option. The technical reasons behind this are not essential for you to look at in detail here.

Exam practice questions

1 An ecosystem can be described as a balanced, self-regulating system because:

A the numbers of plants and animals within it do not change

B there are always more producers than consumers

C photosynthesis will always provide the same input of energy

D natural cycles and feeding relationships normally interact to keep things stable *(1)*

2 In statistical testing, a 5% confidence level indicates:

A there are 5 chances in 100, or less, that the samples might come from the same population

B there are 95 chances in 100, or less, that the samples might come from the same population

C there is a probability of 0.95, or less, that the samples might come from the same population

D there is a probability of 0.05, or less, that the samples are from exactly the same population *(1)*

3 In the nitrogen cycle, ammonium ions are oxidised to nitrates. This process is known as:

A denitrification

B nitrogen fixation

C nitrification

D nitration *(1)*

4 **a)** Decay processes are often brought about by fungi. Describe the process of saprophytic nutrition from the synthesis of enzymes to the absorption of products. *(4)*

b) What will be the end products of the enzyme digestion of:

i) lipids

ii) cellulose

iii) muscle? *(3)*

5 **a)** **i)** Explain what is meant by chemosynthesis. *(2)*

ii) Explain how chemosynthesis is used by soil bacteria to oxidise ammonium ions to nitrate ions. *(4)*

b) Explain how green plants can synthesise amino acids using NADPH $+H^+$ and ATP from the light-dependent stage of photosynthesis and nitrate ions absorbed from soil. *(5)*

Tip

As Questions 1–3 show, multiple-choice questions are not all about simple recall. You need to think carefully, but don't spend too much time puzzling over one mark.

Tip

Question 4 is a straightforward AO1 question, mainly testing knowledge of saprotrophic nutrition. However, it is partially synoptic as you will need to consider the transport of highly active enzymes.

Tip

Question 5 is a typical synoptic question. Whilst part a) is concerned with this chapter, part b) requires you to apply your knowledge of Chapter 2 material on photosynthesis in a slightly different way. You need to be aware that synoptic questions will ask you to switch your concentration from one part of the specification to another quite quickly, and that lots of basic knowledge from Edexcel A level Biology 1 will also be needed to answer many questions.

6 A student carried out an investigation into the effect of different light intensities on the length of bracken (*Pteridium aquilinum*) stems. She selected two different areas of woodland; one area that had been cleared in the past year so there was little foliage to block out the light, and a second area where the trees had been allowed to grow for 10 years and was therefore well-shaded.

She randomly sampled each area and measured the height of 10 plants in each.

a) i) Describe **one** method of random sampling that would be appropriate in this investigation. *(3)*

ii) Name **three** other abiotic factors that would need to be controlled or monitored in this investigation. *(1)*

b) The results of this investigation are shown in the table below.

	Height of plant/cm	
	In shaded woodland	**In unshaded woodland**
	47	59
	56	68
	55	39
	32	55
	49	63
	60	40
	45	67
	39	45
	56	42
	44	51
Mean ($\bar{x}$)	48.3	52.9
Standard deviation (σ)	± 8.72	± 11.13
Variance (S^2)	76.03	123.87

The student decided to test if there was a significant difference in height of the two samples using a *t*-test.

i) Calculate the value of *t* using the formula: *(3)*

$$t = \frac{|\bar{x}_1 - \bar{x}_2|}{\sqrt{\frac{s_1^{\,2}}{n_1} + \frac{s_1^{\,2}}{n_2}}}$$

$\bar{x}$ = mean

s = standard deviation

n = number of entries in a set of data

s^2 = variance

$|\bar{x}_1 - \bar{x}_2|$ = the positive difference between two means

Tip

Question 6 is one of the ways in which your wider practical skills can be tested. You may be asked to make judgments or comment on experimental design for AO3 marks. This is also included to allow you to become familiar with questions involving statistical analysis.

ii) Use your calculated value of *t*, and this extract from a table of critical values of *t*, to explain what conclusions can be drawn from this statistical analysis. *(3)*

Degrees of freedom	Significance level					
	20%	10%	5%	2%	1%	0.1%
17	1.333	1.740	2.110	2.567	2.898	3.965
18	1.330	1.734	2.101	2.552	2.878	3.922
19	1.328	1.729	2.093	2.539	2.861	3.883
20	1.352	1.725	2.086	2.528	2.845	3.850

c) i) Explain what is meant by the term 'standard deviation'. *(2)*

ii) The photograph shows bracken, *Pteridium aquilinum*, growing in a woodland habitat.

Use the information in the photograph and the table of results to comment on the reliability of the conclusions made from these data. *(3)*

Stretch and challenge

7 The nitrogen-fixing bacterium *Rhizobium* sp. is found in the root nodules of leguminous plants.

a) What are root nodules?

b) The relationship between these bacteria and the host plant is known as 'mutualism'. What is meant by 'mutualism' and why is this relationship a good example?

c) The genes involved in nitrogen fixation are known as 'nif' genes.

i) Which enzymes are coded by nif genes?

ii) How do low nitrogen levels affect the expression of nif genes?

iii) The artificial transfer of nif genes into common crop plants is apparently an excellent way of reducing agricultural use of artificial fertilisers, but this has caused great concern. Why might biologists be very concerned about the effect of such crops on ecosystems?

d) Apart from those found in root nodules, what other nitrogen-fixing bacteria are thought to play an important role in nutrient cycles?

14 Changing ecosystems

Prior knowledge

In this chapter you will need to recall that:

- → populations of living organisms are affected by living and non-living factors in their surroundings
- → climate, soil and temperature are important factors in determining the distribution of living things
- → competition between organisms is a constant feature of ecosystems
- → newly established ecosystems change over time to form stable communities
- → global warming is causing significant changes to ecosystems
- → the greenhouse effect is due to increased concentrations of gases such as carbon dioxide in the atmosphere
- → burning of fossil fuels is a major source of atmospheric carbon dioxide
- → melting of polar ice and glaciers are indicators of global warming
- → many countries are attempting to reduce carbon dioxide emissions
- → solar power, wind power and tidal power are important sources of renewable energy
- → over-exploitation of natural resources can cause ecosystems to break down.

Test yourself on prior knowledge

1 Name **two** greenhouse gases other than carbon dioxide.
2 Explain why the presence of a greenhouse gas warms the atmosphere.
3 Describe **two** ways in which temperature can affect living cells.
4 State the major way in which carbon dioxide is removed from the atmosphere.
5 Explain what is meant by 'renewable' energy.
6 State the main difficulty in relying on wind or solar power as a main energy source.
7 Explain why tropical rainforests are such an important ecosystem.
8 Explain why tropical rainforest ecosystems are under threat.

Ecosystems – abiotic and biotic factors

We have defined an ecosystem as a stable unit of nature consisting of a community of organisms interacting between themselves and the physical and chemical environment. The living things known as the **biota** form the biotic environment and the physical, non-living factors form the **abiotic** environment. These aspects of ecosystems are so closely related as to be almost inseparable, as you will see shortly. To learn about the working of the ecosystem, however, we need to look at both aspects in more detail. While we do, keep in mind examples of ecosystems you are familiar with, such as woodland or seashore.

Key terms

Biota All of the living things found within an ecosystem.

Biotic factors All of the living influences within an ecosystem.

Abiotic factors All the non-living aspects of an ecosystem.

Introducing abiotic factors

The physical and chemical components of an ecosystem more-or-less determine the physical conditions in which populations live (Figure 14.1).

Abiotic factors of a terrestrial habitat are of three types, relating to:

- climate – factors such as solar radiation, temperature, rainfall and wind
- soil – factors such as the parent rock, soil water and soil chemistry, and the mineral nutrients available (**edaphic factors**)
- topography – factors such as slope and aspect of the land, and altitude.

Key term

Edaphic factors Factors that influence an ecosystem that are linked to the soil.

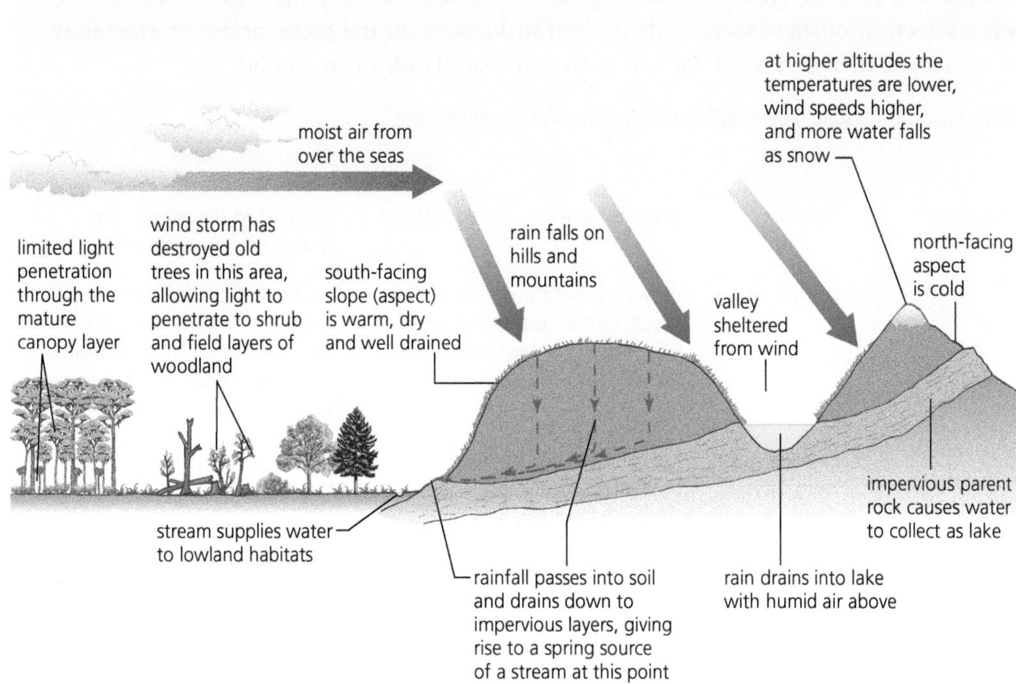

Figure 14.1 Interaction of climate, soil and topography in terrestrial ecosystems

We can illustrate the far-reaching impact of abiotic factors by looking at the effects of solar radiation. Light is the ultimate source of energy for the ecosystem; green plants grow only where there is sufficient light for their autotrophic nutrition. This need for light by green plants has an effect on the structure of plant communities. For example, an area of woodland is stratified into layers from the canopy above to the shrub layer below, the field layer (herbaceous plants) and ground layer (mosses). Each layer has particular plant life adapted to the light regime, and its own fauna. In aquatic habitats, plant life is largely confined to a region close to the surface.

The duration of illumination is the environmental trigger for inducing flowering of many plants, and in the timing of reproduction, migration and hibernation of animals. Light enables animals to see and be seen.

Sunlight is the major source of heat. Very few organisms grow if the temperature of their environment falls outside the range 0–40 °C. The effect of temperature on organisms is direct, as temperature influences the rate of all biochemical reactions. At low temperatures, ice crystals may form in cells, disrupting the cytoplasm. High temperatures denature enzymes, although certain bacteria found in hot springs have evolved tolerance of temperatures above 100 °C.

The length of daily illumination and the intensity of the light are determined by latitude, season, aspect (slope), time of day and the extent of cloud cover. Through its effects on temperature, light intensity also influences humidity.

The measurement of abiotic factors

Which abiotic factors will most strongly influence distribution of particular organisms within a habitat under investigation? We have already noted that the important abiotic factors are climatic factors (light, temperature, water availability and wind) and edaphic factors (the soil, its texture, nutrient status, acidity and moisture content). Topographic factors (angle and aspect of slope, and altitude) operate by their influence on local climate and edaphic factors. In fact, all factors interact to varying degrees. Table 14.1 lists a selection of these factors, their chief influences on the biota, and how they may be measured in a terrestrial habitat, such as a woodland, or in a pond.

Table 14.1 Introducing key abiotic factors and their measurement

Factor		Chief influence on biota	Measurement
Climatic	light	the ultimate source of energy	dedicated digital sensor linked to a sensor meter
	temperature	directly influences rates of biochemical reactions, and has indirect effects on other factors e.g. evaporation	dedicated digital sensor linked to a sensor meter, or a permanently mounted maximum and minimum thermometer
	relative humidity	influences rate of water loss by evaporation, and rate of transpiration in plants	dedicated digital sensor linked to a sensor meter, or a whirling hygrometer
Edaphic	pH	affects availability of nutrient ions that otherwise exist in unavailable forms	dedicated digital sensor linked to a sensor meter, or colorimetrical measurement using a soil test kit
	temperature	affects root growth, microorganism activity, and seed germination	dedicated digital sensor linked to a sensor meter
	texture	relative proportions of different-sized particles affect, among other things, aeration and drainage	hand assessment of soil
Topographical – terrestrial	slope aspect	indirect effects on illumination, temperature, drainage and so on	survey methods, as in profile transect
Topographical – pond	water supply, or flow	ponds with 'spring' source are permanent with relatively stable communities, whereas ponds raised by surface drainage tend to have ephemeral communities	direct observation and measurements of seasonal changes in water levels
	O_2 availability	low or non-existent concentrations of O_2 trigger anaerobic respiration, and favour obligate anaerobic organisms – otherwise, small changes in percentage O_2 concentration in an environment may have little impact on life under generally aerobic conditions	dedicated digital sensor linked to a sensor meter

Introducing biotic factors

The organisms of an ecosystem affect each other (Figure 14.2). Interactions between organisms, known as **biotic factors**, are between members of the same species (intraspecific competition) and between members of different species (interspecific competition). Of course, the impact of biotic factors depends upon the numbers of organisms present in relation to resources available, within a given environment. So we say that biotic factors are **density dependent**. Furthermore, we can recognise that biotic factors must be highly influential since most species occupy only a small part of the environment in which they are equipped to live.

> **Key terms**
>
> **Intraspecific competition** Competition between members of the same species.
>
> **Interspecific competition** Competition between members of different species.

Herbivory – caterpillars of the monarch butterfly feeding on milkweed leaves

Predation – African lion at the moment of capture of prey (kudo – a savannah herbivore)

Parasitism – sheep tick (an ectoparasite) attached to the skin of a cat where it has fed on a blood meal

Mutualism – mushroom of the fly agaric fungus takes sugars and amino acids from the tree's roots in return for essential ions, via its hyphae attached below ground

Figure 14.2 Interactions between species

Competition

Plants compete for space, light and mineral nutrients. Animals compete for food, shelter and a mate. To lose out in competition for resources means the individual grows and reproduces more slowly or, in extreme cases, dies. When the fastest growing competitor eliminates a slower growing competitor, it takes over the area completely. This is known as the principle of competitive exclusion.

Predation, grazing and symbiotic relations

Interactions between individuals of different species may also take the form of predation, grazing or symbiosis.

A predator is an organism that feeds on other living species. Predators are normally larger than their prey, and they tend to kill before they eat. The predator's prey is another animal – the eating of plants by herbivorous animals is a very similar process, but is referred to as grazing or browsing. All food webs show numerous examples of both predation and grazing.

Predator–prey relationships may be studied in the laboratory. Experiments show that both populations oscillate naturally. Predators feed on their prey and the population of prey starts to decline. Meanwhile the well-fed predators breed more so their numbers increase. Eventually their food source starts to become scarce and some die through lack of food. As their numbers decline so the number of prey begins to rise and the cycle begins again.

This simple relationship is much more difficult to demonstrate in natural populations, where individual species are part of more than one food chain so changes are not as clear-cut. The numbers of predator and prey pelts received by the Hudson Bay Trading Company of Canada from trappers over a 100 year period was carefully recorded and appears to be a clear example. (Figure 14.3), but the data were not collected by any scientifically organised random trial and could be affected by other factors.

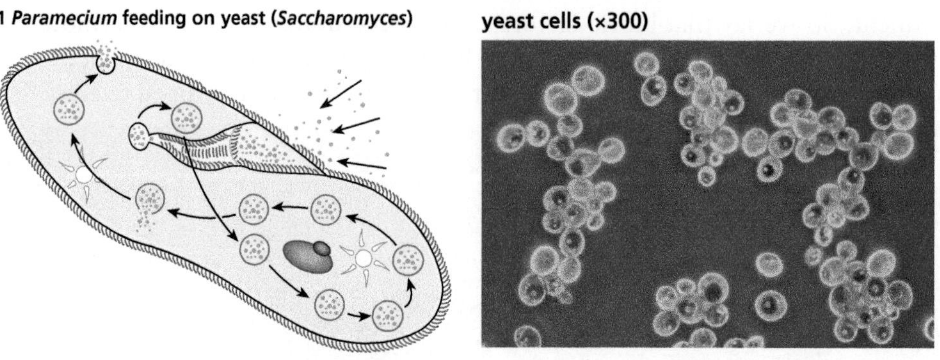

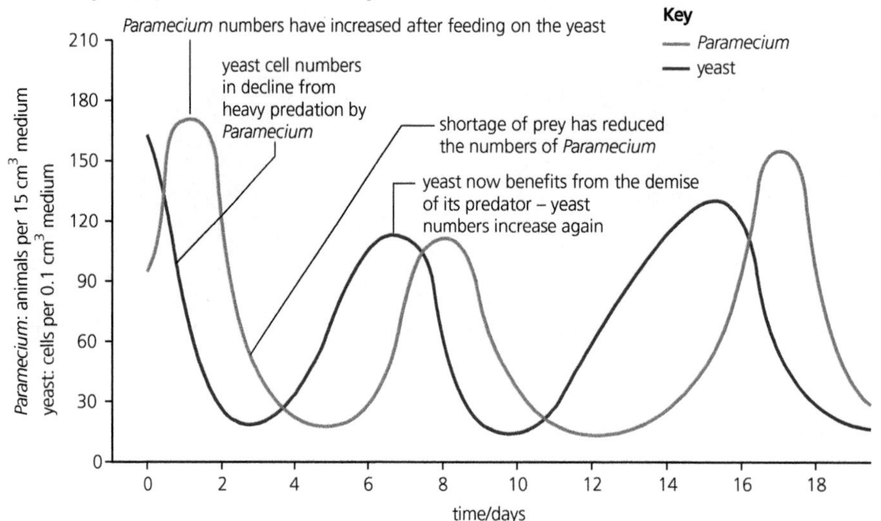

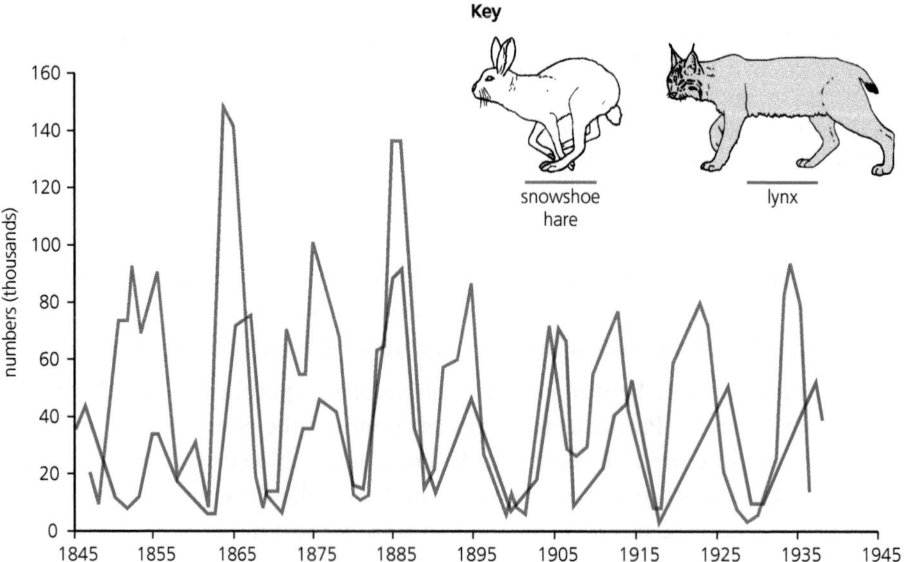

Figure 14.3 Feeding relationship between *Paramecium* and yeast (1) and predator-prey oscillation (2)

Symbiosis (sometimes known as **mutualism**) is the name we give to a relationship between two or more organisms living in intimate association. It simply means 'living together'. Parasitism is one form of symbiosis, in which the parasite lives on or in another organism, the host, for all or much of its life cycle. The parasite depends on the host for food and the host receives no benefit at all. **Endoparasites** live in the body of their host, like the human tapeworm and the malarial parasite. **Ectoparasites** live on the body of their host, like the aphids that tap the phloem sieve tubes of plants.

A parasite may affect the growth and reproduction of its host and in these cases the parasite is a limiting factor for the host species. In fact, parasitic associations show a gradation from those that normally kill the host (like the *Myxoma* virus on the rabbit, causing the fatal disease myxomatosis), to associations so bland that the parasite leaves its host virtually unharmed.

Key terms

Symbiosis A close and often obligatory association between two or more organisms, often, but not necessarily, to mutual advantage.

Parasitism A mutualistic partnership between two living organisms where one partner gains benefit and the other does not.

Test yourself

1 State the name given to abiotic factors related to the soil.
2 Suggest why biotic factors are density dependent.
3 Name **two** factors for which members of the same species may compete.
4 State the difference between a parasite and other symbionts.
5 Explain why many parasites are adapted to do only limited damage to their hosts.
6 In a theoretical predator–prey cycle, suggest why the prey population will peak before the predator population.

Niche – distribution and abundance of species

Competition between organisms is an ecological 'force' resulting in the establishment of distinct niches, and distinct differences in distribution and abundance of related species in particular habitats. The concept of ecological niche is not just some tiny place in which an organism is found, it encompasses everything the organism does and how it interacts with its living and non-living surroundings. In other words, its whole lifestyle that makes it particularly successful in that part of the habitat.

We can illustrate this with two examples:

1 Barnacle distribution

Two species of marine crustaceans, the barnacles *Chthamalus* and *Semibalanus*, are common creatures of seashore habitats. These sedentary animals release their gametes into sea water, where fertilisation occurs. From the fertilised eggs, free-living larvae emerge, which feed and grow before attaching to a surface, thereby adopting the sessile mode of life of the adults. Attachment occurs randomly on firm, submerged surfaces (typically rocks) of the intertidal region of the shore. However, one species (*Chthamalus*) is able to withstand prolonged exposure when the tides recede, whereas these same conditions slow the growth or actually kill off the other species (*Semibalanus*). As a result, as the degree of exposure is experienced by the growing barnacles, *Chthamalus* barnacles become less abundant in lower zones of the shore, and *Semibalanus* barnacles are similarly crowded out from upper (exposed) zones (Figure 14.4 on the next page).

Key term

Niche An ecological term used to describe the exact role that an organism plays within an ecosystem. This includes many features of the lifestyle of the organism including its interaction with the living and non-living environment.

So the distribution of *Chthamalus* in the upper intertidal zones, and of *Semibalanus* in the lower intertidal zones, is a function of differences in their respective niches. That is, behavioural and structural differences in resistance to periodic desiccation and endurance of prolonged submersion have determined where each species survives best.

2 Marine alga distribution

Another illustration of the evolution of distinctive niches between related species that has lead to differences in distribution and abundance is found on the rocky shore among common species of brown algae (Figure 14.5). All seaweeds have a covering of mucilage over their plant 'body' (thallus), and this acts as a reservoir of water. Differences in mucilage production and other differences appear to account for differences in species distribution. The algal species of the upper shore are shown to be better able than others to withstand desiccation – they contain more water when hydrated and they lose it more slowly when exposed.

the degree of exposure determines the distinctive distribution pattern of these two species of barnacle

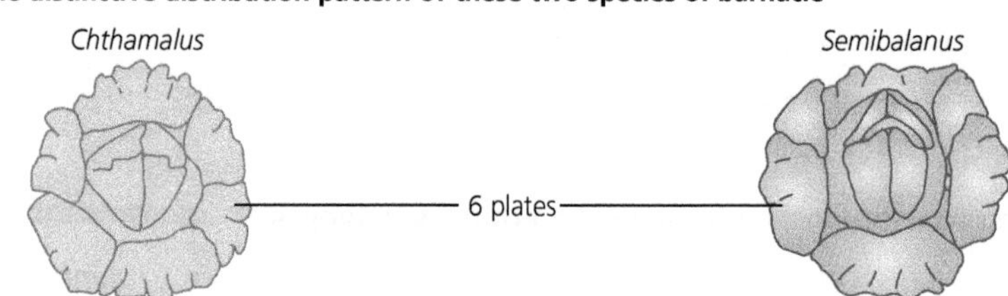

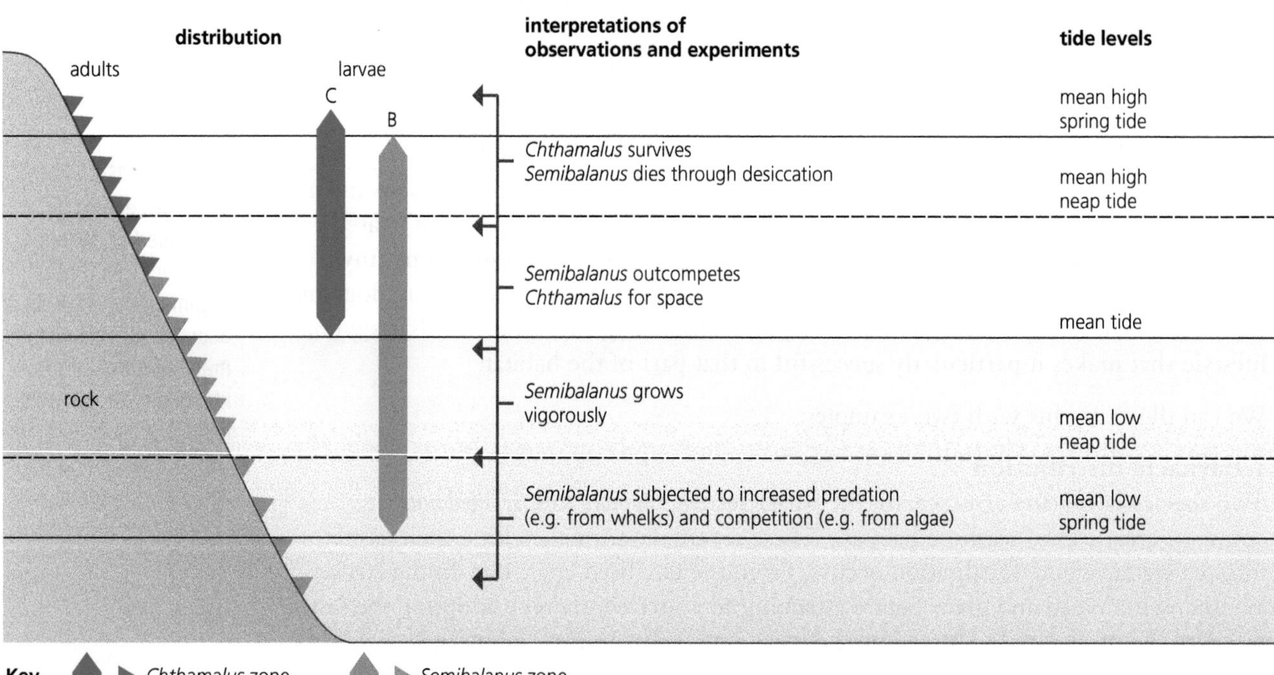

Figure 14.4 The growth of two species of barnacle on the seashore

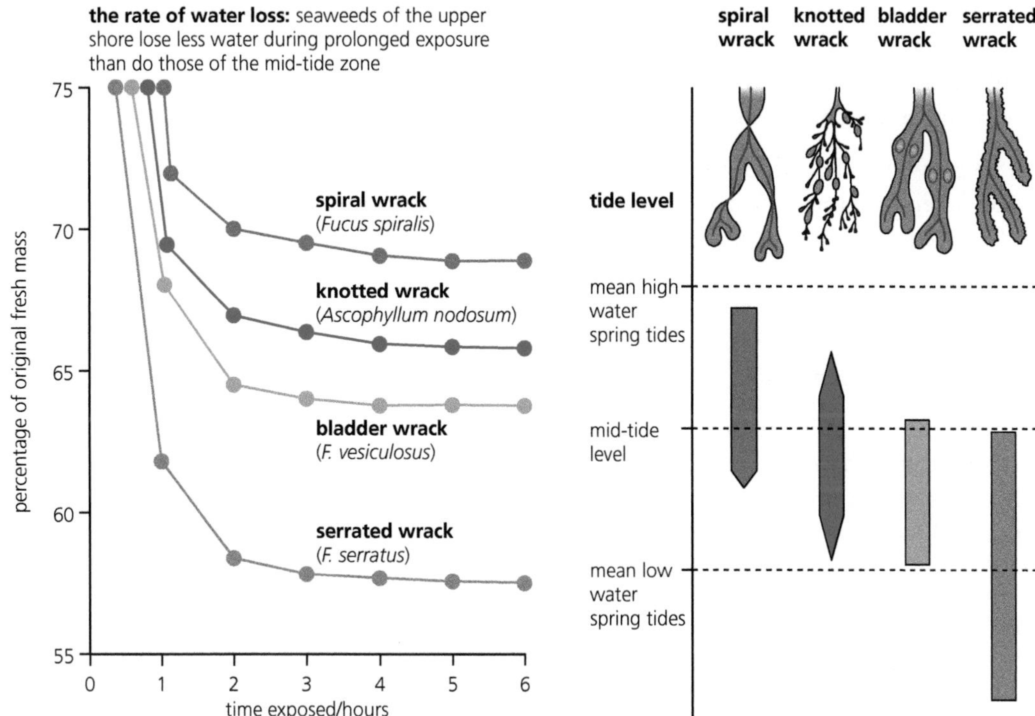

Figure 14.5 The resistance to desiccation of related species of brown algae

Ecological succession

When new land is exposed, it is quickly invaded and colonised by organisms. In fact, a sequence of communities develops with time by a process known as ecological succession. The stages in a succession are a series of plant and animal communities of increasing complexity, which develop as the abiotic factors operating change with succeeding communities. These stages are called **seral stages**, and the whole process is termed a sere. A climax community finally results, characteristic of the area.

Key terms

Ecological succession Sequences of different ecological communities developing in a habitat over time.

Sere All of the stages in a complete seral succession.

Humus The dead and decaying remains of plants and animals, which form an important part of fertile soil structure.

Primary succession

When the succession sequence starts on entirely new land without soil, the process is known as a **primary succession**. New land is formed on the Earth's surface at river deltas, at sand dunes and from cooled volcanic lava, for example. Primary successions also develop in aquatic habitats, such as in a pond formed and fed by a spring.

In all cases, the first significant development is the formation of soil, as at the initial site of a primary succession all that may be present is parent rock, from which the bulk of the soil is formed by erosion. However, mineral particles may also be blown or washed in from elsewhere, and the resulting mineral skeleton is of particles of a wide range of sizes.

Soil, when fully formed, has organic matter called humus, wrapped around the particles of the mineral skeleton. Humus is a substance derived from dead plant and animal remains, together with animal faeces, that have been decomposed by the actions of microorganisms. Humus contributes mineral nutrients as it continues to be decayed, and it also helps soil to hold water and to retain heat. Between mineral particles and humus are innumerable pockets of air.

Key terms

Pioneer species Plants and animals that have adaptations to enable them to become the first colonisers of an empty habitat.

Climax community The final stable stage in an ecological succession.

Humus is first added by plant invaders of the primary succession, known as **pioneer plants**, as shown in Figure 14.6. Since the first-formed soil retains little water, plants able to survive there (called xerophytes) have drought-resistant features. When a sere starts from dry conditions, then the sere is called a xerosere.

The growth and death of the early plant communities continue to add humus, so more soil water is retained. Nutrients are added to the soil when organisms die and the range of nutrients available to plants increases steadily. Different plants now grow – various herbaceous weeds, for example, may start to shade out the pioneers. Herbaceous plants are followed by shrubs and small trees, all growing from seeds carried in by wind, water or the activities of animals. If the site remains undisturbed, a **climax community** such as a stable woodland will result.

So, succession can be seen as a directional change in a community with time. Initially abiotic factors have the greater influence on the survival and growth of organisms. Later, as the numbers of living organisms build up, biotic factors increasingly affect survival too, especially as they come to modify the abiotic factors operating.

An important feature of a succession is the progressive increase in the number of species present. As more species occur in a habitat, the food webs diversify. Now, in the event that one population crashes (such as when a disease sweeps through, or predators have a temporary population explosion), then alternative food chains may be sufficient to supply the higher trophic levels.

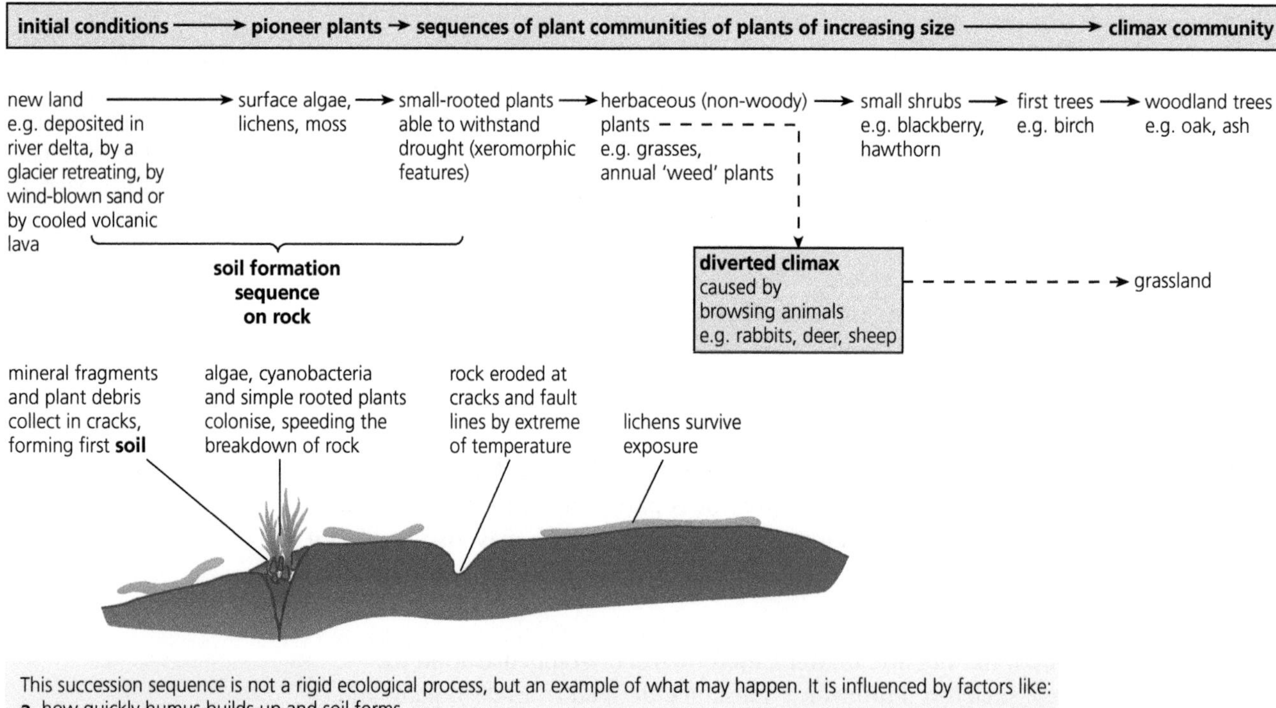

Figure 14.6 A primary succession on dry land – a xerosere

Secondary succession

Sometimes established communities are suddenly disrupted and totally destroyed. This occurs, for example, when fire destroys a large area of vegetation; occasionally it occurs as a result of human activities. In these situations, soil is already formed and present – it is just the existing biota that has been abruptly removed. A succession that starts from existing soil is known as a **secondary succession**.

Secondary successions normally happen quite quickly, since the necessary soil for plant life is already present. Plant communities are established in succession, as spores and seeds are blown in, or carried in by visiting animal life, or as they grow in from the surrounding, unharmed climax communities. After forest fires, for example, the soil is quickly covered by moss species that favour scorched soil habitats. The carpet of moss reduces soil erosion, starts to contribute to the supply of humus to the soil, and provides conditions favourable to the lodging and germination of seeds of higher plants.

Changes in climax communities

There are several types of climax communities.

- **Climatic climax** – a natural climax brought about by prevailing climatic conditions, such as deciduous woodland in temperate wet conditions. In uplands where temperatures are lower and rainfall higher, soils become saturated and more acidic. The low pH inhibits the breakdown of plant material and large areas of peat bog or moorland will become the climax community.
- **Deflected climax** – a change from the climatic climax brought about by the activities of humans, such as farming.

Seral changes do, in theory, lead to stable climax communities. In Great Britain the most likely outcome is the formation of a deciduous woodland ecosystem. Centuries ago this would have meant primarily oak woodland. However, at one time oak was of extreme military importance as the material for building warships and for structural timber. Not surprisingly a great deal of the oak woodland was cut down and today only small, ancient remnants remain.

In the 15th century, wool was an extremely important source of great wealth. This meant that more and more woodland was turned into pasture. The activities of sheep can still be seen today as the green hills of the Lake District are largely due to the effects of their grazing preventing the development of a climatic climax.

Climax communities can change, as we are witnessing in parts of the world where average temperatures are rising. For example, large areas on the fringes of the Sahara desert have become permanent arid desert rather than simply arid scrubland.

Key terms

Climatic climax
The final stage of an ecological succession, which is determined by the prevailing climate.

Deflected climax
The final stage of an ecological succession, which has been determined by human activity rather than natural forces.

Test yourself

7 Explain what is meant by a 'xerosere'.

8 Suggest the most likely climatic climax ecosystem in lowland UK.

9 Explain what changes a climatic climax into a deflected climax.

10 A secondary succession will develop much faster than a primary succession. Explain why.

Core practical 16

Investigate the effect of one abiotic factor on the distribution or morphology of one species, taking into account the safe and ethical use of living organisms

Background information

The exact nature of this investigation will depend upon the habitats available to you and you will need to discuss this with your teacher. In this example we will consider one example but the choice available to you is extremely wide. This practical is closely linked to Core practical 15 in Chapter 13. In particular, you will almost certainly need to consider such aspects as random sampling, which are also covered in Chapter 13. For this example we will look at an investigation that considers a gradual change in an abiotic factor along a transect and we will show how the data may be analysed using a correlation test.

Whatever you choose to investigate it is important that you plan carefully. In this practical you are investigating the effect of one abiotic factor. This will be your main independent variable so naturally it is vital that it should be measured or monitored as precisely and as reliably as possible. This is not always an easy task in field investigations so you must think very carefully. One common abiotic factor is light, and this poses many problems if you are to measure it carefully. Readings will obviously change throughout the day due to weather conditions and movement of the Sun, so if you are considering light and shade, what might be in bright sunlight in the morning will possibly be in deep shade in the afternoon. If the day is cloudy there might be bright, sunny intervals alternating with dull, cloudy conditions. What might be the best measurement in these circumstances? There are no simple answers to this so some thought and trial are needed.

Exactly the same problems occur with the dependent variable. Imagine you were investigating the effect of light intensity on leaf morphology; what would you measure and why - length, width, area, thickness or simply number? All will provide different information, so consider the biology behind your hypothesis and select the parameter which would provide most information. Remember that often a ratio such as width : length will give more information about overall shape, even though each leaf may not be exactly the same size.

The investigation below was carried out on sand dunes, which are a good example of succession in action. At the edge of the sea only pioneer species that can grow in loose, yellow, free-draining sand with little humus and a saline environment are found. Marram grass (*Ammophila arenaria*) is often dominant here. As the dunes become more stable and humus accumulates, the soil darkens to grey dunes. More and more plants are able to colonise the better soils and finally a climax community of woodland develops. Grasses such as red fescue begin to colonise the grey dunes and higher. This investigation is designed to find out if this was because there was a greater humus content.

NOTE: If red fescue is not present on the dunes you are visiting, you can simply choose another plant that is common across the section you study.

If you wish to find out more about dune transects before you read on, then there are a number of excellent photographic sequences showing real transects on the web. These can be found by simply searching for 'virtual sand dune transect' in your chosen browser.

Carrying out the investigation

Aim: To investigate the effect of soil humus content on the distribution of red fescue grass (*Festuca rubra*).

Hypothesis: There will be a significant positive correlation between the humus content of the grey dune and the abundance of red fescue.

Risk assessment: Read the section describing 'safety and ethics in ecological investigations' in Chapter 13 of this book. As with any outdoor activity, do not undertake this in isolated areas unless in a group with good communications. You will be working in exposed areas so check weather forecasts carefully and never swim in the sea without some supervision. Many dune systems have lifeguards and safety equipment nearby. Always wash your hands thoroughly after working outside, and before any hand to mouth contact. If you choose to investigate humus content of soil, you should burn off the humus in a fume cupboard, as the chemicals released into the air are similar to those released from burning cigarettes.

1 First of all make an initial survey of the area you are to work in. Make sure that you can identify red fescue easily. It does have a reddish tinge to the leaves but its typical grass spikelets (flowers) and seeds have a more pronounced red colour, which make it easy to spot.
Find an area of the dunes where red fescue begins to be obvious and continues inland for at least 20 m. You will now need to use some judgment according to what you find.
2 To investigate, use a transect (Chapter 13) and take not less than seven measurements for your correlation test; ideally 10–15. To sample the dune, lay out a tape stretching inland and decide whether to take samples each metre or at longer intervals. Make sure that your transect crosses a typical section of dune but avoid introducing other variables

such as deep dips in the dunes known as 'slacks', where the conditions will change a great deal.

3 The next step is to decide how to record abundance of red fescue. This will depend on what you find on your dune. If the grass is quite sparse and only in small tufts you might simply count these within an open 1 m^2 quadrat. If the tufts are large and merge into each other you might choose a gridded quadrat to estimate percentage cover. If the grass forms a dense turf with other plants then you might choose a point frame. The latter can easily be improvised by drilling 10 holes in a 40 cm piece of wood and using one or two knitting needles (see Figure 13.12).
The process of planning a field investigation in this way is quite common. As habitats vary so much you must adapt your methods to take account of exactly what you find by initial observations, and not expect a rigid idea to be appropriate each time.

4 You now need to sample the soil at your selected sites and determine its humus content. To do this, take a soil sample at the bottom right-hand corner of each of your quadrats or each sample site. To ensure this is the same depth each time, a small empty can or rigid plastic yogurt pot can be pushed firmly into the soil and then removed with the help of a small trowel to prevent the soil spilling out. Place each sample into a polythene bag and mark this with the quadrat number.

5 Back in the laboratory, remove the sample and place it in a small beaker in an oven at 50 °C for several hours to dry it thoroughly. You can check that it is dry by weighing it at intervals until it loses no more mass.

6 The humus content of the dry soil can now be found by weighing a small sample (about 5–10 g) and placing this in a crucible or other heatproof container. Humus can be burned off at high temperature by using a Bunsen burner directly on the sample or using an oven at high temperature (250 °C+) for several hours. The mineral content of the soil will not be affected. Finally, weigh the sample again carefully and calculate the percentage loss in mass. This will be the humus content.

Analysis of the data using Spearman's rank correlation test

The formula you need to calculate your test statistic, known as the Spearman's rank correlation coefficient (r_s), is:

$$r_s = 1 - \frac{6\sum d^2}{n(n^2 - 1)}$$

Where r_s = Spearman's rank correlation coefficient

d = the difference in ranks of the two measurements

n = the number of pairs of measurements.

Here is a typical set of data from an investigation such as yours.

Sample number	% cover of *F. rubra*	% soil humus
1	20	1.9
2	23	2.2
3	18	2.1
4	15	1.6
5	19	1.8
6	25	2.4
7	15	1.7
8	21	2.1

First you have to rank each set of data from highest to lowest and subtract them to find the difference (d).

Then square each difference (d^2) (notice you lose the minus signs by doing this) and find the total Σd^2.

Sample number	% cover of *F. rubra*	Rank of % cover	% soil humus	Rank of % humus	Difference in ranks (d)	d^2
1	20	4	1.9	5	−1	1
2	23	2	2.2	2	0	0
3	18	6	2.1	3.5	2.5	6.25
4	15	7.5	1.6	8	−0.5	0.25
5	19	5	1.8	6	−1	1
6	25	1	2.4	1	0	0
7	15	7.5	1.7	7	0.5	0.25
8	21	3	2.1	3.5	−0.5	0.25
					Sum of d^2 (Σd^2)	9

When two measurements are the same, add together the two ranks and divide by two, for example samples 4 and 7 have 15 per cent cover. The two ranks would be 7 + 8, which you simply divide by two to give rank 7.5 as shown in the table. If there were three of the same value you would add together the three available rankings and divide by three.

For the formula $6\Sigma d^2 = 54$

$n(n^2 - 1) = 8(64 - 1) = 504$

$r_s = 1 - (54 \div 504)$

$r_s = 1 - 0.107 = \mathbf{0.893}$

You now look up the **critical value for eight pairs of measurements** at the 5% confidence level in a statistical table for the Spearman's rank test to find this is 0.738.

Your conclusions would be: There is a significant correlation between percentage cover of *Festuca rubra* and soil humus content at the 5% significance level as our value of r_s is greater than the critical value.

Correlations and causation – a cautionary note

Be very careful when interpreting a correlation, even where this has been shown to be significant by a statistical test.

A correlation does not tell you that one thing causes another. It simply tells you that when one changes so does the other. This might be a strong hint but it does not prove causation.

In the example it shows that as the humus content of the soil increases so does the percentage cover of *Festuca*, but it does not show why or if it causes this effect.

Scientific investigations often come up with this type of result but there may be other linked but very different reasons why this is the case. Your investigation took samples by working inland from the sea. It could be true that the humus content increases, but what if the salinity of the soil also decreased as you moved inland? You would get the same result even though it had nothing at all to do with humus. This problem is very widespread and therefore you need to be very careful when interpreting correlation data.

Questions

1 Why is it necessary to dry the soil before burning off the humus?
2 Why express humus content as a percentage?
3 Why is it important to keep burning the humus until I have a constant final mass?
4 Why is there likely to be more humus as I move inland?
5 How would I display my data?

Human effects on ecosystems

Humans are the dominant animals on the Earth. They have used their intellectual capacity to overcome many of the ecological forces that would bring about a balance in other populations, and to manipulate their environment in unprecedented ways. The effects of their ingenuity and exploitation of resources have influenced the whole of the biosphere.

In this section you will be looking at several aspects of how human activities have affected ecosystems and the difficulties associated with striking a balance between meeting human needs and conserving ecosystems.

Human population growth

Central to the problems of human influences on the biosphere is the enormous growth in human populations throughout the world. This began in earnest following the Industrial Revolution some 200 years ago. Since that time advances in medical knowledge and increased food availability have dramatically decreased death rates and birth rates have increased. The net result is that average life-expectancy has nearly doubled in this time (Figure 14.8). The inevitable result is shown in Figure 14.7. Human pressures on both natural resources and other populations have increased in the same way.

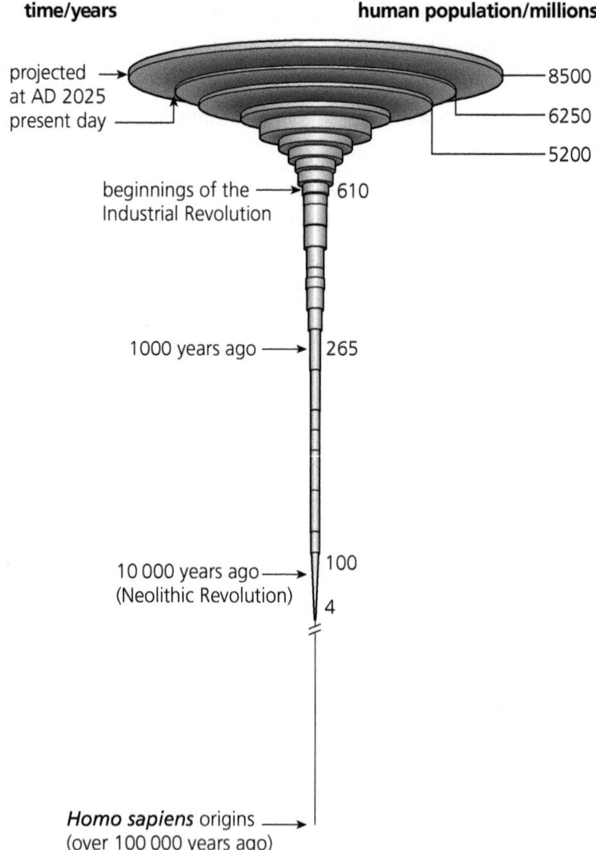

Figure 14.7 The changing pattern of the estimated world human population

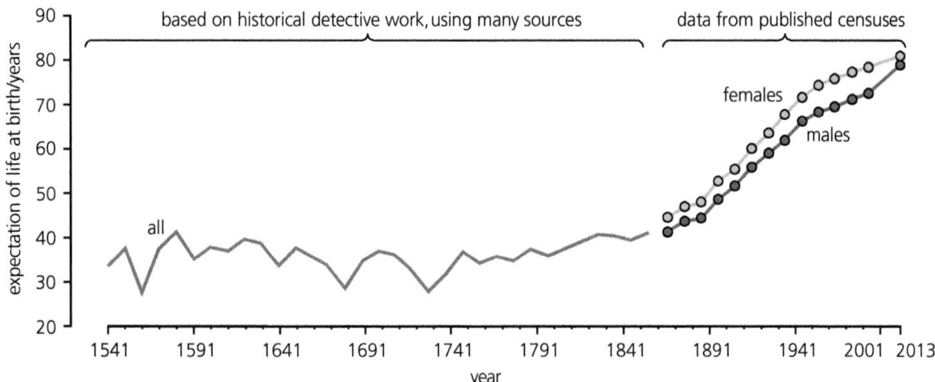

Figure 14.8 Expectation of life at birth in England over approximately the past 500 years

Test yourself

11 Describe the mathematical relationship that illustrates the current trend of human population growth.

12 State **two** factors that have contributed to the increase in human population.

Human pressures on resources – a case study

Abundant supplies of natural resources of food depend upon thriving ecosystems. You have seen that ecosystems have an inbuilt tendency to remain in balance and often a remarkable ability to adjust to changing conditions. However, there are strict limits, which if exceeded result in the complete breakdown of the system.

An understanding of the ecosystem and balancing the needs of humans with good management of the natural environment is the aim of conservation throughout the world. Unfortunately, there are many pressures, which mean that this is not always successful. Conservation of fish stocks is an example of how these efforts have failed in many parts of the world and also provides good lessons for the future.

As you will see in the examples that follow, conservationists need to come to terms with the many different influences on decisions and actions, even though the simple scientific logic may seem obvious.

The cod fisheries of Newfoundland

As scientists, the first thing we need to think about is the undisputed background to the case history. In this example we need to know more about the Atlantic cod (*Gadus morhua)* before looking at the problem.

The Atlantic cod, like many marine fish, reproduces by relying upon strength in numbers. By this we mean that there is almost no parental care of the offspring and only a tiny number will develop into fertile adults, so this drawback is overcome by producing very large numbers of offspring.

At certain times of year cod gather together in large shoals of many hundreds of thousands of individuals. A mature adult female cod will release well over 1 million eggs at this time. Older, larger females may release 10 times this number. These are fertilised by the sperm of the males in open water. Newly fertilised eggs are buoyant and float to the surface, where they are easy prey to many fish species. In fact, only about 1 in a million will develop into an adult cod.

The developing cod larvae use the yolk in the egg (Figure 14.9) to grow into tiny fish, which then feed off zooplankton and small crustaceans. Shallow water and marine algae are ideal places for young cod to escape predators and find enough food to grow. As they increase in size they begin to prey upon larger shrimps and other marine life, until as adults they will feed off larger fish and almost all other smaller species such as crabs. At this stage they have few predators and become one of the top carnivores within their ecosystem.

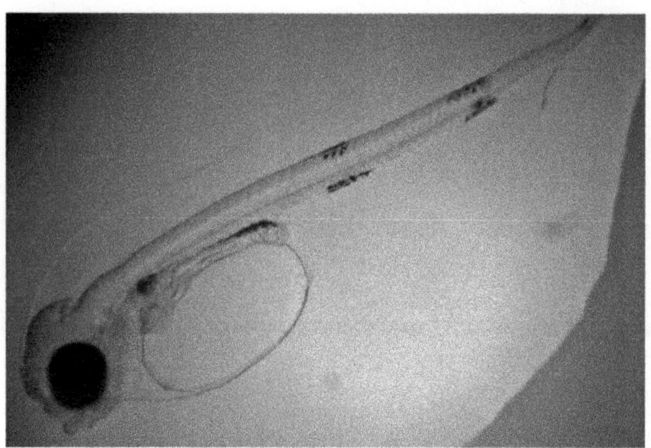

Figure 14.9 Atlantic cod (*Gadus morhua*) larval and adult stages

An adult cod is a large fish about 80 cm in length, weighing about 4 kg and with a life span of at least 10–15 years. Females become fully mature at about 5–6 years of age. However, this average hides the fact that shoals used to contain older cod of more than 1 m in length and a mass of over 50 kg.

Atlantic cod are cold-water species, thriving in Arctic waters. The ideal conditions in the North Western Atlantic, and especially in the shallow waters off the coast of Newfoundland in eastern Canada known as the Grand Banks, mean that historically there were huge populations of fish.

The collapse of the cod stock

The Atlantic cod population on the grand banks of Newfoundland had been recognised for hundreds of years. Native Canadians and early English and French settlers were able to exploit this population without affecting its enormous size. After the Second World War, commercial fishing increased dramatically and catches began to decline. Finally, in 1992 the cod population collapsed completely (Figure 14.10), forcing the Canadian government to ban cod fishing in the area completely. Over 40 000 Canadians in Eastern Canada lost their livelihood and the government introduced special pensions and benefits to prevent mass poverty.

It was hoped that this extreme measure would allow the cod stock to recover in a few years but 20 years later this had not happened.

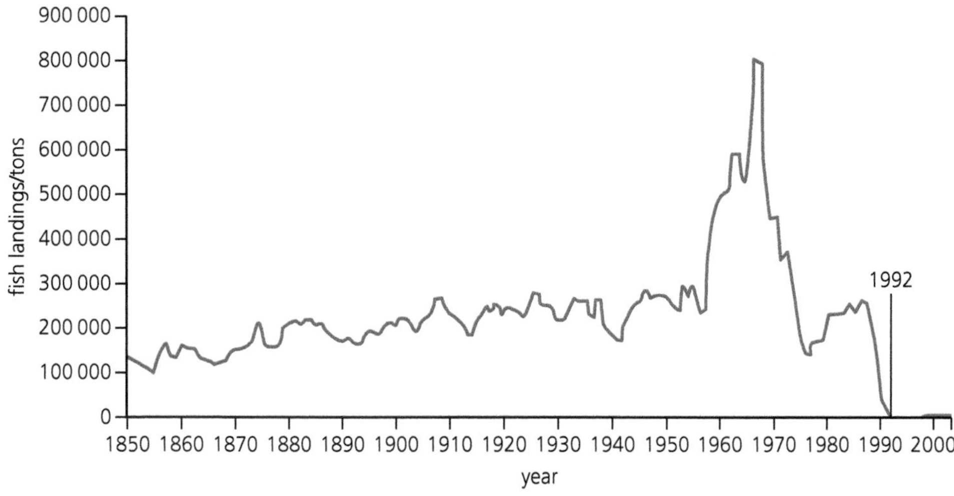

Figure 14.10 Total fish landings in Newfoundland 1850-2005

What caused the collapse of the cod stock?

In hindsight it seems very obvious what happened, but the story also illustrates the problems of reconciling human demands and the conservation of whole ecosystems.

1850–1950

During this time fishing was largely controlled by the limitations of fishing technology and limited international trade in fish. Local fisherman relied on human effort to haul nets and row boats. Even the introduction of sailing boats and the first steam-driven trawlers had little effect on the huge populations of cod.

Figure 14.11 Fishing schooner on the Grand Banks circa 1920

1950–1960

Big international companies from Europe introduced large factory ships (Figure 14.12), which froze their catch on board. These were able to fish day and night for many days using very large trawl nets, which were dragged across the sea bottom where most cod were found. The number of fishing boats increased enormously and their nets caused irreparable damage to the ocean floor and its ecosystem (Figure 14.13 on the next page), which was essential for the young cod to grow. Bottom trawling also results in an enormous 'bycatch', which is a large collection of organisms deemed to be unsaleable and simply dumped back into the sea, destroying many food webs in the process. Amongst this 'bycatch' were large numbers of fish called capelin. These small fish are not commercially valuable but crucially are a very important food source for young cod, so the disruption was compounded. Increasing use of sonar to locate fish meant that, even though the population was declining, the remaining fish were easily found. Taking so many fish from the population also meant that the large female fish were no longer available to produce vast numbers of eggs each breeding season. Between 1955 and 1968 the total catch of fish increased from an average of around 250000 tons to just over 800000 tons per year.

Figure 14.12 A modern factory-fishing vessel

1960–1992

In 1968 international treaties allowed the Canadian government to extend its jurisdiction to 200 miles offshore and it was able to take control of fishing on the Grand Banks. Sadly, what proved to be the last opportunity to conserve the fish stocks was

Figure 14.13 The damaging effects of repeated bottom trawling on the ocean floor

not fully implemented, as Canadian vessels partially replaced international competitors. Total fish landings did reduce dramatically but largely as a result of diminishing stocks as the damage had already been done. Despite a brief recovery in the early 1980s, there was a dramatic collapse in 1992 to such a low level that the government was forced to close the fishery completely.

1992–present day

Despite initial optimism, cod stocks on the Grand Banks have not recovered. Even with government support, the population of Newfoundland suffered severe hardship and many people emigrated. Whilst the original ecosystem has not returned, a new balance has established itself. This is now dominated, not by a large carnivorous fish, but by millions of crustacea such as snow crabs and prawns, which are essentially scavengers. Surprisingly, there is a large international market for such crustaceans and now there is a developing fishing industry again. Worryingly, for many ecologists, this industry is dominated by large industrial-sized fishing vessels.

Conflict and conservation

So what can we learn from the demise of the Atlantic cod fishery of Newfoundland?

First and foremost we can see that the whole story is a complex mixture of scientific prediction, political will, international agreement, national interests, commercial interests and a dash of greed. Above all we can see that any conservation effort is bound to fail unless there is local support and assistance.

Commercial companies, both internationally and in Canada, invested heavily in new ships and technology, obviously to make a profit but at the same time providing jobs and prosperity to the local economy. A large proportion of the Newfoundland population was employed in this way, not only as fishermen, but also whole families working in shore-based factories. Local and national politicians are unlikely to support measures that would severely restrict their constituents' incomes.

Even ecologists employed by the government would be subject to extreme pressures. Predicting the future of natural populations often involves complex computer models, which were far less developed in the 1970s. Even then, some theoretical assumptions had to be made and this means any predictions will have a margin of error. In an attempt to get some actions approved, selecting more optimistic predictions was a much more attractive option.

Sovereign countries have only a limited amount of jurisdiction beyond their coastline and this highlights the need for international agreements.

It is easy to think that these events are far removed from one's home country but nothing is further from the truth. Almost every country in the world has some issues of overfishing. In the UK we have lost the whole of an enormous North Sea herring fishery because of exactly the same problem as that experienced by the cod industry.

There are lessons to be learnt from successes too. There still remains a viable cod-fishing industry in the eastern part of the North Atlantic. This is largely because of strong policies put in place by Norway and Iceland to protect fish stocks as increased fishing threatened to overcome them. Here, too, large populations were dependent upon the fish stocks. This was not achieved without controversy as, at one point, UK and Iceland aggressively confronted each other with naval vessels to protect their fishing fleet from arrest in what became known as the 'cod war'. Eventually, in 1976, agreement was reached and the UK accepted that Iceland could enforce a 200 mile zone around their island in which they alone controlled the fishing. But, just as in Newfoundland, many British trawlermen lost their jobs in the years that followed and the fishing ports of Grimsby and Hull suffered severe decline – a better outcome for conservation but not for diplomatic relations or individual fishermen.

Conserving fish stocks in the 21st century

By far the most influential development in recent years has been the raising of public awareness of conservation issues on a local and international scale. Public support for the implementation of difficult conservation measures has meant that their influence on consumer demand and political decisions has increased.

The problems of overfishing have been highlighted by large international organisations such as Greenpeace and The World Wide Fund for Nature. This has led to independent bodies such as The Marine Stewardship Council (MSC) establishing codes of practice for fisheries.

Figure 14.14 The MSC logo

The MSC standard consists of three core principles that each fishery must demonstrate it meets:

- **Principle 1: Sustainable fish stocks.** The fishing activity must be at a level that is sustainable for the targeted fish population. Any certified fishery must operate so that fishing can continue indefinitely and is not overexploiting the resources.
- **Principle 2: Minimising environmental impact.** Fishing operations should be managed to maintain the structure, productivity, function and diversity of the ecosystem on which the fishery depends.
- **Principle 3: Effective management.** The fishery must meet all local, national and international laws and must have a management system in place to respond to changing circumstances and maintain sustainability.

The crucial point about such standards is that they influence consumers when they purchase fish. There are now very encouraging signs that this is happening. Conservation and sustainable fishing are increasingly in the minds of shoppers, and large retailers are keen to show their 'green' credentials. Logos such as the MSC (Figure 14.14) are now increasingly familiar and have helped to change fishing habits throughout the world.

Increasing awareness amongst consumers of the origins and methods of fish production has also had a strong influence on conservation.

Demand for tuna has increased worldwide but consumer awareness is influencing fishing methods. Tuna are large fish and can be caught in huge nets. Unfortunately this also means that large numbers of other animals such as dolphins are trapped and killed. A glance at supermarket shelves will today show prominent labelling indicating only tuna caught with the less damaging line and pole method is now stocked – a good step forward but there is still concern about the numbers of tuna being harvested worldwide.

Atlantic salmon, like the cod, are a severely threatened species. Here, biologists have devised enclosed fish farms in sheltered bays and inlets where specially bred fish can be produced to meet consumer demand without further pressure on depleted wild stocks. The large majority of salmon found on supermarket shelves is now farmed.

Even though fish farming helps to solve the problem of consumer demand, it also creates important biological problems of its own, such as the escape of farmed fish into wild populations and the increase of pests and diseases in concentrated numbers of near-identical individuals.

Other developments

- Many countries are setting up marine nature reserves. In these reserves all fishing is banned and they are chosen to include areas that will provide protected breeding grounds for fish.
- Satellite imaging means that vast areas of the ocean can be monitored to prevent illegal fishing.

Test yourself

13 Explain why the advent of 'factory ships' meant that more cod were landed.

14 Repeated bottom trawling was extremely damaging to the cod stock. Explain why.

15 Give **two** reasons why it was difficult for the Canadian government to prevent overfishing in the 1960s.

16 Explain what a 'bycatch' is.

17 Pole and line fishing is less damaging to the ecosystem than trawling. Explain why.

Human influences on ecosystems – endangered species

Many ecosystems worldwide are threatened, not only by exploitation of resources for food but also by the removal of key animals and plants from their natural habitats. Animals such as the rhinoceros and tiger are seriously endangered. In part this is due to loss of habitat but they are also under great pressure from poaching. In both cases, misinformed traditional belief in the medicinal power of body parts of these animals, means they command extremely high prices, making poaching a profitable enterprise.

A similar threat is posed by the value of elephant ivory.

Apart from local anti-poaching initiatives and the support of international wildlife charities, one of the most important international agreements on conservation and trafficking in endangered species is the Convention on International Trade in Endangered Species of wild fauna and flora (CITES). This is one of the oldest international agreements on conservation. Established in 1973, it has 180 member-countries.

Key term

CITES The Convention on International Trade in Endangered Species of wild fauna and flora. An important milestone in international cooperation on conservation.

The purpose of CITES is to licence and control the movement of endangered species and their products throughout the world. Each country has a government body responsible for issuing licences for the import and export of any species or product listed on a large agreed database. By restricting such movements, the transfer of plants and animals out of their natural habitat for commercial purposes is prevented and damage to indigenous ecosystems by other countries avoided.

The CITES website and database will give you a clear indication of the wide range of plants and animals that are protected under this scheme.

We shall look at other international agreements later in this chapter.

Climate change

Carbon dioxide is present in the atmosphere at about 0.038 per cent by volume (which represents 0.057 per cent by mass). Atmospheric carbon dioxide is added as a waste product of respiration by all living things, by combustion, and from the decay of organic matter by microorganisms. Much carbon dioxide is removed by fixation during photosynthesis – an interrelationship illustrated in the **carbon cycle** (Figure 14.15). About as much carbon dioxide is withdrawn from the atmosphere during the daylight each day as is released into the air by all the other processes, day and night – or nearly so.

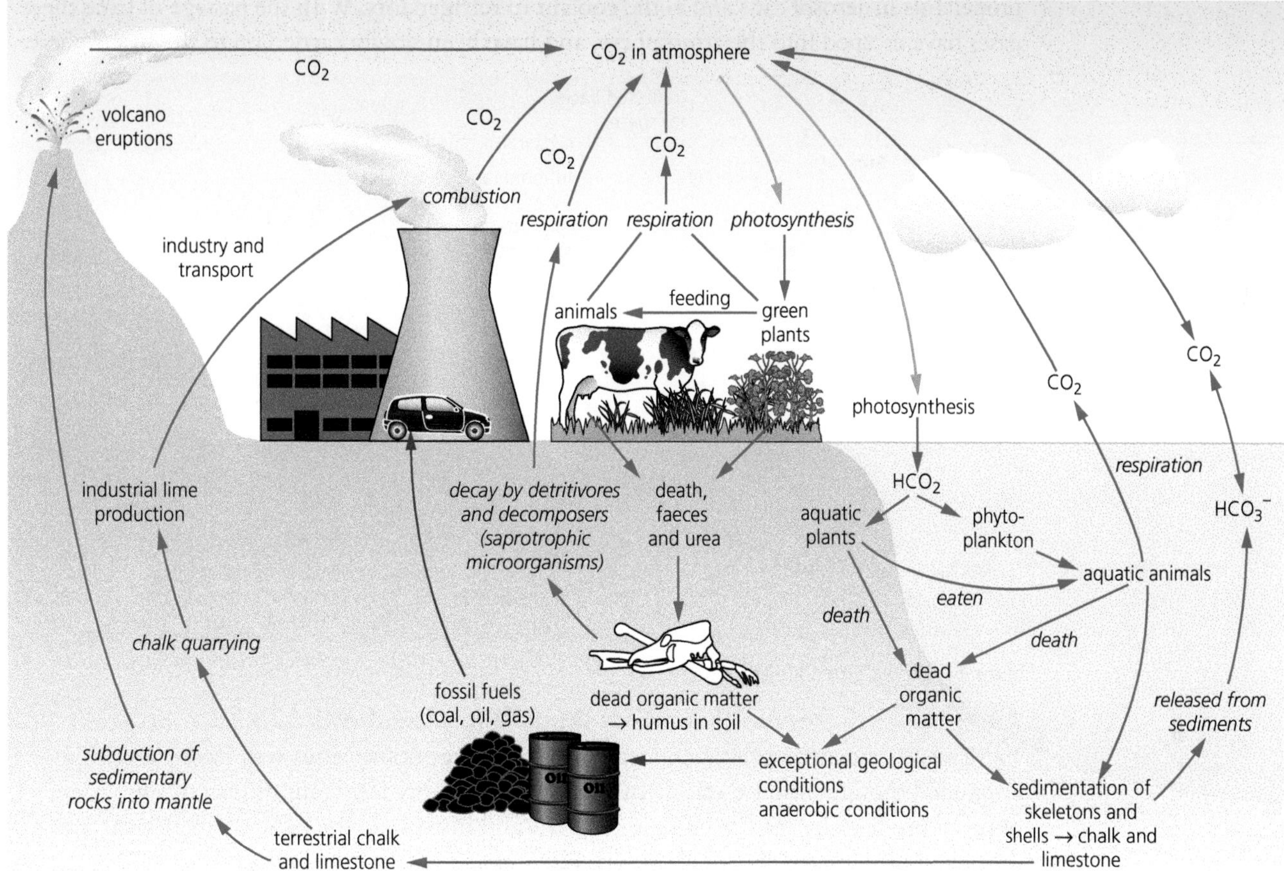

Figure 14.15 The carbon cycle

The effect of a low level of atmospheric carbon dioxide is to maintain a favourable environmental temperature on Earth – a phenomenon known as the **greenhouse effect**. Actually, the level of atmospheric carbon dioxide is now rising, for reasons we will investigate shortly.

> **Key term**
>
> **Greenhouse effect** Atmospheric warming caused by the absorption of re-radiated solar energy by gases such as carbon dioxide and methane.

The mechanism of the greenhouse effect

The radiant energy reaching the Earth from the Sun includes visible light (short wave radiation) and infrared radiation (longer wave radiation – heat), which warms up the sea and land. As it is warmed, the Earth radiates infrared radiation back towards space. However, much of this heat does not escape from our atmosphere. Some is reflected back by clouds and much is absorbed by gases in the atmosphere, which are warmed. In this respect, the atmosphere is working like the glass in a greenhouse, which is why this phenomenon is called a 'greenhouse effect' (Figure 14.16 on the next page). It is vitally important, as without it, surface temperatures would be too cold for life to exist on Earth.

Any gas in the atmosphere that absorbs infrared radiation is referred to as a **greenhouse gas**. Carbon dioxide is neither the only component of our atmosphere with this effect, nor the most 'powerful'. Both water vapour and methane are also naturally occurring greenhouse gases, and the latter is much more efficient at heat retention than carbon dioxide, although not present in the same proportions (so far).

In addition, purely anthropogenic atmospheric pollutants – such as oxides of nitrogen (particularly nitrous oxide), and chlorofluorocarbons (CFCs) – have 'greenhouse' properties, too. Oxides of nitrogen are waste products of the combustion of fossil fuels (oil and coal), and so occur in the exhaust fumes of vehicles. CFCs, on the other hand, are unreactive molecules that were deliberately manufactured by the chemical industry to use as propellants in aerosol cans and as the coolant in refrigerators. With the passage of time these gases have escaped into the atmosphere, and have been slowly carried up to the stratosphere.

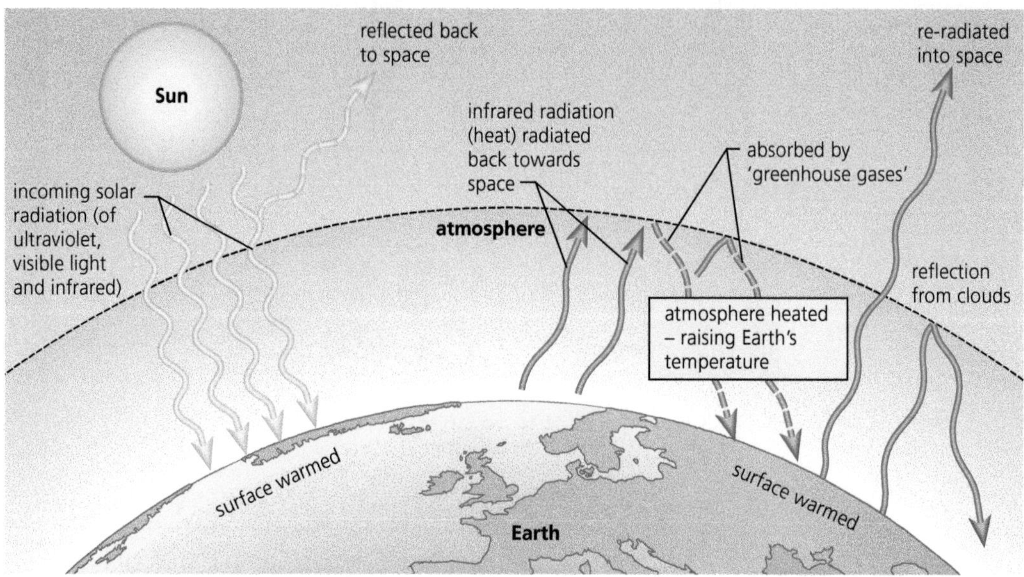

Figure 14.16 The greenhouse effect

An enhanced greenhouse effect leading to global warming?

Increases in the atmospheric concentrations of greenhouse gases will have inevitably enhanced the greenhouse effect. In order to assess how the composition of the atmosphere has changed over time, current and historic levels of atmospheric carbon dioxide and methane must be known. How are such records obtained?

The best long-term records of changing levels of these greenhouse gases (and associated climate change) are based on evidence obtained from ice cores drilled in the Antarctic and Greenland ice sheets. As water freezes, bubbles of air from the surrounding atmosphere become trapped within the ice. For example, data from the Vostok ice core in East Antarctica (that is, the composition of the bubbles of gas obtained from these cores) show us how methane and carbon dioxide levels have varied over no less than 400000 years. Similarly, variations in the concentration of oxygen isotopes from the same source indicate how temperature has changed during the same period (Figure 14.17).

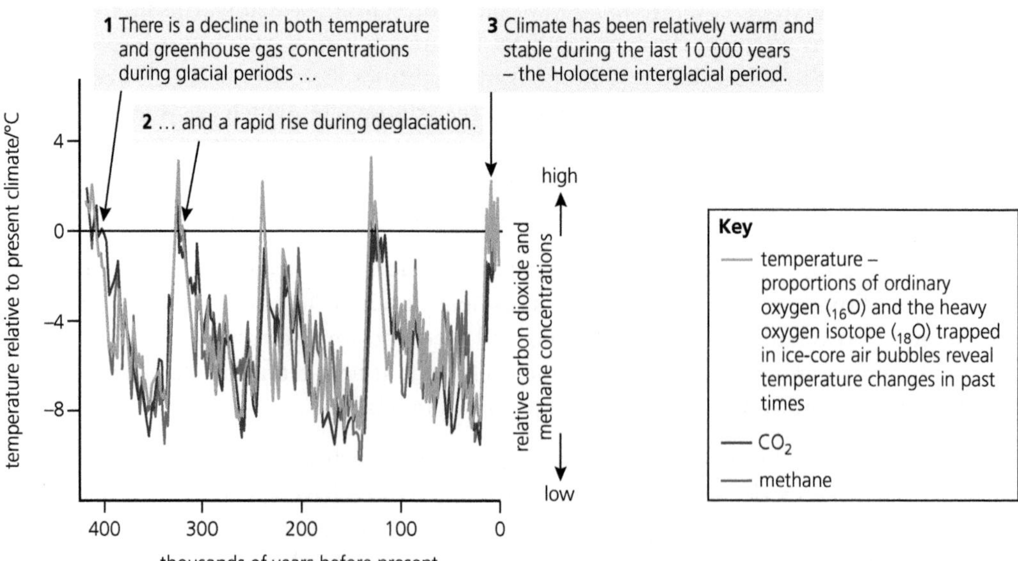

Figure 14.17 Three types of data recovered from the Vostok ice cores over 400 000 years of Earth history

Clearly, here the levels of greenhouse gas in the atmosphere can be closely correlated with global temperature. Environmental conditions on Earth have changed as a consequence. In fact, the Earth's climate has varied greatly over its billions of years of existence. Even over the 100 000 years of human presence, ice ages have come and gone! However, since the beginning of agriculture and the formation of city communities (about 8000 years ago) Earth conditions have been atypically steady. This stability can be correlated with steady atmospheric levels of carbon dioxide and methane (Figure 14.18).

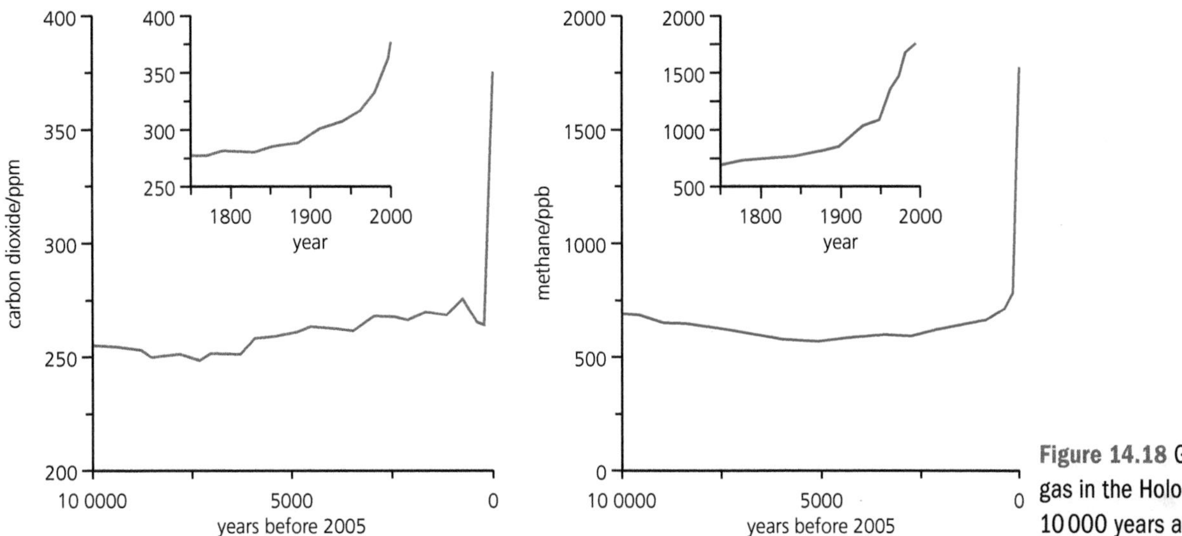

Figure 14.18 Greenhouse gas in the Holocene (from 10 000 years ago)

Since the Industrial Revolution in the developed countries of the world (about 200 years ago), there have been sharp rises in the levels of greenhouse gases, attributed to the burning of coal and oil (Table 14.2). These 'fossil fuels' were mostly laid down in the Carboniferous Period. So, we are now adding to our atmosphere, carbon that has been locked away for about 350 million years. This is an entirely new development in geological history. The effect has been a recent, accelerating rate of global warming. Many climate scientists argue that this development poses a major environmental threat to life as we know it.

Table 14.2 Changing levels of atmospheric CO_2

	CO_2 (in ppm)
pre-Industrial Revolution level	280 (± 10)
by mid 1970s	330
by 1990	360
by 2007	380
by 2050 (if current rate maintained)	**500**

The evidence for global warming – environmental change

The report of the Intergovernmental Panel on Climate Change (IPCC) of 2007 estimated global warming would be in the range 2.4–6.4 °C by 2100. This would generate the warmest period on Earth for at least 100 000 years, with highly significant environmental impacts. The paragraphs that follow describe in outline various theatres of environmental change attributed to global warming. These changes (and others) are already evident in part, and will have disastrous consequences if they continue unabated.

Figure 14.19 The retreat of the famous Fox glacier in New Zealand

Polar ice melt

The Arctic is a highly sensitive region – ice cover varies naturally. However, since 1979 the size of the summer polar ice cap has shrunk more than 20 per cent. In this period, the decline in the ice has been, on average, more than 8 per cent per decade. At this rate, there may be no ice in the summer of 2060. The associated Greenland ice is similarly in decline.

In Antarctica the picture is less clear. The Antarctic Peninsula has warmed, and 13 000 km^2 of ice have been lost in the past 50 years. Also, major sections of the Antarctic ice shelf have broken off. Meanwhile, at times the interior ice has become cooler and thicker – due to circular winds around the land mass preventing warmer air reaching the interior. Warmer seas may be eroding the ice from underneath, but IPCC predicts that the Antarctic's contribution to rising sea levels will be small.

Glacier retreats

Retreat by glaciers is worldwide and rapid – since 1980, glacier retreat has become ubiquitous. Mid-latitude mountain ranges such as the Himalayas, the Tibetan plateau, Alps, Rockies, and the southern Andes, plus the tropical summit of Kilimanjaro, show the greatest losses of glacier ice. Rivers below these mountain ranges are glacier-fed, so the melting of glaciers will have increasing impact on the water supplies for a great many people. The photographs in Figure 14.19 show the rapid retreat of the Fox Glacier in New Zealand from 2005 to 2015.

Rising sea levels

The impact of global warming on sea levels is due to thermal expansion of sea water and the widespread melting of ice. The global average sea level rose at an average rate of 1.8 mm per year in the period 1960–2003, but during the later part of that period the rate was far higher than at the beginning. If this acceleration continues at the current rate, sea levels could rise by at least 30 cm in this century. This phenomenon will threaten low-lying islands and countries (including Bangladesh), and major city communities such as London, Shanghai, New York and Tokyo.

Changing weather and ocean current patterns

At the poles, cold, salty water sinks and is replaced by surface water warmed in the tropics. Now, melting ice decreases ocean salinity, which then slows the great ocean currents that convey heat energy from warmer to colder regions through their pattern of convection. So, for example, as the Gulf stream (which to date, keeps temperatures in Europe relatively warmer than in Canada) slows down, more heat is retained in the Gulf of Mexico. Here, hurricanes get their energy from hot water, and become more frequent and more severe.

Also, alteration in the patterns of heat and rainfall distribution over continental land masses are predicted to cause Russia and Canada to experience the largest mean temperature rises, followed by several Asian countries and already drought-ridden countries in West Africa. Least warming is anticipated in Ireland and Britain in the northern hemisphere, and New Zealand, Chile, Uruguay and Argentina in the south. The most immediately vulnerable populations are already impoverished communities in parts of Africa; the least vulnerable is the wealthy population of Luxembourg.

Coral bleaching

Microscopic algae live symbiotically in the cells of corals, giving them their distinctive colouration, but when under environmental stress (for example, high water temperatures), the algae are expelled (causing loss of colour) and the coral starts to die. Mass bleaching events occurred in the Great Barrier Reef in 1999 and 2002. The effects from thermal stress are likely to be exacerbated under future climate scenarios, threatening biodiversity in coral communities. A photo of bleached corals can be seen in Figure 14.20.

Figure 14.20 Bleached corals

Other bio-indicators of climate change

Evidence of past climate changes may also be deduced from the remains of organisms that survived in habitats in the past. For example, in ancient peat deposits are found the 'fossilised' pollen grains of the once dominant vegetation. Alternatively, the study of tree rings (known as **dendrochronology**) indicates climatically favourable and less favourable years for tree growth, for as far back as preserved ancient timbers go.

The evidence for global warming – the debate continues

Global warming as an issue is seldom out of the news. Many scientists believe that the trends shown by average temperatures and the evidence from polar ice caps and glaciers make a very strong argument that this is an important and possibly long-term effect. However, there are many others who claim that this is simply natural climatic variation, which the Earth has seen many times before. One problem with the argument is that many predictions are based upon **extrapolation** of existing data. This means simply analysing the current trend and assuming that this is going to continue in the same pattern for many years to come. All scientists are very wary of extrapolation along with its assumptions, however the data keeps building and it could be that current estimates may even underestimate the effect.

Key term

Dendrochronology
The dating of wooden artefacts by analysis of their unique pattern of tree rings formed from xylem vessels.

Extrapolation The practice of extending conclusions, or graphical presentation, beyond that shown by data to make predictions about future trends and patterns.

Even if we accept the fact of global warming, there are problems in explaining its causes. Figure 14.21 shows the correlation between rising temperature and rising atmospheric CO_2.

Even simple observation shows a clear correlation between the two and it is possible to show that this is statistically significant. But, as we have discussed earlier in this chapter, showing a correlation does not demonstrate causation. You cannot automatically say that the rise in carbon dioxide levels is the main cause of global warming. Many vested interests who are opposed to limiting carbon emissions are quick to seize upon this dilemma. As scientists, you must also look carefully at the data. For example, carbon dioxide levels in the atmosphere are extremely low, so just where and how accurately were they measured in 1860? What other secondary causes might give a false correlation?

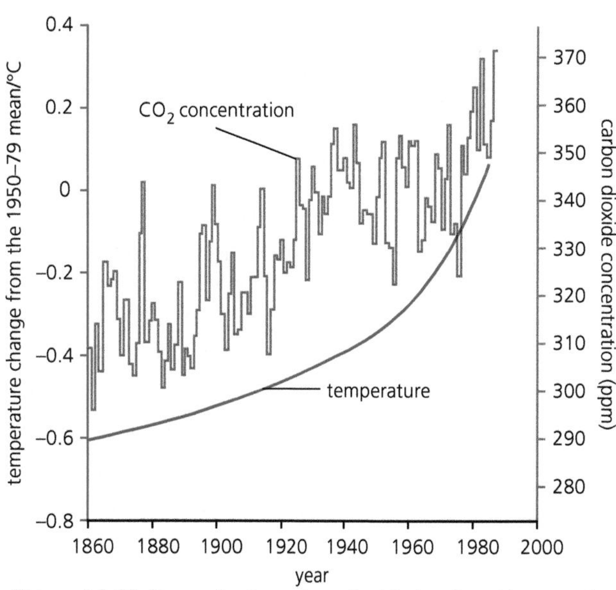

Figure 14.21 Atmospheric carbon dioxide level and temperature

Test yourself

18 Name the type of radiation that is absorbed by greenhouse gases.

19 Describe how evidence of ancient atmospheric composition is provided by ice cores.

20 Scientists are very wary of making conclusions based on extrapolation. Explain why.

To prevent the most severe impacts of climate change, the international community agreed, in 2014, that global warming should be kept below 2 °C compared to the temperature in pre-industrial times. That means a temperature increase of no more than around 1.2 °C above today's level.

To stay within this ceiling, the scientific evidence shows that the world must stop the growth in global greenhouse gas emissions by 2020 at the latest, reduce them by at least half of 1990 levels by the middle of this century and continue cutting them thereafter.

At the present time there is a strong consensus amongst scientists that rising carbon dioxide levels are at least very strongly implicated as the causative agent of global warming. So how do scientists come to a consensus and discuss the various lines of evidence pointing to the effect of carbon dioxide?

Communication and validation of evidence by scientists

At this point it is a good idea to refresh your memory on how scientists communicate and discuss their findings to build a body of reliable evidence. You have met this before in *Edexcel A level Biology 1*, Chapter 7 when discussing the debate concerning the three-domain classification model. The main methods of communication and validation are listed below.

1 Scientific journals

All scientists publish full details of their investigations in well-known scientific journals. Their reports must contain full details of their methodology, the original data and an analysis of their findings, following some strict rules. These journals are available to scientists worldwide who can read about the work of others and the latest developments in their field.

2 Peer review

Before a scientific journal will accept work for publication it must be verified by senior scientists in the place where it was carried out. It is then scrutinised by an independent panel of scientists who are experts in the same field. They check the details of the method, the data collected and the validity of conclusions. Peer reviewers often ask for more details or a revision of conclusions before approving its publications. The process of peer review and publication of scientific papers is quite strict and therefore ensures that the information contained in the papers is very reliable. In this way a large body of scientific knowledge and understanding has been built up over many years.

3 Conferences (symposia)

Most important fields of research are carried out by many scientists in many parts of the world. Universities and other institutions often host meetings of scientists from around the world specialising in one particular area of research. At these meetings invited participants often present their latest findings before they have been published. However, the most important function of these meetings is to allow individuals to share ideas, discuss common problems and argue their case where different models are proposed.

Finally, it is important to realise that scientists are also human. Debates on the merits of different models can become very heated as proponents defend their ideas.

Action to limit carbon dioxide emissions – the need for international cooperation

Agreeing that carbon dioxide might be the problem is one thing; doing something about it is quite another. Global warming is a problem of the Earth's atmosphere. The atmosphere knows no national boundaries and the activities of one country affect all. Any action that is not agreed by most countries of the world is bound to have very limited effect.

The first stages of an attempt to gain international agreement on reducing emissions of carbon dioxide and other greenhouse gases were begun in 1992 by the United Nations and resulted in what is known as the **Kyoto Protocol**. What followed is shown in the timeline below.

- **1992:** The UN Conference on the Environment and Development is held in Rio de Janeiro. It results in the Framework Convention on Climate Change (FCCC or UNFCCC) among other agreements.
- **1995:** Parties to the UNFCCC meet in Berlin (the 1st Conference of Parties (COP) to the UNFCCC) to outline specific targets on emissions.
- **1997:** In December the parties conclude the Kyoto Protocol in Kyoto, Japan, in which they agree to the broad outlines of emissions targets.
- **2002:** Russia and Canada ratify the Kyoto Protocol to the UNFCCC, bringing the treaty into effect on 16 February 2005.
- **2011:** Canada becomes the first signatory to announce its withdrawal from the Kyoto Protocol.
- **2012:** On 31 December the protocol expires.

Key term

Kyoto protocol An international treaty signed by 180 countries to reduce carbon emissions between 2008 and 2012.

The principles of the Kyoto Protocol are simple. Each country would sign the agreement and commit themselves to reducing greenhouse gas emissions to about 8 per cent lower than those of their country in 1990. This agreement was to run from 2008 to 2012 and took over 5 years to agree and a further 5 years to begin implementation.

To illustrate the enormous problems that need to be overcome in achieving agreement with so many countries, we need to look at one of the major obstacles. Poorer underdeveloped countries felt that imposing targets would be unfair as richer countries had already developed their industries and economies without restraining carbon emissions. These developed countries, in turn, felt that restrictions on them would give unfair advantage to others. To meet both these arguments a compromise scheme was devised. The carbon credit was introduced. A carbon credit is simply a licence to emit 1 tonne equivalent of carbon dioxide. All countries were assigned a fixed quota of emissions; if measures taken meant that less than this was emitted then they would gain credits. If they exceeded their target then they would need to purchase credits from another country. As most of the underdeveloped countries were unlikely to breach their target they would have credits to sell, to finance their changes with the money coming from the richer countries.

Like all compromises this had drawbacks. Principally the poorer countries were never going to reach their limit, so by selling off their carbon credits they allowed others to emit more, with the result that there would be no reduction overall. Nevertheless the protocol was up and running and action to reduce emissions was put in place on a large scale.

Despite the slow progress, the Kyoto Protocol was the very first large-scale international treaty to address global warming. In all, more than 180 countries were involved and it was a great testimony to patience and diplomatic skill. It has resulted in many countries implementing far-reaching carbon-emission policies and setting challenging targets.

Sadly, although initially signing the agreement, the USA never ratified their acceptance and, as the country that produces about a third of global emissions, this was a major weakness. Its neighbour Canada also withdrew its participation in 2011.

There is hope that a new Protocol will be agreed to cover the period 2013–2020 with a major conference in 2015 but the signs are not promising. Three of the world's largest countries – China, India and the USA – have already declared that they will not enter into any legally binding agreement but at least they are open to voluntary reduction measures and there is hope for some positive outcome.

Test yourself

21 Give **two** reasons why it is so difficult to reach international agreement on climate control.

22 Describe what the Kyoto protocol aimed to achieve.

Exam practice questions

1 The exact role that an organism plays within an ecosystem is known as its:

A habitat **C** niche

B community **D** sere *(1)*

2 Greenhouse gases only absorb radiation that has been reflected from the Earth's surface because:

A the wavelength of the radiation is changed on reflection

B the energy of direct radiation from sunlight is too high

C the destruction of the ozone layer has allowed more radiation into the atmosphere

D greenhouse gases are only found in lower levels of the atmosphere *(1)*

3 Which of the following would be described as a secondary succession?

A The colonisation of a newly-formed volcanic island.

B The activities of sheep maintaining a grassland ecosystem rather than woodland.

C The development of new communities following a severe forest fire.

D The build up of non-decayed mosses to form a peat bog. *(1)*

4 The coastal waters of Peru are home to very large shoals of small fish called anchovies (*Engraulis* sp.). Anchovies are an extremely important part of the marine ecosystem as primary consumers, and different species are common throughout the world's oceans. Whilst anchovies are consumed directly because of their strong flavours, most of the catch is processed into fish oils and dried as animal feed.

Overfishing has meant that in recent years the stocks of anchovies have shown periods of significant decline followed by partial recovery. The Peruvian government has been compelled to take action to try to prevent a total collapse of the anchovy population.

a) Explain how overfishing can result in long-term damage to the whole marine ecosystem. *(3)*

b) Explain why using anchovies as food for intensively reared animals such as pigs is an inefficient use of animal protein. *(3)*

c) What action might the Peruvian government take to protect fish stocks? *(3)*

5 Scientists are able to study the temperature and the carbon dioxide content of the Antarctic atmosphere, in the Vostok area, by analysis of air trapped in the ice over many thousands of years. Long cores of ice are extracted from the ice by drilling. The deeper the core, the older the air trapped within it. The nature of the analysis meant that there were far fewer readings for carbon dioxide concentration than there were for temperature.

Tip

Although in Question 4 you will need to apply your knowledge to a new situation to some extent, this is a very straightforward question based on specification material.

Tip

Question 5 has a mixture of demands. Part c) is quite simple AO1 material, which you will have met both in *Edexcel A level Biology 1* and in this chapter. However, part a) requires you to look very carefully at the data and understand clearly what conclusions might be drawn from it.

An extract from the data obtained in this way is shown in the graph.

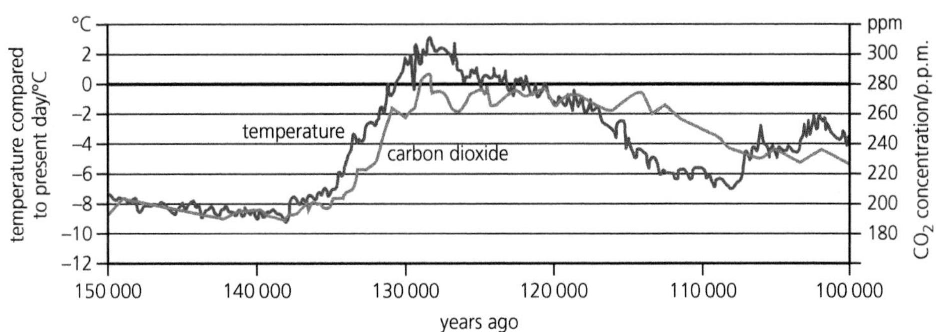

a) Analyse the data in the graph between 110 000 and 105 000 years ago to explain why some scientists might argue that this evidence does not support the idea that a rise in carbon dioxide levels actually causes a rise in atmospheric temperature. *(3)*

b) Other scientists have argued that the overall pattern of changes between 150 000 and 130 000 years ago displays such a strong correlation that this must show that carbon dioxide levels were the cause of the temperature rise. Evaluate this conclusion. *(4)*

c) Describe how these data would be communicated to the scientific community, and debated within it. *(4)*

6 A student carried out an investigation to test the hypothesis that there would be a correlation between light intensity and the number of different plant species (species richness) growing in an area of grassland. He placed a belt transect across a region of grassland that covered a shaded area and a sunlit area. He then measured the light intensity using a light meter at three random points in each of ten 0.5 m × 0.5 m quadrats along the transect and calculated the mean. Finally, the number of different plant species in each quadrat was counted.

a) i) State **two** precautions that should be taken when measuring light intensity using a light meter in this way. *(2)*

ii) Describe **two** ways in which measurement of light intensity could be made more representative of the actual exposure of the plants to light. *(2)*

b) The student decided to test the strength of the correlation between light intensity and number of species found by applying a Spearman's rank correlation test.

i) State a suitable null hypothesis for this investigation. *(1)*

ii) Complete the table of results shown on the next page and calculate the value of Σd^2. *(2)*

Tip

Question 6 has been extended to include 16 mark points. This has been done to give examples of how you may be asked to deal with statistical testing and practical skills. It is unlikely that you will meet such a long question in an examination paper but you will need to be able to deal with all the types of question that it contains.

Mean Light Intensity (lux)	Number of plant species found	Rank Light intensity	Rank number of plants	Difference in ranks (*d*)	d^2
6350	0	1	1	0	0
7033	2	2	5.5	−3.5	12.25
7560	2	3	5.5	−2.5	6.25
7780	2	4	5.5	−1.5	2.25
10276	1	5	2.5	2.5	6.25
10773	3	6			
12090	1	7	2.5	4.5	20.25
13010	3	8			
13596	2	9	5.5	3.5	12.25
17566	4	10	10	0	0
				$\Sigma d^2 =$	

iii) Calculate Spearman's rank correlation coefficient (r_s) for these data using the formula

$$r_s = 1 - \frac{6\sum d^2}{n(n^2 - 1)}$$

Where n= the number of pairs of data in the sample. *(3)*

iv) The critical value of r_s for 10 pairs of measurements at the 5% confidence level is 0.648. Using this information and your calculated value of r_s, what conclusions can be drawn about this relationship? *(2)*

v) Use the information on methods and the collected data to explain why this conclusion could be regarded as unreliable. *(4)*

Stretch and challenge

7 The data from the Vostock ice cores, such as that shown in Question 5 on the previous page, has caused some debate amongst scientists investigating the model that it is atmospheric carbon dioxide that is causing global warming.

a) Research these data further and find out exactly how the information on temperature and atmospheric composition were found.

b) Explain how scientists argue that it can be used to both support and challenge this model.

8 The European Common Fisheries Policy is typical of many inter-governmental attempts to conserve fish stocks.

One recent example of the conflict that this can cause was the restriction of fishing in an area known as the 'Irish Box'. Make a list of the typical arguments that might be put forward by

a) fishermen and

b) conservationists

to argue for and against such restrictions.

Index

D

E

F

G

H

I

Free online resources

Answers for the following features found in this book are available online:

- Test yourself questions
- Activities

You'll also find an Extended glossary for each chapter to help you learn the key terms and formulae you'll need in your exam.

Scan the QR codes below for each chapter.

Alternatively, you can browse through all resources at:

www.hoddereducation.co.uk/EdexABiology2

How to use the QR codes

To use the QR codes you will need a QR code reader for your smartphone/tablet. There are many free readers available, depending on the smartphone/tablet you are using. We have supplied some suggestions below, but this is not an exhaustive list and you should only download software compatible with your device and operating system. We do not endorse any of the third-party products listed below and downloading them is at your own risk.

- for iPhone/iPad, search the App store for Qrafter
- for Android, search the Play store for QR Droid
- for Blackberry, search Blackberry World for QR Scanner Pro
- for Windows/Symbian, search the Store for Upcode

Once you have downloaded a QR code reader, simply open the reader app and use it to take a photo of the code. You will then see a menu of the free resources available for that topic.

1 Cellular respiration

3 Microbial techniques

2 Photosynthesis

4 Pathogens and antibiotics

5 Response to infection

11 Chemical control

6 Cell control

12 Coordination and control in action

7 DNA profiling, gene sequencing and gene technology

13 Ecosystems

8 Genetics

14 Changing ecosystems

9 Gene pools

15 Mathematics for biology

10 Nervous systems

16 Preparing for the exams